MW00528462

THE GARDENER'S BOTANICAL

Published in North America in 2020 by Princeton University Press,
41 William Street, Princeton, New Jersey 08540
press.princeton.edu

ISBN: 978-0-691-20017-0

Library of Congress Control Number: 2019942331

British Library Cataloging-in-Publication Data is available

This book has been composed in Effra Regular, Garamond Premier Pro,
GT Super Display, and GT Walsheim Pro
Printed on acid-free paper

Conceived, designed, and produced by
The Bright Press, an imprint of The Quarto Group.
The Old Brewery, 6 Blundell Street
London N7 9BH
United Kingdom
(0)20 7700 6700
www.QuartoKnows.com

Designed by Greg Stalley
Layout by Ginny Zeal
Editorial consultant: Rosalyn Marshall

Printed in China

1 3 5 7 9 10 8 6 4 2

THE GARDENER'S BOTANICAL

An Encyclopedia of Latin Plant Names

ROSS BAYTON

WITH MORE THAN 5,000 ENTRIES
AND 350 BOTANICAL ILLUSTRATIONS

PRINCETON UNIVERSITY PRESS

PRINCETON AND OXFORD

Contents

Introduction

For any gardener, knowing the names of the plants in your care is crucial. Not only does it let you share what you are growing with friends and family, it unlocks a world of useful information, both in books and online, helping you to make the best of the plants you grow. But why do the names have to be in Latin? Many garden plants have English names that are much easier to remember. The trouble is, those names vary depending on who you talk to. In the United States, bluebell is a common name for plants in the *Mertensia* genus, but in Scotland a bluebell is *Campanula rotundifolia* and in England it is *Hyacinthoides non-scripta*. Not surprisingly, there are other bluebells, including desert bluebell (*Phacelia campanularia*) in California and royal bluebell (*Wahlenbergia gloriosa*) in Australia. There are even bluebell creepers, *Clitoria ternatea* in the United States and Asia as well as *Billardiera heterophylla* in Australia. With "bluebells" occurring seemingly worldwide, it is essential for each to have its own unique name, because the cultivation requirements for an English bluebell differ widely from those of its Antipodean namesakes.

The decision to use Latin for this naming system stems largely from its historical use by scholars. A university education in the eighteenth century required familiarity with both Latin and Greek, allowing for students to access the works of classical writers from ancient Greece and Rome. Much of the "Latin" in use in horticulture is actually derived from Greek. How botanical Latin has changed over time, and botanical art alongside, is discussed on pages 12 and 17.

The enduring benefits of using this botanical tongue are that scientists, growers, and gardeners can communicate about plants with accuracy, no matter what language they speak. So whether it is bluebell (Scotland), harebell (England), bluebell-bellflower (United States), *campanilla* (Spanish), *campanule à feuilles rondes* (French), or *rundblättrige glockenblume* (German), it is universally *Campanula rotundifolia*. Latin names are not static, however, and new information is always being discovered—on page 20 we look at how DNA is revealing old and new relationships between plants.

Understanding Latin names brings other benefits. Most obviously, they can reveal useful information that is helpful to the gardener. *Armeria maritima* thrives by the sea, while *Geum rivale* does not mind getting its feet wet. *Toxicodendron* is best avoided, while silky soft *Alchemilla mollis* is a pleasure to touch. Botanical Latin has other practical uses, too, which are discussed on page 22.

Latin names also reveal fascinating stories, such as the tale of French diplomat Jean Nicot, who popularized tobacco, which now bears his name (*Nicotiana*), or the moving account of a promising Scottish botanist who fell during World War I and is commemorated for evermore (*Briggsia*). Delve into our dictionary of botanical Latin and you will find interest, inspiration, and intrigue in equal measure.

LEFT: French diplomat Jean Nicot is immortalized in the Latin name for tobacco, *Nicotiana tabacum*.

How to Use This Book

The following section of the book contains about 5,000 Latin botanical terms arranged alphabetically. These are terms that you might come across either for the genus or the species name of a plant. The term appears first, then a guide to pronunciation. The following line provides the feminine and neuter versions of the name, if applicable, followed by the etymology of the word. This will explain the meaning (or meanings) behind the Latin term. For the genus entries, the family name is also included in parentheses at the end. An example of a plant name that features the term is also supplied.

Where variations in spelling occur, they are grouped together, as in:

---

cashmerianus
kash-meer-ee-AH-nus
cashmeriana, cashmerianum

–

cashmirianus
kash-meer-ee-AH-nus
cashmiriana, cashmirianum

–

cashmiriensis
kash-meer-ee-EN-sis
cashmiriensis, cashmiriense
From or of Kashmir, as in *Cupressus cashmeriana*

Where a genus term can also be used for a species, it is included in parentheses afterward.

At the end of the book is an index of 2,000 of some of the most frequently used common names. Here, you can look up a plant that you know by its common name and find the Latin name. You can then look up what the Latin term means in the alphabetical listings section.

Pronunciation of botanical Latin can vary, so the examples provided here are a guide rather than a definitive directive. Capital letters indicate where the stress should fall. Where gender variations occur, pronunciation is given for only the masculine version.

Most gardeners will have encountered plants that seem to have not only numerous common names but also variations in their Latin appellations. Some epithets, which were once in wide usage, have now become obsolete, perhaps due to reclassification of the plants to which they were once applied. However, because these names may still sometimes be found in old horticultural works, as synonyms in modern texts, or when browsing the varied sources of the Internet, they have been included for completeness. The *RHS Plant Finder* by the Royal Horticultural Society in England provides a general guide to current nomenclatural opinion, if required, that is often used internationally as a reliable up-to-date source, but you can also check the USDA plants database for plants by their scientific or common names.

Helped by an informed understanding of botanical nomenclature, it is hoped that gardeners will be able to make better gardens filled with better plants; plants that sit well in their site, thrive in the conditions provided for them, and have the form, habit, and color that make for the most aesthetically pleasing association with their neighbors.

Genus name in Latin (if the name can also be used as a species name, this appears in parentheses afterward).

Alyogyne
al-ee-oh-GY-nee
From Greek *alytos*, meaning "undivided," and *gyne*, meaning "female," because the styles are undivided (*Malvaceae*)

A guide to pronunciation is provided and capital letters indicate where the emphasis should fall.

Origin of the term is explained and Latin meaning.

Family name

Genus spotlight focuses on interesting details about key genera.

Species name in Latin. The masculine version appears first with other versions listed below.

amphibius
am-FIB-ee-us
amphibia, amphibium
Growing both on land and in water, as in *Persicaria amphibia*

When appropriate, an example is given of a plant name that features the Latin term.

Lines link entries that are different spelling variations with the same meaning.

Fact spotlight provides fascinating insights into botanical Latin.

Extensive index of common names allows readers to find the Latin name for common plants, which can then be cross-referenced to the alphabetical listings.

A Guide to the Use of Latin

When writing Latin names, it is important to order the various elements in the correct sequence and observe typographical conventions. This is a simplified outline of the binomial system, so there are exceptions to these rules, along with further complexities of structure. Because this is a book aimed at gardeners, not botanists, only the broad principles are dealt with here.

Family

(for example, *Sapindaceae*)
This name appears as uppercase and lowercase, and the International Code of Botanical Nomenclature also recommends italics. Family names are easily recognized, because they end in *-aceae*.

Genus

(for example, *Acer*)
This appears in italics with an uppercase initial letter. It is a noun and has a gender: masculine, feminine, or neuter. The plural term is genera. When listing several species of the same genus together, the genus name is often abbreviated, for example: *Acer amoenum*, *A. barbinerve*, and *A. calcaratum*.

Species

(for example, *Acer palmatum*)
The species is a specific unit within the genus and the term is often referred to as the specific epithet. It appears in lowercase italics. The species name is mostly an adjective, but it can sometimes be a noun (for example, *Agave potatorum*, where the epithet means "of drinkers"). Adjectives always agree in gender with the genus name they follow, but nouns used as specific epithets are invariable. It is the combination of the generic and specific epithet that gives us the species name in the binomial, or two-word, system.

Subspecies

(for example, *Acer negundo* subsp. *mexicanum*)
This part of a plant's name appears as lowercase italics and is preceded by the abbreviated form subsp. (or occasionally ssp.), which appears as lowercase roman type. It is a distinct variant of the main species.

Varietas

(for example, *Acer palmatum* var. *coreanum*)
This appears as lowercase italics and is preceded by the abbreviated form var., which appears as lowercase roman type. Also known as the variety, it is used to recognize slight variations in botanical structure.

Forma

(for example, *Acer pseudoplatanus* f. *purpureum*)
When used, it appears as lowercase italics and is preceded by the abbreviated form f., which appears in roman type. Also known as the form, it distinguishes minor variations, such as the color of the flower.

Cultivar

(for example, *Acer forrestii* 'Alice')
This part of the name appears as uppercase and lowercase roman type with single quotation marks. Also known as a named variety, it is applied to artificially maintained plants. Modern cultivar names (for example, those after 1959) should not include Latin or Latinized words.

Acer palmatum

Hybrid

(for example, *Hamamelis* × *intermedia*)

It appears as uppercase and lowercase italics and is denoted by an × that is not italicized (note that this is a multiplication sign, not the letter x). It may be applied to plants that are the product of a cross between species of the same genus. If a hybrid results from the crossing of species from different genera, then the hybrid generic name is preceded by a multiplication sign. However, if a hybrid results from the grafting of species from different genera, it is indicated by an addition sign instead of a multiplication sign.

Synonyms

(for example, *Plumbago indica*, syn. *P. rosea*)

Within a classification a plant has only one correct name, but it may have several incorrect ones. These are known as synonyms (abbreviated as syn.) and may have arisen due to two or more botanists giving the same plant different names or from a plant being classified in different ways.

Common Names

Where common names are used, they appear as lowercase roman type except where they derive from a proper name, such as that of a person or a place. (For example, soapwort but Carolina violet.) Note that the Latinized versions of proper names do not have capital letters, for example, *forrestii* or *freemanii*.

Many Latin genus names are in ordinary use as common names, for example, fuchsia; in this context, they appear in lowercase roman type, and they can also be used in the plural (fuchsias, rhododendrons).

Gender

In Latin, adjectives must agree with the gender of the noun that they qualify; therefore in botanical names, the species must agree with the genus. An exception to this rule is made if the species name is a noun (for example, *forrestii*, "of Forrest"); in this instance, there is no genus and species agreement. To help familiarize the reader with the different gender forms, where a specific epithet appears, the masculine, feminine, and neuter versions are usually listed—for example, *grandiflorus* (*grandiflora*, *grandiflorum*).

Plumbago indica

Why Botanical Latin Is Used for Plant Names

John David, Head of Horticultural Taxonomy, Royal Horticultural Society (RHS)

The system of plant names we use today dates back to the plan first devised by the Swedish botanist Carolus Linnaeus (Carl von Linné) in his *Species Plantarum* of 1753, but the names he used in many cases were far older. Linnaeus drew heavily on earlier works, particularly the many herbals published in the previous two centuries that, in turn, are ultimately derived from the *De Materia Medica* of Dioscorides (around AD 77). Although he mined these early sources for names, Linnaeus did not always choose to use the name for the same plant as it was originally intended.

Linnaeus's innovation, known as the binomial system, employed a unique combination of two words to refer to each plant and was developed as a kind of *aide-memoire*. By the Renaissance, plants had become known by long phrase names that could include more than twenty words, but usually between five and ten. These could be difficult to remember, so a shorthand of just two words was a godsend. Linnaeus's system was quickly picked up by other botanists; one of the earliest works in Great Britain to fully adopt the system is the eighth edition of Miller's *Gardener's Dictionary* (1768). Linnaeus developed this system for animals too.

The two elements of a Latin name are a combination of the name of a genus (plural genera) and the name of a species, such that closely related species share the same genus (generic) name. The genera were included by Linnaeus in classes, and this is where his so-called sexual system comes in. His classes were defined by the number and combination of male and female parts in the flower. An example being Hexandria Monogyna, for flowers with six stamens (male parts) and only one style (female part). This was more controversial than it might seem today, because Linnaeus described these combinations in terms of marriages. For a conservative and deeply religious society, the thought of six husbands in bed with one wife—or in the case of Polyandria, twenty or more husbands in one marriage—was shocking. It was felt at the time that Linnaeus's system was not something for polite conversation.

ABOVE: Swedish botanist Carolus Linnaeus, in the traditional dress of Lapland, is shown here carrying his favorite flower, *Linnaea borealis*.

RIGHT: The system of classification created by Linnaeus emphasized the sexual anatomy of flowers.

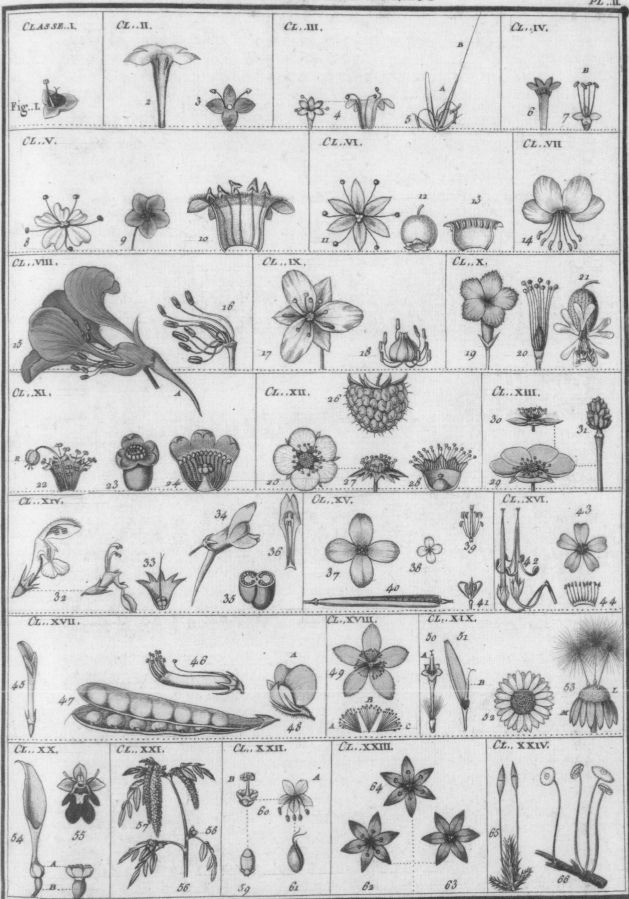

Botanical Cabinet

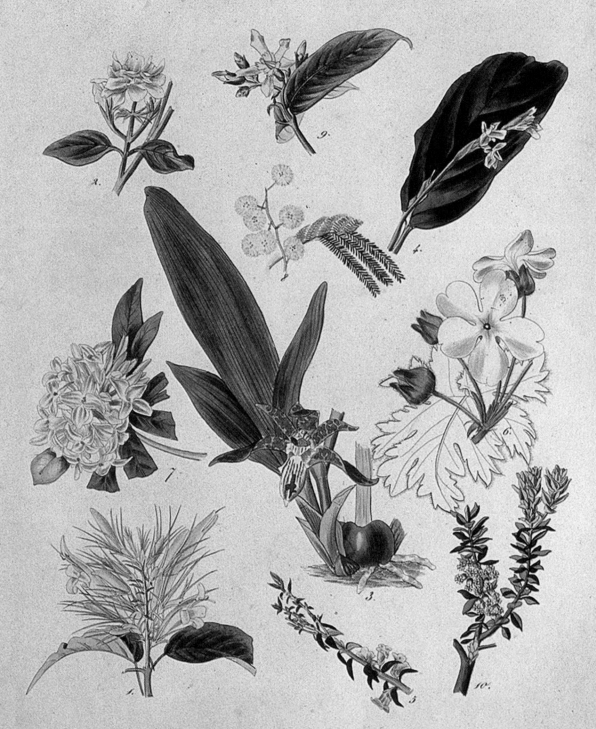

14

Another reason why Linnaeus's system of naming plants was so important is that until the era of exploration of the Americas, Africa, and Asia, Europeans had needed names for only European plants, which numbered in the low thousands. By the early eighteenth century, however, plants were flowing in from all over the world and needed to be described and named quickly. With so many plants being discovered and described, and the establishment of a more sophisticated understanding of how they were related, came the need for more ranks in the classification, which were inserted between genus and class. These were the family (ending with -*aceae*) and order (ending with -*ales*) in the nineteenth century, with the class (-*opsida*) being redefined and the rank of phylum (-*phyta*) added above the class in the twentieth century. Like the genus and species names, these are all in Latin form, but are plural adjectives. In addition, there is provision for ranks in between: variety, tribe, section, series, and any number of sub- or super- ranks, as well as hybrids. Although this complex arrangement of ranks is important for botanists looking at the plant kingdom as a whole, it does not detract from the simplicity of Linnaeus's binomial system for other users of plant names.

The adoption of Latin for naming and describing plants comes from the status of that language as the common language of learned discourse for centuries, letting students access the works of classical writers from ancient Greece and Rome. However, botanical Latin would hardly be recognized by a Roman, because it is an end point of the development, through medieval Church Latin, of a language that from the seventh century onward has not been anyone's native tongue. The formal publication of a new plant required a description in Latin up until 2012, and even now it is still one of only two alternatives, the other being English. However, despite its advantage in removing almost all ambiguity, the technicalities of case, number, gender, tense, mood, and declension, grammatical features almost nonexistent in many of the world's languages, meant that the use of Latin presented an almost insurmountable challenge for botanists around the globe. While it is no longer required for describing plants, it is retained for plant names, and an understanding of Latin is helpful for the creation of new names of plants, an appreciation of their meaning, and their pronunciation.

LEFT: The binomial system of Linnaeus was timely, because it provided a system for naming the many plants found during the age of discovery.

BELOW: Plant hunter Frederick Burbidge (1847–1905) discovered several new species while collecting for the Veitch Nurseries in Asia.

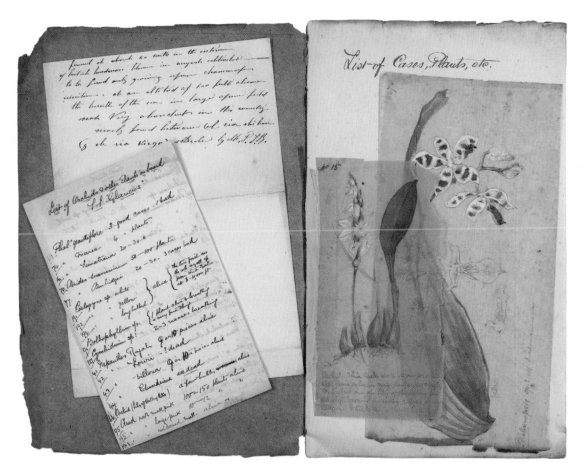

Tab. XII.

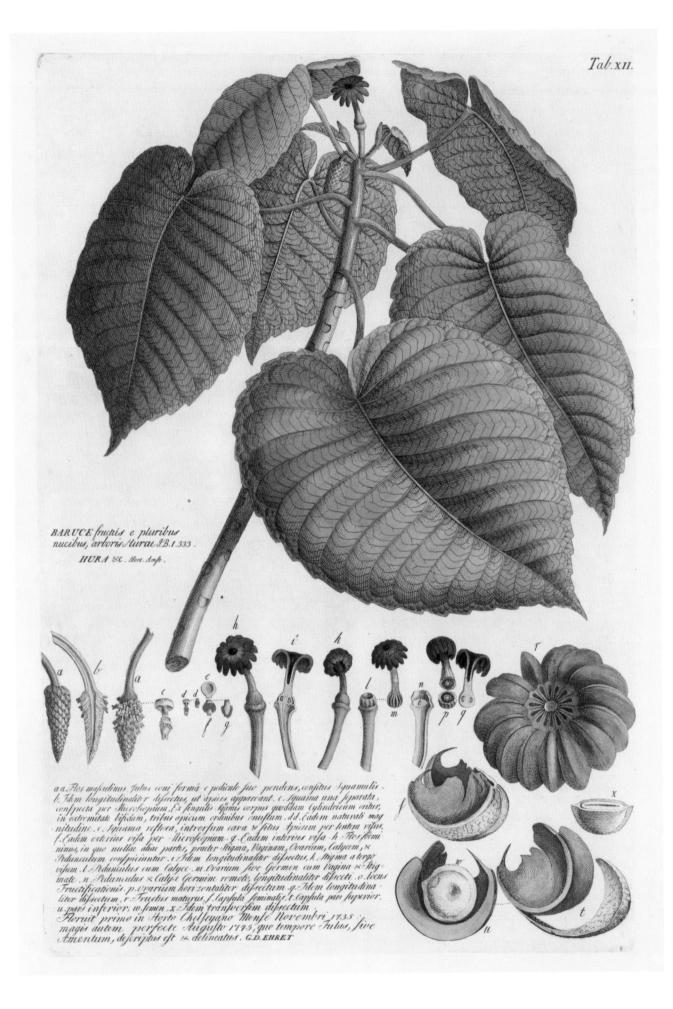

BARUCE fructûs e pluribus
nucibus, arboris Hurae I.B.1.333.

HURA &C. Hort. Amst.

a a. Flos masculinus Julus coni formâ e pediculo suo pendens, consitus squamulis.
b. Idem longitudinaliter dissectus, ut apices appareant. c. Squama una separata,
conspecta per Microscopium. C. x singulis squamis corpus quoddam Cylindricum oritur,
in extremitate bifidum, tribus apicum ordinibus onustum. d d. Eadem naturali mag-
nitudine. e. Squama reflexa, introrsum cava & situs Apicum per totum visus.
f. Eadem exterius visa per Microscopium. g. Eadem interius visa. h. Flos foemi-
ninus, in quo nullae aliae partes, praeter Stigma, Vaginam, Ovarium, Calycem, &
Pedunculum conspiciuntur. x. Idem longitudinaliter dissectus. k. Stigma a tergo
visum. l. Pedunculus cum Calyce. m. Ovarium sive Germen cum Vagina & Stig-
mate. n. Pedunculus & Calyx Germine remoto, longitudinaliter dissecti. o. Locus
Fructificationis. p. Ovarium horizontaliter dissectum. q. Idem longitudinaliter dissectum. r. Fructus maturus. s. Capsula seminalis. t. Capsula pars superior.
u. pars inferior. w. semen. x. Idem transversim dissectum.

Floruit primo in Horto Chelseyano Mense Novembri 1735.
magis autem perfecte Augusto 1743, quo tempore Julus, sive
Amentum, descriptus est & delineatus. G. D. EHRET

A Brief History of Botanical Art

Brent Elliott, Royal Horticultural Society Librarian (retired)

Botanical art is art that serves the purposes of botanists, and throughout history it has usually facilitated the accurate identification of plants. To be worthwhile, illustrations need to be available in multiple copies; a picture that exists in a single copy may be an excellent portrait of a plant, but it can be of little use other than to local botanists. We know from Pliny the Elder, writing in the first century AD, that scholars had tried to produce illustrated books about plants; but the only way of distributing books then was to copy by hand, and illustrations copied by hand over and over again gradually ceased to resemble the originals sufficiently to be reliable identification guides. So the attempt to illustrate books on plants was abandoned, and scholars had to rely on written descriptions alone.

The printing press changed all that. The first printed botanical books simply copied the poor schematic illustrations found in medieval manuscripts, but in 1530 Otto Brunfels published his *Herbarum vivae eicones*, with the first printed images of plants drawn from actual plant specimens, made by an artist named Hans Weiditz. This was followed in 1542 by the *De historia stirpium commentarii insignes* of Leonhart Fuchs, the first textbook on plants where both text and pictures were based on studying living specimens. Albrecht Meyer's illustrations for Fuchs established the conventions of botanical art for the next century: plants were depicted at life size where possible; the complete plant was shown, including the root system; the plants were shown in outline, with little attempt at illustrating surface texture.

Fuchs's reasoning for the use of outlines was that readers might want to color the pictures. Amateur coloring remained a problem for generations. How could you ensure that people using the book in different places actually saw the same pictures if in some cases they had been colored, perhaps by

LEFT: Botanical illustrations from the mid-1700s (here, *Hura crepitans* in *Euphorbiaceae*) show detailed dissections of the flowering parts.

ABOVE: Early botanical illustrations, such as this *Capsicum annuum* from Fuchs's 1543 German edition, rely on outlines instead of texture.

Pl. CCXIX.

RICINUS, *foliis peltatis inæqualiter serratis, capfulis hispidis.*

G. D. Ehret delin.

I. Miller Sculp.

Publishd according to Act of Parliament by J. Miller March 26. 1758.

18

someone not familiar with the plants? In the late seventeenth century, artists for prestigious publications began to depict the surface texture of leaves and stems in as much detail as possible to deter people from coloring the illustrations. The first experiments in color printing took place in the eighteenth century, but it would not be until the mid-nineteenth century that satisfactory results were obtained.

Most botanical illustration in the sixteenth and seventeenth centuries was made for herbals—books on the plants used in medicine. During the eighteenth century, the emphasis shifted to floras of particular parts of the world, and monographs on particular plant genera or families. In the 1730s, the botanist Carl Linnaeus popularized a system of classification based on the number and arrangement of the parts of the flower, and artists began to add detailed dissections of the flower to their works. These dissections came to be presented in rows, usually below the main figure. Leaf structure was considered subordinate, and an outline of the leaf was often shown as a background to the main image.

Linnaeus's sexual system of classification was superseded in the nineteenth century by systems that no longer focused exclusively on the flower, and artists returned to depicting all

LEFT: The illustrations to Philip Miller's *Gardeners Dictionary* included this one by G. D. Ehret, showing *Ricinus communis.*

ABOVE: This picture of *Hibiscus mutabilis* from Rheede's *Hortus Malabaricus* (1678-1703) shows how such illustrations spread knowledge of the world's flora.

parts of the plant. By this time, they had a new tool to help them: the microscope. The first plant drawings to be created with the assistance of a microscope had been made as long ago as the 1620s, but these were not published at the time, and the power of the microscope as an aid to botany was widely recognized only in the 1830s, with the publication of Franz Bauer's drawings of orchid anatomy. A new type of botanical illustration appeared, one that used anatomical details to compare different species in the same genus or family.

In the twentieth century, despite the increasing use of photography to portray plants, botanical works continued to use artists' drawings. It would take several photographs to provide all the information that can be conveyed in a single good drawing. This situation may be changing in the twenty-first century, with the advent of digital photography and the use of computer software to assemble multiple photographs into a single image. The coming decades promise to be an exceedingly interesting period in the history of botanical art.

How the Study of DNA Is Changing Plant Names

Alastair Culham, Curator of the University of Reading Herbarium

As both a botanist and a gardener, I often have to justify my use of two languages that are not natively spoken today: Latin, the international language of scientific communication for many centuries, which forms the basis for plant names, and the language of DNA sequences, the code that tells most living organisms what they are, how they should interact with the environment, and even when to die. Latin causes people to ask me why plant names are so difficult and DNA data evokes questions about why plant names change so often.

DNA structure has been understood since the 1950s, but only since the 1990s has it been possible to read long stretches of DNA in a fairly cheap and efficient way. DNA provides a complex and highly detailed story of how things are related to each other, written by DNA from parents combining in offspring and the accumulation and stabilization of chance mutations in the DNA bases. Plants have a particularly complex pattern of inheritance of DNA. Most DNA is in the cell nucleus, arranged in chromosomes and inherited 50/50 from each parent. However, there are two much smaller sets of DNA: one is in the mitochondrion, which provides energy for cell functions; the other, unique to plants, is in the plastid (including chloroplasts) with genes linked to photosynthesis and other functions. Both mitochondrial and plastid DNA are usually inherited from only the female (seed) parent. The result is that a plant cell has DNA with a mix of different histories, therefore different parts of DNA can tell different stories about how plants are related.

To get from DNA to a theory of how species are related to each other, to form genera, families, and orders, involves the comparison of DNA sequences through alignment and analysis of the changes in the DNA between samples. This is called phylogenetic analysis, which is about building a theory

ABOVE: The acornlike structures on *Fucus vesciculosus* (right) once led to its inclusion in the oak genus *Quercus*, along with *Quercus robur* (left).

of what is related to what through reconstruction of changes in DNA sequence over time. The resulting tree of relationships, the phylogeny, is hierarchical and the tips are generally species. These in turn form larger groups, genera, and these yet larger families and so on.

There are many hidden levels of complexity. It is not easy to align one strand of DNA against another, because they can change in the DNA sequence and can both extend and contract in length. Over millions of years, DNA strands can become so different it is almost impossible to tell which part from one species is derived from which part of another. As plants become more closely related, the differences become smaller and alignment is easier. This goes on until the DNA sequences are effectively the same, at which point species cannot be told apart. Until about 2015, it was expensive to read large portions of DNA in a single plant (it took ten years to read 90 percent of the smallest plant genome), but new

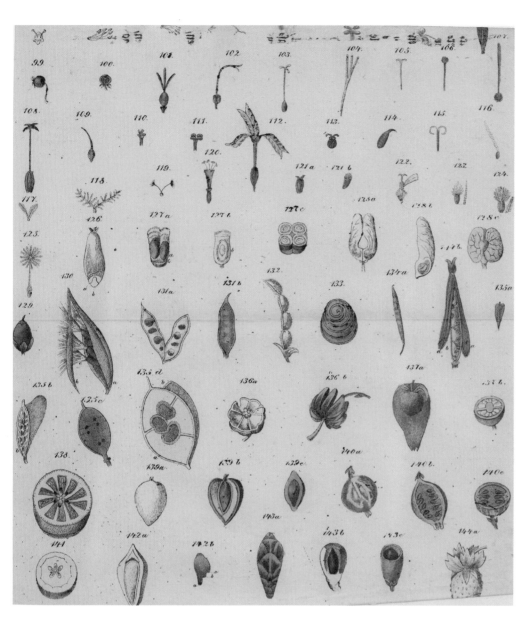

technology has made reading a whole genome more cost effective. Now it is possible to uniquely identify most individuals based on DNA sequence.

The phylogenies constructed for plants were originally based on one gene, and gradually more genes have been added. However, the easy sources of data have been only the chloroplast DNA and the DNA that codes for ribosomes (protein manufactures). When these limited data sets were analyzed and used to build phylogenetic trees, some startling new relationships among plants were found, resulting in many changes of plant names. Sin more DNA data have been read, some of those early ideas of relationship have been replaced with newer, more robust ideas, leading to additional name changes.

Most botanists try to make sure that plant names are informative of the relationships among those plants. This has big practical advantages; for instance, most gardeners would know that plants in the mint family, *Lamiaceae*, will probably be aromatic, or that *Capsicum* species will have spicy hot fruit. My favorite way to illustrate how name changes have helped us understand what we see is to look in a botany book by Clusius, published in 1601. He included bladderwrack (the black rockweed seaweed) in his genus *Quercus* (the oak) due to the common presence of acornlike structures. Imagine the disappointment if you bought a *Quercus* from a catalog and got brown seaweed. DNA data helps improve the way classification reflects relationships among plants. Try to think of every name change as an improved model instead of a nuisance.

Practical Uses of Botanical Latin

James Armitage, Editor of *The Plantsman*

Many gardeners will have thought at some point that Latin names just get in the way. Even an eager plant enthusiast will be familiar with the experience of trying to recall some impossibly antediluvian epithet, clutching his or her forehead and muttering feebly, "I know it begins with a D..." Common names are so charming and memorable—foxglove, Jack-in-the-pulpit, dove tree—why not just use these?

There are several excellent reasons why we need a fixed scientific set of names constructed from a dead language. One great advantage of Latin is its universality. No nation can claim it as its mother tongue, so it is at once everyone's and no one's, neither advantaging nor disadvantaging any one user. In the vast, shifting diversity of language and script found across the globe, Latin stands as a constant, unchanging point, so a plant name is the same in New Zealand as it is in Korea, Sweden, Morocco, Brazil, or Canada.

Latin names are also predictive in a way common names tend not to be. There is nothing in the colloquialisms skunkbush and staghorn sumach to suggest any particular relationship or similarity. However, once you know that they both refer to species of the genus *Rhus*, you are able to predict that each plant will probably have pinnate leaves and good autumnal color.

Aside from other considerations, an important justification for the use of Latin names is that they have so much to tell us that is of practical value. Color, form, size, ethnobotany, habitat, geography, history—all these things and many more are referred to in the names of plants, and all impart practical information that helps to make us better gardeners.

It is always helpful to gain some idea of a plant's appearance from a mere glance at its name. Those developing a color palette in blue and white, for instance, might want to keep an eye open for plants bearing the epithets *cyanus* and *caeruleus* (blue) or *albus* and *lacteus* (white). More fiery colors are represented by *aurantiacum* (orange), *coccineus* (scarlet), and *ruber* (red), among many others.

It is not just characteristics as obvious as color that are indicated in plant names. Epithets with the suffix *-folius* refer to details of the leaves. So we have *angustifolius* (narrow leaves), *brevifolius* (short leaves), and *grandifolius* (big leaves).

Stature might be indicated in names such as *grandis* (big), *minima* (smallest), and *nana* (dwarf), and growth habit in names such as *fastigiata* (erect, upright branches), *pendula* (hanging), and *scandens* (climbing). For plants that satisfy with their aroma, the epithet *odorata* (with a fragrant scent) might be considered, while the name *foetida* (with a bad smell) forewarns of a less agreeable experience.

Once acquaintance is made with the richness and diverse derivation of botanical Latin, scientific names can become every bit as romantic, evocative, and charming as common names—and nearly as easy to remember.

RIGHT: Although exhibiting many different colors, *Digitalis* flowers are usually tubular, much like the thimble from which their Latin name derives.

Genre Digitale

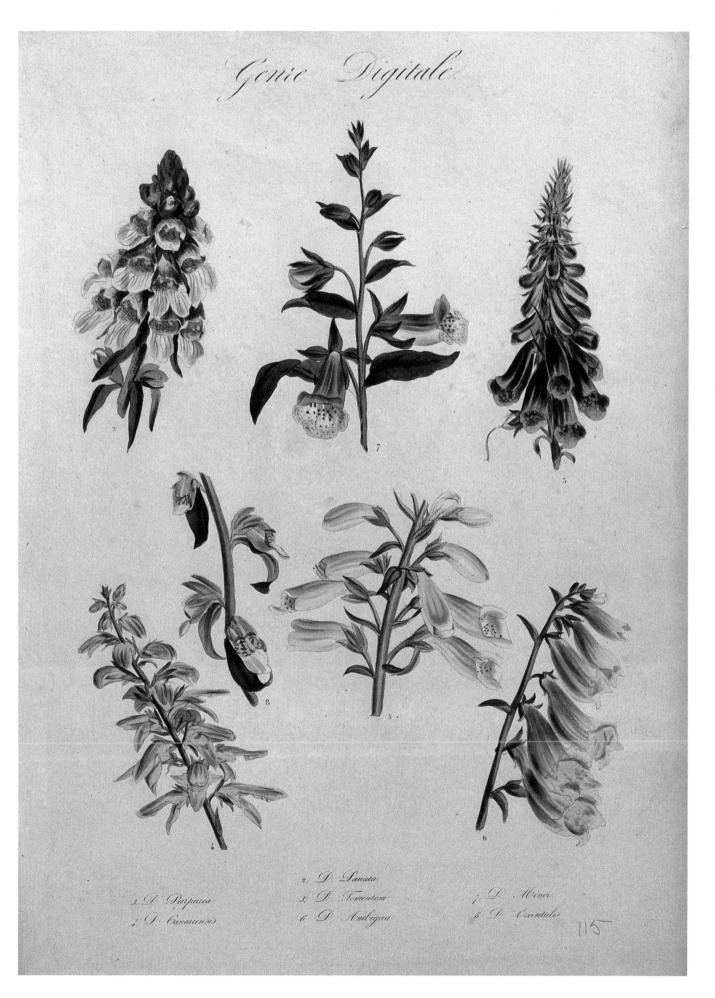

2. D. Lanata

3. D. Purpurea
4. D. Canariensis

5. D. Tomentosa
6. D. Ambigua

7. D. Minor
8. D. Orientalis

115

a-

Used in compound words to denote without or contrary to

abbreviatus

ab-bree-vee-AH-tus

abbreviata, abbreviatum

Shortened; abbreviated, as in *Buddleja abbreviata*

Abelia

uh-BEE-lee-uh

Named after Clarke Abel (1780–1826), British surgeon and naturalist (*Caprifoliaceae*)

Abeliophyllum

uh-BEE-lee-oh-fil-um

Leaves resembling *Abelia*, from Greek *phyllon*, meaning "leaf" (*Oleaceae*)

Abies (also abies)

A-bees

From Latin name for fir tree, published by Pliny the Elder in AD 77 (*Pinaceae*)

—

abietinus

ay-bee-TEE-nus

abietina, abietinum

Like fir tree (*Abies*), as in *Campanula patula* subsp. *abietina*

abortivus

a-bor-TEE-vus

abortiva, abortivum

Incomplete; with parts missing, as in *Oncidium abortivum*

abrotanifolius

ab-ro-tan-ih-FOH-lee-us

abrotanifolia, abrotanifolium

With leaves like southernwood (*Artemisia abrotanum*), as in *Euryops abrotanifolius*

Abutilon

uh-BEW-ti-lon

From Latin, derived from Arabic (*abu-tilun*) for Indian mallow (*Malvaceae*)

abyssinicus

a-biss-IN-ih-kus

abyssinica, abyssinicum

Connected with Abyssinia (Ethiopia), as in *Aponogeton abyssinicus*

Acacia [1]

uh-KAY-sha

From Greek *akantha*, meaning "thorn." Many species bear prominent spines (*Fabaceae*)

Acaena

uh-SEE-nuh

From Greek *akantha*, meaning "thorn," because the fruit is spiny (*Rosaceae*)

Acalypha

uh-KA-lee-fuh

From Greek *akaluphe*, meaning "nettle," to which some species resemble (*Euphorbiaceae*)

acanth-

Used in compound words to denote spiny, spiky, or thorny properties

acanthifolius

uh-kan-thi-FOH-lee-us

acanthifolia, acanthifolium

With leaves like *Acanthus*, as in *Carlina acanthifolia*

Acantholimon

uh-KAN-tho-lim-on

From Greek *akantha*, meaning "thorn," and *Limonium*, a related genus (*Plumbaginaceae*)

Acanthostachys

uh-KAN-tho-stak-is

From Greek *akantha*, meaning "thorn," and *stachys*, an "ear of grain," because the inflorescences are spiny and similar in shape to a cereal seed head (*Bromeliaceae*)

Acacia acinacea

1

Acanthus

uh-KAN-thus

From Greek *akantha*, meaning "thorn,"
because floral bracts and leaves may be spiny
(*Acanthaceae*)

acaulis

a-KAW-lis

acaulis, acaule

Short-stemmed: without a stem, as in
Gentiana acaulis

Acca

AK-uh

From the Peruvian indigenous name *aca,*
as in *A. macrostemma* (*Myrtaceae*)

-aceae

Denoting the rank of family

Acer

AY-sa

From the Latin name for maple tree, derived
from the Latin for sharp or pointed, because
maple wood was used in manufacturing
spears (*Sapindaceae*)

acerifolius

a-ser-ih-FOH-lee-us

acerifolia, acerifolium

With leaves like maple (*Acer*), as in
Quercus acerifolia

acerosus

a-seh-ROH-sus

acerosa, acerosum

Like a needle, as in *Melaleuca acerosa*

acetosella

a-kee-TOE-sell-uh

With slightly sour leaves, as in *Oxalis
acetosella*

Achillea

uh-KEY-lee-ah

Named after the Greek warrior Achilles, who
used the plant to heal wounds (*Asteraceae*)

achilleifolius

ah-key-lee-FOH-lee-us

achilleifolia, achilleifolium

With leaves like common yarrow (*Achillea
millefolium*), as in *Tanacetum achilleifolium*

Achimenes

uh-KIM-en-eez

From Greek *cheimanos*, meaning "tender,"
because plants are not frost hardy, or from
Achaemenes, the Greek mythical ancestor of
a Persian royal house (*Gesneriaceae*)

Achlys

AK-liss

Named after the Greek goddess of mist and
darkness, because plants grow in shade
(*Berberidaceae*)

acicularis

ass-ik-yew-LAH-ris

acicularis, aciculare

Shaped like a needle, as in *Rosa acicularis*

acinaceus

a-sin-AY-see-us

acinacea, acinaceum

In the shape of a curved sword or scimitar,
as in *Acacia acinacea*

Acanthus

The Greek word *akantha*, meaning "thorn" or "thorny,"
is a common component of plant Latin names, as in
Acanthostachys or *Pyracantha*, warning that the plant may
be spiny to touch. *Acanthus* itself can have spiny leaves, as
in *A. spinosus* and *A. sennii*, but some species have softer
foliage (for example, *A. mollis*) and the spines are restricted
to the flower spikes, where spine-tipped bracts enclose
the buds. Consider carefully before planting herbaceous
Acanthus, because their deep roots are difficult to extract
and can resprout if cut, turning unwanted plants into
pernicious weeds.

Acanthus mollis

Acinos
uh-SEE-nos
From Greek *akinos*, used by Dioscorides for a small fragrant plant (*Lamiaceae*)

Aciphylla
a-si-FILL-uh
From Greek *akis*, meaning "point," and *phyllon*, meaning "leaf," because plants have sharply pointed foliage (*Apiaceae*)

Acis
A-kis
Named after the Greek mythological character Acis, killed by a cyclops and transformed into a Sicilian river of the same name (*Amaryllidaceae*)

Acmena
AK-mee-na
From Greek *acmene*, a beautiful wood nymph (*Myrtaceae*)

acmopetala
ak-mo-PET-uh-la
With pointed petals, as in *Fritillaria acmopetala*

Acoelorrhaphe
a-see-lo-RAY-fee
From Greek *a* ("without"), *koilos* ("hollow"), and *rhaphis* ("seam"), because the seed lacks the characteristic ridge or depression commonly found in related palms (*Arecaceae*)

Acokanthera
ah-ko-KAN-ther-ah
From Greek *akoke*, meaning "point," and *anthera*, referring to the pointed anthers (*Apocynaceae*)

aconitifolius
a-kon-eye-tee-FOH-lee-us
aconitifolia, aconitifolium
With leaves like aconite (*Aconitum*), as in *Ranunculus aconitifolius*

Aconitum
ah-kon-EYE-tum
From Greek *akoniton*, used by Theophrastus to refer to a poisonous plant, possibly derived from *akon*, a poison-tipped dart or javelin (*Ranunculaceae*)

Acorus
a-CORE-us
From Greek *coreon*, meaning "pupil," as used to treat ailments of the eye (*Acoraceae*)

Acradenia
ak-ra-DEEN-ee-uh
From Greek *akros*, meaning "tip," and *aden*, meaning "gland," referring to the glands on top of the ovary (*Rutaceae*)

acraeus
ak-ra-EE-us
acraea, acraeum
Dwelling on high ground, as in *Euryops acraeus*

GENUS SPOTLIGHT

Achillea

This plant is named after a hero in Greek mythology, Achilles, who is said to have used an extract of yarrow (*A. millefolium*) to staunch the flow of blood from wounds. The inflorescences in many ways resemble those of the carrot family (*Apiaceae*) in that many flowers cluster together to form an umbrella-like platform that is readily accessed by butterflies, hover flies, and other pollinating insects. However, yarrow is in the daisy family (*Asteraceae*), and each of the small "flowers" is itself an inflorescence, composed of several much smaller florets. By gathering together so many blooms, yarrows produce a long-lasting and nectar-rich display, much loved by gardeners and insects alike.

Achillea millefolium

Actaea

ak-TAY-uh
From the Greek word for elder (*Sambucus*), because the foliage is similar (*Ranunculaceae*)

Actinidia

ak-tin-ID-ee-uh
From Greek *aktin*, meaning "ray," because the floral styles radiate outward from the center of the flower (*Actinidiaceae*)

actinophyllus

ak-ten-oh-FIL-us
actinophylla, actinophyllum
With radiating leaves, as in *Schefflera actinophylla*

acu-

Used in compound words to denote sharply pointed

aculeatus

a-kew-lee-AH-tus
aculeata, aculeatum
Prickly, as in *Polystichum aculeatum*

aculeolatus

a-kew-lee-oh-LAH-tus
aculeolata, aculeolatum
With small prickles, as in *Arabis aculeolata*

acuminatifolius

a-kew-min-at-ih-FOH-lee-us
acuminatifolia, acuminatifolium
With leaves that taper sharply to long narrow points, as in *Polygonatum acuminatifolium*

acuminatus

ah-kew-min-AH-tus
acuminata, acuminatum
Tapering to a long, narrow point, as in *Magnolia acuminata*

acutifolius

a-kew-ti-FOH-lee-us
acutifolia, acutifolium
With leaves that taper quickly to sharp points, as in *Begonia acutifolia*

acutilobus

a-KEW-ti-low-bus
acutiloba, acutilobum
With sharply pointed lobes, as in *Hepatica acutiloba*

acutissimus

ak-yoo-TISS-ee-mus
acutissima, acutissimum
With an acute point, as in *Ligustrum acutissimum*

acutus

a-KEW-tus
acuta, acutum
With a sharp but not tapering point, as in *Cynanchum acutum*

1

Adansonia digitata

ad-

Used in compound words to denote to

Ada

A-duh
Named after Ada, Queen of Caria and adopted mother of Alexander the Great. Could also refer to the Hebrew *adah*, meaning "beauty" (*Orchidaceae*)

Adansonia [1]

ad-an-SOWN-ee-ah
Named after Michel Adanson (1727–1806), French explorer and biologist (*Malvaceae*)

aden-

Used in compound words to denote that a part of the plant has glands

Adenia

ah-DEE-nee-uh
From Arabic *aden*, the vernacular name for the outwardly similar *Adenia venenata* and *Adenium obesum* in Arabia (*Passifloraceae*)

Adenium

ah-DEE-nee-um
From Arabic *aden*, the vernacular name for the outwardly similar *Adenia venenata* and *Adenium obesum* in Arabia (*Apocynaceae*)

Adiantum capillus-veneris

Adenocarpus

ad-en-oh-CARP-us

From Greek *aden* meaning "gland," and *karpos*, meaning "fruit," because the seedpods are glandular (*Fabaceae*)

Adenophora

ah-den-OFF-or-uh

From the Greek *aden*, meaning "gland," and *phoros*, meaning "to bear," because the nectaries are glandular (*Campanulaceae*)

adenophorus

ad-eh-NO-for-us

adenophora, adenophorum

With glands, usually in reference to nectar, as in *Salvia adenophora*

adenophyllus

ad-en-oh-FIL-us

adenophylla, adenophyllum

With sticky (gland-bearing) leaves, as in *Oxalis adenophylla*

adenopodus

a-den-OH-poh-dus

adenopoda, adenopodum

With sticky pedicels (small stalks), as in *Begonia adenopoda*

adiantifolius

ad-ee-an-tee-FOH-lee-us

adiantifolia, adiantifolium

With leaves like maidenhair fern (*Adiantum*), as in *Anemia adiantifolia*

Adiantum [2]

ad-ee-AN-tum

From Greek *adiantos*, meaning "unwettable," because the fronds shed water (*Pteridaceae*)

adlamii

ad-LAM-ee-eye

Named after Richard Wills Adlam (1853–1903), a British collector who supplied plants to London's Royal Botanic Gardens, Kew, in the 1890s

Adlumia

ad-LOOM-ee-uh

Named after John Adlum (1759–1836), American surveyor, judge, and viticulturist (*Papaveraceae*)

admirabilis

ad-mir-AH-bil-is

admirabilis, admirabile

Of note, as in *Drosera admirabilis*

adnatus

ad-NAH-tus

adnata, adnatum

Joined together, as in *Sambucus adnata*

Adonis

ad-OWN-iss

Named after the Greek mythological character Adonis, a lover of the goddess Aphrodite (*Ranunculaceae*)

adpressus

ad-PRESS-us

adpressa, adpressum

Pressed close to; refers to the way hairs (for example) press against a stem, as in *Cotoneaster adpressus*

Adromischus

ad-roh-MISS-kus

From Greek *hadros*, meaning "stout," and *mischos*, meaning "stalk," referring to the thick flower stems (*Crassulaceae*)

adscendens

ad-SEN-denz

Ascending; rising, as in *Aster adscendens*

adsurgens

ad-SER-jenz

Rising upward, as in *Phlox adsurgens*

aduncus

ad-UN-kus

adunca, aduncum

Hooked, as in *Viola adunca*

Aechmea

EK-mee-uh

From Greek *aichme*, meaning "point," because the floral bracts have sharp tips (*Bromeliaceae*)

Aegopodium

ie-go-POH-dee-um

From Greek *aigos*, meaning "goat," and *podion*, meaning "foot," because the leaves are said to resemble goat hooves (*Apiaceae*)

aegyptiacus

eh-jip-tee-AH-kus

aegyptiaca, aegyptiacum

—

aegypticus

eh-JIP-tih-kus

aegyptica, aegypticum

—

aegyptius

eh-JIP-tee-us

aegyptica, aegyptium

Connected with Egypt, as in *Achillea aegyptiaca*

aemulans

EM-yoo-lanz

—

aemulus

EM-yoo-lus

aemula, aemulum

Imitating; rivaling, as in *Scaevola aemula*

Aeonium

ie-OH-nee-um

From Greek *aionios*, meaning "everlasting," because these succulents are evergreen (*Crassulaceae*)

aequalis

ee-KWA-lis

aequalis, aequale

Equal, as in *Phygelius aequalis*

aequinoctialis

eek-wee-nok-tee-AH-lis

aequinoctialis, aequinoctiale

Connected with the equatorial regions, as in *Cydista aequinoctialis*

aequitrilobus

eek-wee-try-LOH-bus

aequitriloba, aequitrilobum

With three equal lobes, as in *Cymbalaria aequitriloba*

Aerangis

air-AN-gis

From Greek *aer*, meaning "air," and *angos*, meaning "vessel," because the flower has a pendulous spur that holds nectar (*Orchidaceae*)

Aerides

air-id-EEZ

From Greek *aer*, meaning "air," referring to the epiphytic habit (*Orchidaceae*)

aerius

ER-re-us

aeria, aerium

From high altitudes, as in *Crocus aerius*

aeruginosus

air-oo-jin-OH-sus

aeruginosa, aeruginosum

The color of rust, as in *Curcuma aeruginosa*

Aeschynanthus

ay-shi-NAN-thus

From Greek *aischyne*, meaning "shame," and *anthos*, meaning "flower," possibly alluding to the shocking red color of the flowers of many species (*Gesneriaceae*)

aesculifolius

es-kew-li-FOH-lee-us

aesculifolia, aesculifolium

With leaves like horse chestnut (*Aesculus*), as in *Rodgersia aesculifolia*

Aesculus [1]

ES-kew-lus

From the Latin name for edible acorn (*Sapindaceae*)

aestivalis

ee-stiv-AH-lis

aestivalis, aestivale

Relating to summer, as in *Vitis aestivalis*

aestivus

EE-stiv-us

aestiva, aestivum

Developing or ripening in the summer months, as in *Leucojum aestivum*

Aethionema

ee-thee-oh-NEEM-uh

From Greek *aethes*, meaning "irregular," or *aitho*, meaning "burned," plus *nema*, meaning "filament," referring to some characteristic of the stamens (*Brassicaceae*)

aethiopicus

ee-thee-OH-pih-kus

aethiopica, aethiopicum

Connected with Africa, as in *Zantedeschia aethiopica*

aethusifolius

e-thu-si-FOH-lee-us

aethusifolia, aethusifolium

With pungent leaves like *Aethusa*, as in *Aruncus aethusifolius*

aetnensis

eet-NEN-sis

aetnensis, aetnense

From Mount Etna, Italy, as in *Genista aetnensis*

aetolicus

eet-OH-lih-kus

aetolica, aetolicum

Connected with Aetolia, Greece, as in *Viola aetolica*

afer

A-fer

afra, afrum

Specifically connected with North African coastal countries, such as Algeria and Tunisia, as in *Lycium afrum*

affinis

uh-FEE-nis

affinis, affine

Related or similar to, as in *Dryopteris affinis*

afghanicus

af-GAN-ih-kus

afghanica, afghanicum

—

afghanistanica

af-gan-is-STAN-ee-ka

Connected with Afghanistan, as in *Corydalis afghanica*

aflatunensis

a-flat-u-NEN-sis

aflatunensis, aflatunense

From Aflatun, Kyrgyzstan, as in *Allium aflatunense*

africanus

af-ri-KAHN-us

africana, africanum

Connected with Africa, as in *Sparrmannia africana*

Agapanthus

a-guh-PAN-thus

From Greek *agape*, meaning "love," and *anthos*, meaning "flower," but the reason for this name is unknown (*Amaryllidaceae*)

Agapetes

ag-ah-PEET-ees

From Greek *agapetos*, meaning "beloved" or "desirable," due to their great beauty (*Ericaceae*)

Agastache

ag-AH-sta-kee

From Greek *agan*, meaning "very much," and *stachys*, an "ear of grain," an allusion to the plentiful flower heads (*Lamiaceae*)

agastus

ag-AS-tus

agasta, agastum

With great charm, as in *Rhododendron* × *agastum*

Agathis

AG-ah-thiss

From Greek *agathis*, meaning "ball of thread," because the seed cones are spherical (*Araucariaceae*)

Agathosma

ag-uth-OZ-ma

From Greek *agathos*, meaning "good," and *osme*, meaning "fragrance" (*Rutaceae*)

Agave

uh-GAH-vee

From Greek *agavos*, meaning "great," in reference to the impressive inflorescence of many species (*Asparagaceae*)

agavoides

ah-gav-OY-deez

Resembling *Agave*, as in *Echeveria agavoides*

ageratifolius

ad-jur-rat-ih-FOH-lee-us

ageratifolia, ageratifolium

With leaves like *Ageratum*, as in *Achillea ageratifolia*

Ageratina

aj-err-ah-TEE-na

Resembling the related genus *Ageratum* (*Asteraceae*)

ageratoides

ad-jur-rat-OY-deez

Resembling *Ageratum*, as in *Aster ageratoides*

Ageratum

aj-err-AH-toom

From Greek *ageraton*, meaning "not growing old," in reference to the long-lasting flowers (*Asteraceae*)

aggregatus

ag-gre-GAH-tus

aggregata, aggregatum

Denotes aggregate flowers or fruits, such as raspberry or strawberry, as in *Eucalyptus aggregata*

Aglaomorpha

uh-glay-oh-MOR-fa

From Greek *aglaos*, meaning "beautiful," and *morpha*, meaning "shape" or "form" (*Polypodiaceae*)

Aglaonema

uh-glay-oh NEE-ma

From Greek *aglaos*, meaning "beautiful," and *nema*, meaning "filament," possibly referring to the stamens, although these are typically small in this genus (*Araceae*)

agnus-castus

AG-nus KAS-tus

From *agnos*, the Greek name for *Vitex agnus-castus*, and *castus*, "chaste," as in *Vitex agnus-castus*

Agonis

ah-GO-niss

From Greek *agon*, meaning a "gathering," in reference to the clustered flowers or copious seed (*Myrtaceae*)

Agoseris

ag-OSS-er-iss

From Greek *aigos*, meaning "goat," and *seris*, meaning "lettuce" (*Asteraceae*)

agrarius

ag-RAH-ree-us

agraria, agrarium

From fields and cultivated land, as in *Fumaria agraria*

Aesculus glabra

1

agrestis

ag-RES-tis

agrestis, agreste

Found growing in fields, as in *Fritillaria agrestis*

agrifolius

ag-rih-FOH-lee-us

agrifolia, agrifolium

With leaves with a rough or scabby texture, as in *Quercus agrifolia*

Agrimonia

ah-gree-MOH-nee-ah

From Greek *argemone*, meaning "poppy," although several alternative meanings are also possible, because this plant does not closely resemble a poppy (*Rosaceae*)

agrippinum

ag-rip-EE-num

Named after Agrippina, mother of the Roman emperor Nero, as in *Colchicum agrippinum*

Agrostemma [1]

ah-gro-STEM-uh

From Greek *agros*, meaning "field," and *stemma*, meaning "wreath" (*Caryophyllaceae*)

Agrostis

ah-GROS-tis

The Latin and Greek name for "grass," "weed," or "couch grass" (*Poaceae*)

Aichryson

ie-CRY-son

From Greek *aei*, meaning "always," and *khrysos*, meaning "gold," because the flowers are yellow (*Crassulaceae*)

Ailanthus [2]

ay-LAN-thoos

From Ambonese (Indonesia) *ailanto*, meaning "reach for the sky" or "tree of heaven" (*Simaroubaceae*)

aitchisonii

EYE-chi-soh-nee-eye

Named after Dr. J. E. T. Aitchison (1836–98), a British doctor and botanist who collected plant material in Asia, as in *Corydalis aitchisonii*

aizoides

ay-ZOY-deez

Like the genus *Aizoon*, as in *Saxifraga aizoides*

ajacis

a-JAY-sis

A species name that honors the Greek hero Ajax, as in *Consolida ajacis*

ajanensis

ah-yah-NEN-sis

ajanensis, ajanense

From Ajan on the Siberian coast, as in *Dryas ajanensis*

1

Agrostemma githago

2

Ailanthus altissima

Ajania

ah-JAH-nee-uh

Named for the city of Ajan (now Ayan) on the Pacific coast of Russia (*Asteraceae*)

Ajuga [3, 4]

ah-JOO-ga

Origins uncertain, although could derive from Greek *a*, meaning "without," and *iugum*, a "yoke," because the calyx is undivided (*Lamiaceae*)

Akebia

ah-KEE-bee-uh

From Japanese *akebi*, the vernacular name for *A. quinata* (*Lardizabalaceae*)

alabamensis

al-uh-bam-EN-sis

alabamensis, alabamense

—

alabamicus

al-a-BAM-ih-kus

alabamica, alabamicum

From or of Alabama State, as in *Rhododendron alabamense*

Alangium

al-ANJ-ee-um

From Malayalam *alangi*, the vernacular name used in Kerala, India, for *A. salviifolium* (*Cornaceae*)

alaternus

a-la-TER-nus

The Roman name for *Rhamnus alaternus*

alatus

a-LAH-tus

alata, alatum

Winged, as in *Euonymus alatus*

albanensis

al-ba-NEN-sis

albanensis, albanense

From St Albans, Hertfordshire, England, as in *Coelogyne* × *albanense*

alberti

al-BER-tee

—

albertianus

al-ber-tee-AH-nus

albertiana, albertianum

—

albertii

al-BER-tee-eye

Named after various people called Albert, such as Albert von Regel (1845–1908), plant collector, as in *Tulipa albertii*

albescens

al-BES-enz

Becoming white, as in *Kniphofia albescens*

3

Ajuga reptans

4

Ajuga genevensis

Iris albicans

Alcea rosea

Alchemilla fissa

albicans [1]
AL-bih-kanz
Off-white, as in *Hebe albicans*

albicaulis
al-bih-KAW-lis
albicaulis, albicaule
With white stems, as in *Lupinus albicaulis*

albidus
AL-bi-dus
albida, albidum
White, as in *Trillium albidum*

albiflorus
al-BIH-flor-us
albiflora, albiflorum
With white flowers, as in *Buddleja albiflora*

albifrons
AL-by-fronz
With white fronds, as in *Cyathea albifrons*

Albizia
al-BITZ-ee-uh
Named after Filippo degli Albizzi
(eighteenth century, dates uncertain), Italian
naturalist and nobleman (*Fabaceae*)

albomaculatus
al-boh-mak-yoo-LAH-tus
albomaculata, albomaculatum
With white spots, as in *Asarum
albomaculatum*

albomarginatus
AL-bow-mar-gin-AH-tus
albomarginata, albomarginatum
With white margins, as in *Agave
albomarginata*

albopictus
al-boh-PIK-tus
albopicta, albopictum
With white hairs, as in *Begonia albopicta*

albosinensis
al-bo-sy-NEN-sis
albosinensis, albosinense
Meaning white and from China, as in *Betula
albosinensis*

albovariegatus
al-bo-var-ee-GAH-tus
albovariegata, albovariegatum
Variegated with white, as in *Holcus mollis*
'Albovariegatus'

Albuca
all-BOO-ka
From Latin *albicans*, meaning "becoming
white," in reference to the flower color,
although most species have yellow flowers
(*Asparagaceae*)

albulus
ALB-yoo-lus
albula, albulum
Whitish in color, as in *Carex albula*

albus
AL-bus
alba, album
White, as in *Veratrum album*

Alcea [2]
al-SEE-uh
From Greek *alkea*, meaning "mallow"
(*Malvaceae*)

Alchemilla [3]
all-ke-MILL-uh
From Arabic *al-kemelih*, meaning "alchemy,"
because the hairy leaves repel water in an
almost miraculous fashion (*Rosaceae*)

alcicornis
al-kee-KOR-nis
alcicornis, alcicorne
Palmate leaves that resemble the horns of
the North American moose (European elk),
as in *Platycerium alcicorne*

aleppensis
a-le-PEN-sis
aleppensis aleppense
—

aleppicus
a-LEP-ih-kus
aleppica, aleppicum
From Aleppo, Syria, as in *Adonis aleppica*

aleuticus

a-LEW-tih-kus

aleutica, aleuticum

Connected with the Aleutian Islands, Alaska, as in *Adiantum aleuticum*

alexandrae

al-ex-AN-dry

Named after Queen Alexandra (1844–1925), wife of Edward VII of England, as in *Archontophoenix alexandrae*

alexandrinus

al-ex-an-DREE-nus

alexandrina, alexandrinum

Connected with Alexandria, Egypt, as in *Senna alexandrina*

algeriensis

al-jir-ee-EN-sis

algeriensis, algeriense

From Algeria, as in *Ornithogalum algeriense*

algidus

AL-gee-dus

algida, algidum

Cold; of high mountain regions, as in *Olearia algida*

alienus

a-LY-en-us

aliena, alienum

A plant of foreign origin, as in *Heterolepis aliena*

Alisma

ah-LIZ-mah

From the ancient Greek name for a water plant (*Alismataceae*)

Alkanna

al-KA-na

From Arabic *al-hinna*, meaning "henna," because the plant produces a dye (*Boraginaceae*)

alkekengi

al-KEK-en-jee

From the Arabic for bladder cherry, as in *Physalis alkekengi*

Allamanda [4]

al-ah-MAN-duh

Named after Jean Frédéric-Louis Allamand (1736–1803), Swiss botanist (*Apocynaceae*)

alleghaniensis

al-leh-gay-nee-EN-sis

alleghaniensis, alleghaniense

From the Allegheny Mountains, as in *Betula alleghaniensis*

alliaceus

al-lee-AY-see-us

alliacea, alliaceum

Like *Allium* (onion or garlic), as in *Tulbaghia alliacea*

Alliaria

al-ee-AIR-ee-uh

From Latin *allium*, meaning "garlic," due to the scent of crushed leaves (*Brassicaceae*)

alliariifolius

al-ee-ar-ee-FOH-lee-us

alliariifolia, alliariifolium

With leaves like *Alliaria*, as in *Valeriana alliariifolia*

allionii

al-ee-OH-nee-eye

Named after Carlo Allioni (1728–1804), Italian botanist, as in *Primula allionii*

Allium [5]

al-EE-um

From Latin *allium*, meaning "garlic," as in *A. sativum* (*Amaryllidaceae*)

Alluaudia

al-AWED-ee-uh

Named after François Alluaud (1778–1866), French scientist and author (*Didiereaceae*)

alnifolius

al-nee-FOH-lee-us

alnifolia, alnifolium

With leaves like alder (*Alnus*), as in *Sorbus alnifolia*

Alnus [6] (also alnus)

ALL-noos

From the Latin for alder tree (*Betulaceae*)

4

Allamanda cathartica

5

Allium caeruleum

6

Alnus incana

Alpinia nutans

Alocasia
al-oh-KAY-see-uh
From Greek *a*, meaning "without," and *Colocasia*, a related genus (*Araceae*)

Aloe
A-loh-ee
From Arabic *alloeh*, referring to the bitter juice extracted from the leaves (*Asphodelaceae*)

aloides
al-OY-deez
Resembling *Aloe*, as in *Lachenalia aloides*

aloifolius
al-oh-ih-FOH-lee-us
aloifolia, aloifolium
With leaves like *Aloe*, as in *Yucca aloifolia*

Aloinopsis
a-loh-in-OP-sis
Resembling the genus *Aloe* (*Aizoaceae*)

Alonsoa
al-ON-so-uh
Named after Zenón de Alonso Acosta (dates unknown), Spanish soldier in Bogotá, Colombia (*Scrophulariaceae*)

alopecuroides
al-oh-pek-yur-OY-deez
Like the genus *Alopecurus* (foxtail), as in *Pennisetum alopecuroides*

Alopecurus
al-oh-pek-URE-us
From Greek *alopex*, meaning "fox," and *oura*, meaning "tail," referring to the shape of the inflorescence (*Poaceae*)

Aloysia
al-OY-see-uh
Named after Maria Luisa of Parma (1751–1819), wife of King Charles IV of Spain (*Verbenaceae*)

alpestris
al-PES-tris
alpestris, alpestre
Of lower, usually wooded, mountain habitats, such as in *Narcissus alpestris*

alpicola
al-PIH-koh-luh
Of high mountain habitats, as in *Primula alpicola*

alpigenus
AL-pi-GEE-nus
alpigena, alpigenum
Of a mountainous region, as in *Saxifraga alpigena*

Alpinia
al-PIN-ee-uh
Named after Prospero Alpini (1553–1617), Italian physician and botanist (*Zingiberaceae*)

alpinus
al-PEE-nus
alpina, alpinum
Of high, often rocky regions; from the Alps region of Europe, as in *Pulsatilla alpina*

Alsobia
al-SOH-bee-uh
From Greek *alsos*, meaning "forest," and *bios*, meaning "life," alluding to the epiphytic habit (*Gesneriaceae*)

Alstroemeria
al-stro-MEE-ree-uh
Named after Clas Alströmer (1736–94), Swedish nobleman (*Alstroemeriaceae*)

altaclerensis
al-ta-cler-EN-sis
altaclerensis, altaclerense
From Highclere Castle, Hampshire, England, as in *Ilex × altaclerensis*

altaicus
al-TAY-ih-kus
altaica, altaicum
Connected with the Altai Mountains, Central Asia, as in *Tulipa altaica*

alternans
al-TER-nans
Alternating, as in *Chamaedorea alternans*

Alternanthera
awl-ter-NAN-thu-rah
From Latin *alternans* and *anthera*, because the anthers alternate with staminodes in the flowers (*Amaranthaceae*)

alternifolius
al-tern-ee-FOH-lee-us
alternifolia, alternifolium
With leaves that grow from alternating points of a stem instead of opposite each other, as in *Buddleja alternifolia*

Althaea
al-THEE-uh
From Greek *althos*, meaning "healing"; also Althaea, Queen of Calydon in Greek mythology (*Malvaceae*)

althaeoides
al-thay-OY-deez
Resembling hollyhock (formerly *Althaea*), as in *Convolvulus althaeoides*

altissimus
al-TISS-ih-mus
altissima, altissimum
Very tall; the tallest, as in *Ailanthus altissima*

altus
AHL-tus
alta, altum
Tall, as in *Sempervivum altum*

MIXING IT UP

When creating a new genus, the usual route is to choose a name that describes the plant, or failing that, name it after someone. But with so many names already in use, it can be hard to come up with something new. A novel approach is to create an anagram of an existing genus, as in *Tellima* (from *Mitella*), *Docynia* (from *Cydonia*), *Leymus* (from *Elymus*), and *Saruma* (from *Asarum*). Using similar names not only makes them easier to remember, but also helps to reinforce the familial relationship between the pair.

Alyogyne
al-ee-oh-GY-nee
From Greek *alytos*, meaning "undivided," and *gyne*, meaning "female," because the styles are undivided (*Malvaceae*)

Alyssum
A-liss-um
From Greek *a*, meaning "without," and *lyssa*, meaning "rabies" or "madness," a medicinal herb (*Brassicaceae*)

amabilis
am-AH-bih-lis
amabilis, amabile
Lovely, as in *Cynoglossum amabile*

amanus
a-MAH-nus
amana, amanum
Of the Amanus Mountains, Turkey, as in *Origanum amanum*

amaranthoides
am-ar-anth-OY-deez
Resembling amaranth (*Amaranthus*), as in *Calomeria amaranthoides*

Amaranthus
am-uh-RANTH-us
From Greek *amarantos*, meaning "unfading," because the flowers are long-lasting (*Amaranthaceae*)

amarellus
a-mar-ELL-us
amarella, amarellum
—

amarus
a-MAH-rus
amara, amarum
Bitter, as in *Ribes amarum*

amaricaulis
am-ar-ee-KAW-lis
amaricaulis, amaricaule
With a bitter-tasting stem, as in *Hyophorbe amaricaulis*

Amaryllis
am-uh-RIL-us
From Greek *amarysso*, meaning "to sparkle," the name of a shepherdess in Virgil's *Eclogues* (*Amaryllidaceae*)

amazonicus
am-uh-ZOH-nih-kus
amazonica, amazonicum
Connected with the Amazon River, South America, as in *Victoria amazonica*

ambi-
Used in compound words to denote around

ambiguus
am-big-YOO-us
ambigua, ambiguum
Uncertain or doubtful, as in *Digitalis ambigua*

amblyanthus
am-blee-AN-thus
amblyantha, amblyanthum
With a blunt flower, as in *Indigofera amblyantha*

Ambrosia
am-BRO-zee-uh
From Greek *ambrosia*, meaning "food of the gods," although the allusion to this often weedy genus is unclear (*Asteraceae*)

GENUS SPOTLIGHT

Alyssum

For many gardeners, alyssum is a white-flowered bedding plant, often grown together with blue lobelia in containers and park bedding displays. However, that plant is now classified as *Lobularia maritima*; true *Alyssum* species are often known as "madworts," because they are supposedly endowed with medicinal properties. In particular, they are said to fend off rabies, the Latin name *Alyssum* means "without madness." Another popular alyssum is no longer in *Alyssum* (*A. saxatile* =*Aurinia saxatilis*), but several species remain and they are often useful for growing in rock or crevice gardens.

Aurinia (formerly *Alyssum*) *saxatilis*

ambrosioides

am-bro-zhee-OY-deez

Resembling *Ambrosia*, as in *Cephalaria ambrosioides*

Amelanchier

am-uh-LANG-kee-uh

From Provençal (France) *amalenquièr*, the vernacular name for *A. ovalis* (*Rosaceae*)

amelloides

am-el-OY-deez

Resembling *Aster amellus* (from its Roman name), as in *Felicia amelloides*

americanus

a-mer-ih-KAH-nus

americana, americanum

Connected with North or South America, as in *Lysichiton americanus*

amesianus

ame-see-AH-nus

amesiana, amesianum

Named after Frederick Lothrop Ames (1835–93), horticulturist and orchid grower, and Oakes Ames (1874–1950), supervisor of the Arnold Arboretum and professor of botany, Harvard, Massachusetts, as in *Cirrhopetalum amesianum*

amethystinus

am-eth-ih-STEE-nus

amethystina, amethystinum

Violet, as in *Brimeura amethystina*

Amherstia [1]

am-HURR-stee-ah

Named after Sarah, Countess Amherst (1762–1838), British naturalist and botanist (*Fabaceae*)

Amicia

am-ISS-ee-uh

Named after Giovanni Battista Amici (1786–1863), Italian astronomer and botanist (*Fabaceae*)

Ammi

AM-ee

Ancient name for a related plant, possibly derived from Greek *ammos*, meaning "sand," an allusion to habitat (*Apiaceae*)

Ammobium

am-MOW-bee-um

From Greek *ammo*, meaning "sand," and *bio*, meaning "to live"; plants prefer a sandy habitat (*Asteraceae*)

1

Amherstia nobilis

Ammophila

ah-MOFF-il-uh

From Greek *ammo*, meaning "sand," and *philos*, meaning "loving"; these are dune grasses (*Poaceae*)

ammophilus

am-oh-FIL-us

ammophila, ammophilum

Of sandy places, as in *Oenothera ammophila*

amoenus

am-oh-EN-us

amoena, amoenum

Pleasant; delightful, as in *Lilium amoenum*

Amomyrtus

AM-oh-mer-tus

A combination of the names of two related genera, *Amomis* (*Pimenta*) and *Myrtus* (*Myrtaceae*)

Amorpha

a-MOR-fuh

From Greek *amorphos*, meaning "shapeless"; each flower has only a single petal (*Fabaceae*)

Amorphophallus

ah-mor-foh-FA-loos

From Greek *amorphos*, meaning "shapeless" or "misshapen," and *phallos*, meaning "penis," referring to the shape of the floral spadix (*Araceae*)

Ampelopsis

am-pel-OP-sis

From Greek *ampelos*, meaning "a vine," and *opsis*, meaning "resembling" (*Vitaceae*)

amphibius

am-FIB-ee-us

amphibia, amphibium

Growing both on land and in water, as in *Persicaria amphibia*

amplexicaulis

am-pleks-ih-KAW-lis

amplexicaulis, amplexicaule

Clasping the stem, as in *Persicaria amplexicaulis*

amplexifolius

am-pleks-ih-FOH-lee-us

amplexifolia, amplexifolium

Clasping the leaf, as in *Streptopus amplexifolius*

ampliatus

am-pli-AH-tus

ampliata, ampliatum

Enlarged, as in *Oncidium ampliatum*

amplissimus

am-PLIS-ih-mus

amplissima, amplissimum

Very large, as in *Chelonistele amplissima*

amplus

AMP-lus

ampla, amplum

Large, as in *Epigeneium amplum*

Amsonia

am-SOW-nee-uh

Named after John Amson (1698–1765?), British physician, botanist, and alderman in Colonial Virginia (*Apocynaceae*)

amurensis

am-or-EN-sis

amurensis, amurense

From the Amur River region in Asia, as in *Sorbus amurensis*

amygdaliformis

am-mig-dal-ih-FOR-mis

amygdaliformis, amygdaliforme

Shaped like an almond, as in *Pyrus amygdaliformis*

amygdalinus

am-mig-duh-LEE-nus

amygdalina, amygdalinum

Relating to the almond, as in *Eucalyptus amygdalina*

amygdaloides

am-ig-duh-LOY-deez

Resembling the almond, as in *Euphorbia amygdaloides*

Anacampseros

ah-nah-KAMP-sir-os

From Greek *anakampseros*, the name for a herb that restores passion (*Anacampserotaceae*)

Anacardium

ah-nah-KAR-dee-um

From Greek *ana*, meaning "upward," and *kardia*, or "heart," referring to the cashew seed, which is not at the heart of the fruit but on its tip (*Anacardiaceae*)

Anacyclus

ah-nah-SIKE-loos

From Greek *a* ("without"), *anthos* ("flower"), and *kyklos* ("ring"), because the outer flowers in the capitulae lack petals in the type species (*Asteraceae*)

Anagallis [1]

an-uh-GAH-lis

From Greek *ana*, meaning "again," and *agallo* "to bloom," because the flowers open and close depending on the weather (*Primulaceae*)

Ananas [2]

ah-NAN-oos

From the indigenous South American name for the pineapple (*Bromeliaceae*)

1

Anagallis arvensis

2

Ananas comosus

ananassa
a-NAN-ass-uh

—

ananassae
a-NAN-ass-uh-ee

With a fragrance like pineapple, as in *Fragaria × ananassa*

Anaphalis
ah-NA-fah-lus

From the classical Greek name for this or a closely related plant (*Asteraceae*)

anatolicus
an-ah-TOH-lih-kus

anatolica, anatolicum

Connected with Anatolia, Turkey, as in *Muscari anatolicum*

anceps
AN-seps

Two-sided, ambiguous, as in *Laelia anceps*

3

Anemone obtusiloba

Anchusa
an-CHOO-sah

From Greek *ankousa*, "cosmetic paint," because the roots contain dye (*Boraginaceae*)

ancyrensis
an-syr-EN-sis

ancyrensis, ancyrense

From Ankara, Turkey, as in *Crocus ancyrensis*

andersonianus
an-der-soh-nee-AH-nus

andersoniana, andersonianum

—

andersonii
an-der-SON-ee-eye

Named after Dr. Charles Lewis Anderson (1827–1910), American botanist, as in *Arctostaphylos andersonii*

andicola
an-DIH-koh-luh

—

andinus
an-DEE-nus

andina, andinum

Connected with the Andes, South America, as in *Calceolaria andina*

andrachne
an-DRAK-nee

—

andrachnoides
an-drak-NOY-deez

From Greek *andrachne* (strawberry tree), as in *Arbutus × andrachnoides*

andraeanus
an-dree-AH-nus

andraeana, andraeanum

—

andreanus
an-dree-AH-nus

andreana, andreanum

Named after Édouard François André (1840–1911), French explorer, as in *Anthurium andraeanum*

androgynus
an-DROG-in-us

androgyna, androgynum

With separate male and female flowers growing on the same spike, as in *Semele androgyna*

Andromeda
an-DRAW-mee-dah

From Greek mythology, Andromeda was the daughter of Aethiopian king Cepheus and his wife Cassiopeia, rescued from a sea monster by Perseus (*Ericaceae*)

Andropogon
an-DRO-poh-gon

From Greek *andros*, meaning "man," and *pogon*, meaning "beard," to which the flower clusters resemble (*Poaceae*)

Androsace
an-DROS-uh-see

From Greek *andros* ("man") and *sakos* ("shield"), referring to the shape of the leaves or stamens (*Primulaceae*)

androsaemifolius
an-dro-say-MEE-fol-ee-us

androsaemifolia, androsaemifolium

With leaves like *Androsaemum*, as in *Apocynum androsaemifolium* (*Androsaemum* is now listed under *Hypericum*)

androsaemus
an-dro-SAY-mus

androsaema, androsaemum

With sap the color of blood, as in *Hypericum androsaemum*

Anemanthele
ah-nem-AN-thuh-lee

From Greek *anemos*, meaning "wind," and *anthele*, meaning "plume," from the vernacular name wind plume-grass (*Poaceae*)

Anemone [3]
ah-NEM-oh-nee

From Greek *anemos*, meaning "wind," because the seed is blown in the wind; also, in Greek mythology, Aphrodite's lover, Adonis, was killed by a boar and scarlet anemones sprang up from each drop of blood shed, although the flowers may have been from the related genus *Adonis* (*Ranunculaceae*)

Anemonopsis
ah-nem-uh-NOP-sis

Resembling the related genus *Anemone* (*Ranunculaceae*)

Anemopaegma
ah-nem-oh-PEG-mah

From Greek *anemos*, meaning "wind," and *paigma*, meaning "play"; these climbers are said to play in the wind (*Bignoniaceae*)

Anemopsis
ah-nem-OP-sis
Resembling the genus *Anemone* (*Saururaceae*)

Anethum
ah-NEE-thum
From Greek *anethon*, meaning "anise" or "dill" (*Apiaceae*)

Angelica
an-JEL-ih-kuh
From Greek *angelos*, a "messanger" or "angel," referring to its medicinal properties (*Apiaceae*)

Angelonia
an-je-LOH-nee-uh
From the indigenous South American name (*Plantaginaceae*)

anglicus
AN-glih-kus
anglica, anglicum
Connected with England, as in *Sedum anglicum*

Angophora
an-go-FOUR-uh
From Greek *angos*, meaning a "vessel," and *phero*, meaning "to bear," because the vessel-like fruit bears seed (*Myrtaceae*)

Angraecum
an-GREE-koom
From Malay *angrek*, the name of a similar epiphytic orchid (*Orchidaceae*)

angularis
ang-yoo-LAH-ris
angularis, angulare
—

angulatus
ang-yoo-LAH-tus
angulata, angulatum
Angular in shape or form, as in *Jasminum angulare*

Anguloa
an-goo-LOW-uh
Named after Francisco de Angulo (?–1815), Spanish director-general of mines and botany student (*Orchidaceae*)

angulosus
an-gew-LOH-sus
angulosa, angulosum
With several corners or angles, as in *Bupleurum angulosum*

angustatus
an-gus-TAH-tus
angustata, angustatum
Narrow, as in *Arisaema angustatum*

angustifolius
an-gus-tee-FOH-lee-us
angustifolia, angustifolium
With narrow leaves, as in *Pulmonaria angustifolia*

angustus
an-GUS-tus
angusta, angustum
Narrow, as in *Rhodiola angusta*

Anigozanthos
an-ig-oh-ZAN-thoos
From Greek *anisos*, meaning "unequal" and *anthos*, meaning "flower," because the petals are of differing sizes (*Haemodoraceae*)

anisatus
an-ee-SAH-tus
anisata, anisatum
—

Anisodontea
an-iss-oh-DON-tee-uh
From Greek *anisos*, meaning "unequal," and *odontos*, meaning "tooth," because the leaf edges are irregularly toothed (*Malvaceae*)

anisodorus
an-ee-so-DOR-us
anisodora, anisodorum
With the scent of anise (*Pimpinella anisum*), as in *Illicium anisatum*

anisophyllus
an-ee-so-FIL-us
anisophylla, anisophyllum
With leaves of unequal size, as in *Strobilanthes anisophylla*

Anisotome
an-iss-oh-TOE-mee
From Greek *anisos*, meaning "unequal," and *tome*, meaning "division," because the leaves are irregularly divided (*Apiaceae*)

annamensis
an-a-MEN-sis
annamensis, annamense
From Annam, Asia (part of modern-day Vietnam), as in *Viburnum annamensis*

Annona
ah-NOH-na
From Taíno *annon*, a Caribbean indigenous name, or from Latin *annona*, meaning "yearly produce" (*Annonaceae*)

annulatus
an-yoo-LAH-tus
annulata, annulatum
With rings, as in *Begonia annulata*

annuus
AN-yoo-us
annua, annuum
Annual, as in *Helianthus annuus*

anomalus
ah-NOM-uh-lus
anomala, anomalum
Unlike the norm found in a genus, as in *Hydrangea anomala*

Anopterus
ah-NOP-ter-oos
From Greek *ano*, meaning "upward," and *pteron*, meaning "wing," because the seed has wings at its tips (*Escalloniaceae*)

anosmus
an-OS-mus
anosma, anosmum
Without scent, as in *Dendrobium anosmum*

Anredera
an-RED-er-uh
Possibly from Spanish *enredadera*, meaning "climbing plant" (*Basellaceae*)

antarcticus
ant-ARK-tih-kus
antarctica, antarcticum
Connected with the Antarctic region, as in *Dicksonia antarctica*

Antennaria
an-ten-AIR-ree-uh
From Latin *antenna*, because the pappus bristles resemble insect antennae (*Asteraceae*)

Anthemis
AN-tha-miss
From Greek *anthemis*, meaning "flower" (*Asteraceae*)

anthemoides
an-them-OY-deez
Resembling chamomile (Greek *anthemis*), as in *Rhodanthe anthemoides*

Chamaenerion angustifolium

Anthericum

an-ther-i-koom

From Greek *antherikos*, a name for the flowering stem of asphodel (*Asparagaceae*)

Anthoxanthum

an-thox-ANTH-um

From Greek *anthos*, meaning "flower," and *xanthos*, meaning "yellow" (*Poaceae*)

Anthriscus

an-THRIS-koos

From Greek *anthriskon*, the classical name for a similar plant (*Apiaceae*)

Anthurium [1]

an-THOO-ree-um

From Greek *anthos*, meaning "flower," and *oura*, meaning "tail," because the inflorescence has a tail-like spadix (*Araceae*)

Anthyllis

an-THIL-us

From Greek *anthyllis*, the classical name for a similar plant (*Fabaceae*)

Antigonon

an-tee-GOH-non

From Greek *anti,* meaning "against," and *gonia*, meaning "knee," in reference to the flexible stems; alternatively, could combine Greek *anti* with related genus *Polygonum* (*Polygonaceae*)

antipodus

an-te-PO-dus

antipoda, antipodum

—

antipodeum

an-te-PO-dee-um

Connected with the Antipodes, as in *Gaultheria antipoda*

antiquorum

an-ti-KWOR-um

Of the ancients, as in *Helleborus antiquorum*

antiquus

an-TIK-yoo-us

antiqua, antiquum

Ancient; antique, as in *Asplenium antiquum*

antirrhiniflorus

an-tee-rin-IF-lor-us

antirrhiniflora, antirrhiniflorum

With flowers like snapdragon (*Antirrhinum*), as in *Maurandella antirrhiniflora*

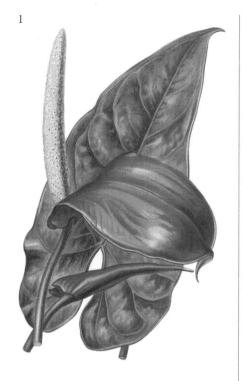

Anthurium × crombezianum

antirrhinoides

an-tee-ry-NOY-deez

Resembling snapdragon (*Antirrhinum*), as in *Keckiella antirrhinoides*

Antirrhinum [2]

an-tee-RYE-num

From Greek *antirrhinon*, meaning "calf's snout," perhaps referring to the shape of the petals or fruit (*Plantaginaceae*)

apenninus

ap-en-NEE-nus

apennina, apenninum

Connected with the Apennine Mountains, Italy, as in *Anemone apennina*

apertus

AP-ert-us

aperta, apertum

Open; exposed, as in *Nomocharis aperta*

apetalus

a-PET-uh-lus

apetala, apetalum

Without petals, as in *Sagina apetala*

Aphanes

A-fan-ees

From Greek *aphanes,* meaning "inconspicuous," as is this plant (*Rosaceae*)

Antirrhinum majus

Aphelandra

aff-ell-AND-rah

From Greek *apheles*, meaning "simple," and *andros*, meaning "male," because the anthers are only one-celled (*Acanthaceae*)

aphyllus

a-FIL-us

aphylla, aphyllum

Having, or appearing to have, no leaves, as in *Asparagus aphyllus*

apiculatus

uh-pik-yoo-LAH-tus

apiculata, apiculatum

With a short, sharp point, as in *Luma apiculata*

apiferus

a-PIH-fer-us

apifera, apiferum

Bearing bees, as in *Ophrys apifera*

apiifolius

ap-ee-FOH-lee-us

apiifolia, apiifolium

With leaves like celery (*Apium*), as in *Clematis apiifolia*

Apios

A-pee-oss

From Greek *apios*, meaning "pear," referring to the shape of the tubers (*Fabaceae*)

Apium
A-pee-um
From Latin *apium*, the classical name for celery (*Apiaceae*)

Apocynum
ah-poh-SINE-um
From Greek *apokynon*, meaning "away from dogs," in reference to its toxicity; commonly known as "dogbane" (*Apocynaceae*)

apodus
a-POH-dus
apoda, apodum
Without stalks, as in *Selaginella apoda*

Aponogeton
ap-oh-no-GEE-ton
Combines the Latin name of a healing spring *Aquae Aponi* with the Greek *geiton*, meaning "neighbor," a reference to its aquatic habitat (*Aponogetonaceae*)

appendiculatus
ap-pen-dik-yoo-LAH-tus
appendiculata, appendiculatum
With appendages such as hairs, as in *Caltha appendiculata*

applanatus
ap-PLAN-a-tus
applanata, applanatum
Flattened, as in *Sanguisorba applanata*

appressus
a-PRESS-us
appressa, appressum
Pressed close against, as in *Carex appressa*

apricus
AP-rih-kus
aprica, apricum
Open to the sun, as in *Silene aprica*

Aptenia
ap-TEE-nee-uh
From Greek *a*, meaning "without," and *pteron*, meaning "wing," because the fruit lacks the wings characteristic of some related genera (*Aizoaceae*)

apterus
AP-ter-us
aptera, apterum
Without wings, as in *Odontoglossum apterum*

aquaticus
a-KWA-tih-kus
aquatica, aquaticum
—

aquatalis
ak-wa-TIL-is
aquatalis, aquatale
Growing in or near water, as in *Mentha aquatica*

aquifolius
a-kwee-FOH-lee-us
aquifolia, aquifolium
Holly-leaved (from the Latin name for holly, *aquifolium*), as in *Mahonia aquifolium*

Aquilegia [3]
ak-wi-LEE-juh
From Latin *aquila*, meaning "eagle," in reference to the clawed petals; the common name "columbine" is from the Latin *columba*, meaning "dove," because each flower resembles five doves clustered together head to head (*Ranunculaceae*)

aquilegiifolius
ak-wil-egg-ee-FOH-lee-us
aquilegiifolia, aquilegiifolium
With leaves like columbine (*Aquilegia*), as in *Thalictrum aquilegiifolium*

aquilinus
ak-will-LEE-nus
aquilina, aquilinum
Like an eagle; aquiline, as in *Pteridium aquilinum*

arabicus
a-RAB-ih-kus
arabica, arabicum
Connected with Arabia, as in *Coffea arabica*

Arabidopsis
ar-ab-ih-DOP-sis
Resembling the related genus *Arabis* (*Brassicaeae*)

Aquilegia alpina

1

2

Arachis hypogaea

Araucaria araucana

Arabis
A-ruh-bis
From Greek *arabis*, meaning "Arabian" (*Brassicaceae*)

Arachis [1]
ah-RAH-kis
From Greek *arakhos*, meaning "chickling vetch" (*Lathyrus sativus*), a legume with edible seed; this is the peanut or monkey-nut (*Fabaceae*)

Arachniodes
ah-RAK-nee-oh-deez
From Greek *arachnion*, meaning "spider," and *odes*, meaning "resembling," perhaps because some species have weblike dissected fronds (*Dryopteridaceae*)

arachnoides
a-rak-NOY-deez
—

arachnoideus
a-rak-NOY-dee-us
arachnoidea, arachnoideum
Like a spider web, as in *Sempervivum arachnoideum*

Aralia
uh-RAY-lee-uh
From French Canadian *aralie*, the name for a North American species, possibly derived from an indigenous name (*Araliaceae*)

aralioides
a-ray-lee-OY-deez
Like *Aralia*, as in *Trochodendron aralioides*

araucana
air-ah-KAY-nuh
Relating to the Arauco region in Chile, as in *Araucaria araucana*

Araucaria [2]
a-raw-KAH-ree-uh
Named for the Mapuche, an indigenous people of Chile, referred to by the Spanish as *araucanos*; also, the Araucanía region of Chile, inhabited by *Araucaria araucana* (*Araucariaceae*)

Araujia
ah-ROW-jee-uh
Named after António de Araújo e Azevedo (1754–1817), a Portuguese statesman and botanist (*Apocynaceae*)

arbor-tristis
ar-bor-TRIS-tis
Latin for sad tree, as in *Nyctanthes arbor-tristis*

arborescens
ar-bo-RES-senz
—

arboreus
ar-BOR-ee-us
arborea, arboreum
A woody or treelike plant, as in *Erica arborea*

arboricola
ar-bor-IH-koh-luh
Living on trees, as in *Schefflera arboricola*

arbusculus
ar-BUS-kyoo-lus
arbuscula, arbusculum
Like a small tree, as in *Daphne arbuscula*

arbutifolius
ar-bew-tih-FOH-lee-us
arbutifolia, arbutifolium
With leaves like a strawberry tree (*Arbutus*), as in *Aronia arbutifolia*

Arbutus

ar-BEW-tus

From classical Latin name for *A. unedo*
(*Ericaceae*)

archangelica

ark-an-JEL-ih-kuh

In reference to the archangel Raphael, as in
Angelica archangelica

archeri

ARCH-er-eye

Named after William Archer (1820–74),
Australian botanist, as in *Eucalyptus archeri*

Archontophoenix

are-KON-to-fee-nix

From Greek *archon*, meaning "ruler," and
Phoenix, a related genus, in reference to its
stature (*Arecaceae*)

arcticus

ARK-tih-kus

arctica, arcticum

Connected with the Arctic region, as in
Lupinus arcticus

Arctium

ARK-tee-um

From Greek *arktos*, meaning "bear," perhaps
in reference to the spiny bracts surrounding
the flowers, which are shaggy like a bear
(*Asteraceae*)

Arctostaphylos

ark-toh-STAF-ih-loss

From Greek *arktos*, meaning "bear," and
staphyle, a "bunch of grapes," because bears
feed on the berries, especially those of
A. uva-ursi (*Ericaceae*)

Arctotheca

ark-toh-THEE-kuh

From Greek *arktos*, meaning "bear," and
theke, meaning "capsule," because the fruit of
some species is covered in woolly hairs
(*Asteraceae*)

Arctotis

ark-TOH-tiss

From Greek *arktos*, meaning "bear," and *otos*,
meaning "ear," because the overlapping floral
bracts resemble bear ears (*Asteraceae*)

arcuatus

ark-yoo-AH-tus

arcuata, arcuatum

In the shape of a bow or arc, as in *Blechnum
arcuatum*

Ardisia

arr-DIZ-ee-uh

From Greek *ardis*, the point of a spear or
arrow, because the anthers are pointed
(*Primulaceae*)

Areca

ah-REE-kuh

From the Malabar (India) name for betel
nuts, as in *A. catechu* (*Arecaceae*)

Arenaria

a-ruh-NAIR-ee-uh

From Latin *arena*, meaning "sand,"
a common habitat for sandworts
(*Caryophyllaceae*)

arenarius

ar-en-AH-ree-us

arenaria, arenarium

—

Arenga

ah-REN-guh

From Javanese *aren*, the name for *A. pinnata*
(*Arecaceae*)

arendsii

ar-END-see-eye

Named after Georg Arends (1862–1952),
German nurseryman, as in *Astilbe* × *arendsii*

arenicola

ar-en-IH-koh-luh

—

arenosus

ar-en-OH-sus

arenosa, arenosum

Growing in sandy places, as in *Stapelia
arenosa*

areolatus

ar-ee-oh-LAH-tus

areolata, areolatum

Areolate, with surface divided into small
areas, as in *Coprosma areolata*

Argemone

ar-GEM-oh-nee

From the classical Greek name for a
poppylike plant (*Papaveraceae*)

argent-

Used in compound words to denote silver

argentatus

ar-jen-TAH-tus

argentata, argentatum

—

argenteus

ar-JEN-tee-us

argentea, argenteum

Silver in color, as in *Salvia argentea*

argenteomarginatus

ar-gent-eoh-mar-gin-AH-tus

argenteomarginata, argenteomarginatum

With silver edges, as in *Begonia
argenteomarginata*

argentinus

ar-jen-TEE-nus

argentina, argentinum

Connected with Argentina, as in *Tillandsia
argentina*

A
B
C
D
E
F
G
H
I
J
K
L
M
N
O
P
Q
R
S
T
U
V
W
X
Y
Z

i

THE GENDER OF TREES

As in Spanish and French,
Latin nouns are attributed
to gender: masculine,
feminine, or neuter.
Botanical Latin names
also follow this pattern;
masculine genera often end
in *-us*, for example, *Acorus*;
feminine genera typically
end in *-a*, for example,
Hosta; while neuter genera
often end in *-um*, for
example, *Delphinium*. The
species name that follows
must agree in gender, so
Acorus gramineus, *Hosta
lancifolia*, and *Delphinium
elatum*. A major exception is
trees, many of which have
masculine genera (*Pinus*,
Quercus, *Malus*, *Prunus*),
but are treated as feminine
(*Pinus nigra*, *Quercus alba*),
a quirk of history, because
fruit-bearing trees were
considered feminine by
the Romans.

argophyllus
ar-go-FIL-us
argophylla, argophyllum
With silver leaves, as in *Eriogonum argophyllum*

argutifolius
ar-gew-tih-FOH-lee-us
argutifolia, argutifolium
With sharp-toothed leaves, as in *Helleborus argutifolius*

argutus
ar-GOO-tus
arguta, argutum
With notched edges, as in *Rubus argutus*

argyraeus
ar-jy-RAY-us
argyraea, argyraeum
—

argyreus
ar-JY-ree-us
argyrea, argyreum
Silvery in color, as in *Dierama argyreum*

Argyranthemum
ar-gi-RAN-thu-mum
From Greek *argyros*, meaning "silver," and *anthemis*, meaning "flower" (*Asteraceae*)

argyro-
Used in compound words to denote silver

argyrocomus
ar-gy-roh-KOH-mus
argyrocoma, argyrocomum
With silver hairs, as in *Astelia argyrocoma*

Argyrocytisus
ar-GYRO-sigh-tiss-us
From Greek *argyros*, meaning "silver," and *Cytisus,* a related genus (*Fabaceae*)

Argyroderma
ar-GYRO-der-muh
From Greek *argyros*, meaning "silver," and *dermis*, meaning "skin," because these succulents have silvery-gray foliage (*Aizoaceae*)

argyroneurus
ar-ji-roh-NOOR-us
argyroneura, argyroneurum
With silver veins, as in *Fittonia argyroneura*

argyrophyllus
ar-ger-o-FIL-us
argyrophylla, argyrophyllum
With silver leaves, as in *Rhododendron argyrophyllum*

aria
AR-ee-a
From Greek *aria*, probably whitebeam, as in *Sorbus aria*

aridus
AR-id-us
arida, aridum
Growing in dry places, as in *Mimulus aridus*

arietinus
ar-ee-eh-TEEN-us
arietina, arietinum
In the shape of a ram's head; horned, as in *Cypripedium arietinum*

arifolius
air-ih-FOH-lee-us
arifolia, arifolium
With leaves like *Arum*, as in *Persicaria arifolia*

Ariocarpus
a-ree-oh-CAR-pus
From the genus *Aria* (=*Sorbus aria*) and Greek *karpos*, meaning "fruit," because their fruit is said to be similar (*Cactaceae*)

Arisaema [1]
a-ri-SEE-muh
From Greek *aris*, a type of arum, and *haema*, meaning "blood," because some species have red-spotted leaves (*Araceae*)

Arisarum
a-ri-SAR-um
From Greek *arisaron*, the classical name for *A. vulgare* (*Araceae*)

aristatus
a-ris-TAH-tus
aristata, aristatum
Bearded, as in *Aloe aristata*

Aristea
a-RIS-tee-uh
From Greek *arista*, meaning "awn," in reference to the pointed bracts (*Iridaceae*)

Arisaema triphyllum

Aristolochia
a-ris-tuh-LOKE-ee-uh
From Greek *aristos*, meaning "best" and *lochia*, meaning "delivery," in reference to the medicinal use of this plant in childbirth (*Aristolochiaceae*)

aristolochioides
a-ris-toh-loh-kee-OY-deez
Resembling *Aristolochia*, as in *Nepenthes aristolochioides*

Aristotelia
a-ris-toe-TEE-lee-uh
Named after Aristotle (384–322 BC), Greek philosopher and scientist (*Elaeocarpaceae*)

arizonicus
ar-ih-ZON-ih-kus
arizonica, arizonicum
Connected with Arizona, as in *Yucca arizonica*

armandii
ar-MOND-ee-eye
Named after Armand David (1826–1900), French naturalist and missionary, as in *Pinus armandii*

armatus

arm-AH-tus

armata, armatum

With thorns, spines, or spikes, as in *Dryandra armata*

armeniacus

ar-men-ee-AH-kus

armeniaca, armeniacum

Connected with Armenia, as in *Muscari armeniacum*

armenus

ar-MEE-nus

armena, armenum

Connected with Armenia, as in *Fritillaria armena*

Armeria (also armeria)

ar-MEER-ree-uh

From Celtic *arm or*, meaning "by the sea," in reference to their coastal habitat; also Latin for outwardly similar carnation, as in *Dianthus armeria* (*Plumbaginaceae*)

Arum maculatum

2

armillaris

arm-il-LAH-ris

armillaris, armillare

Like a bracelet, as in *Melaleuca armillaris*

Armoracia

ar-mo-RAY-see-uh

From the classical Greek name for horseradish, as in *A. rusticana* (*Brassicaceae*)

Arnebia

ar-NEE-bee-uh

From Arabic *shajaret el arneb*, the name for the plant (*Boraginaceae*)

Arnica

AR-ni-kuh

From Greek *arni*, meaning "lamb," due to the soft, silky leaves (*Asteraceae*)

arnoldianus

ar-nold-ee-AH-nus

arnoldiana, arnoldianum

Connected with the Arnold Arboretum in Boston, Massachusetts, as in *Abies × arnoldiana*

aromaticus

ar-oh-MAT-ih-kus

aromatica, aromaticum

With a fragrant, aromatic scent, as in *Lycaste aromatica*

Aronia

uh-ROH-nee-uh

From the genus *Aria* (=*Sorbus aria*), due to the similar fruit (*Rosaceae*)

Arrhenatherum

a-ruh-NATH-uh-room

From Greek *arrhen*, meaning "male," and *ather*, meaning "barb," because there are awns on the male flowers (*Poaceae*)

Artemisia

ar-te-MIZ-ee-uh

Named after Artemis, the Greek goddess of the hunt, or her namesake, Artemisia II of Caria, sister/wife of King Mausolus, and a botanist (*Asteraceae*)

artemisioides

ar-tem-iss-ee-OY-deez

Resembling *Artemisia*, as in *Senna artemisioides*

Arthropodium

arth-ro-POH-dee-um

From Greek *arthron*, meaning "joint," and *podion*, meaning "foot," because the pedicels (flower stalks) are jointed (*Asparagaceae*)

articulatus

ar-tik-oo-LAH-tus

articulata, articulatum

With a jointed stem, as in *Senecio articulatus*

Arum [2]

AIR-um

From Greek *aron*, the classical name for the plant (*Araceae*)

Aruncus

ah-RUN-kus

From Greek *aryngos*, the classical name for the plant (*Rosaceae*)

arundinaceus

a-run-din-uh-KEE-us

arundinacea, arundinaceum

Like a reed, as in *Phalaris arundinacea*

Arundinaria

ah-run-di-NAIR-ree-uh

From Latin *arundo*, meaning "reed," and *aria*, "pertaining to" (*Poaceae*)

Arundo
ah-RUN-doh
From Latin *arundo*, meaning "reed"
(*Poaceae*)

arvensis
ar-VEN-sis
arvensis, arvense
Growing in cultivated fields, as in *Rosa arvensis*

asarifolius
as-ah-rih-FOH-lee-us
asarifolia, asarifolium
With leaves like wild ginger (*Asarum*), as in *Cardamine asarifolia*

Asarina
ah-suh-REE-nuh
From Spanish *asarina*, vernacular name for related *Antirrhinum* (*Plantaginaceae*)

Asarum
ah-SAR-um
From Greek *asaron*, classical name of unknown plant (*Aristolochiaceae*)

ascendens
as-SEN-denz
Rising upward, as in *Calamintha ascendens*

asclepiadeus
ass-cle-pee-AD-ee-us
asclepiadea, asclepiadeum
Like milkweed (*Asclepias*), as in *Gentiana asclepiadea*

Asclepias [1]
ah-SKLEE-pee-uhs
Named after Asklepios, the Greek god of medicine, because some species have medicinal properties (*Apocynaceae*)

aselliformis
ass-el-ee-FOR-mis
aselliformis, aselliforme
Shaped like a wood louse, as in *Pelecyphora aselliformis*

asiaticus
a-see-AT-ih-kus
asiatica, asiaticum
Connected with Asia, as in *Trachelospermum asiaticum*

Asimina
ah-si-MEE-nuh
From the Native American name for the plant, *assimin* (*Annonaceae*)

asparagoides
as-par-a-GOY-deez
Like asparagus, as in *Acacia asparagoides*

Asparagus
ah-SPA-ra-gus
From Greek *asparasso*, meaning "to rip," because some species are spiny (*Asparagaceae*)

asper
AS-per
aspera, asperum
—

asperatus
as-per-AH-tus
asperata, asperatum
With a rough texture, as in *Hydrangea aspera*

asperifolius
as-per-ih-FOH-lee-us
asperifolia, asperifolium
With rough leaves, as in *Cornus asperifolia*

asperrimus
as-PER-rih-mus
asperrima, asperrimum
With a very rough texture, as in *Agave asperrima*

Asperula
ah-speh-ROO-lah
From Latin *asper*, meaning "rough," because many species have coarse stems (*Rubiaceae*)

Asphodeline
as-fod-ah-LEE-nee
A modified version of the related genus *Asphodelus* (*Asphodelaceae*)

1

Asclepias curassavica

2

Astrantia major

asphodeloides

ass-FOD-el-oy-deez

Like *Asphodelus*, as in *Geranium asphodeloides*

Asphodelus

as-FOD-uh-lus

The classical Greek name for *A. ramosus* (*Asphodelaceae*)

asplenifolius

ass-plee-ni-FOH-lee-us

asplenifolia, asplenifolium

—

aspleniifolius

ass-plee-ni-eye-FOH-lee-us

aspleniifolia, aspleniifolium

With fine, feathery, fernlike leaves, as in *Phyllocladus aspleniifolia*

Aspidistra

as-pi-DIS-truh

From Greek *aspidion*, meaning "shield," an allusion to the shieldlike stigma (*Asparagaceae*)

Asplenium

as-PLEE-nee-um

From Greek *splen*, meaning "spleen," because spleenworts were used medicinally to treat ailments of this organ (*Aspleniaceae*)

assa-foetida

ass-uh-FET-uh-duh

From Persian *aza*, "mastic," and Latin *foetidus*, "stinking," as in *Ferula assa-foetida*

assimilis

as-SIM-il-is

assimilis, assimile

Similar; alike, as in *Camellia assimilis*

assurgentiflorus

as-sur-jen-tih-FLOR-us

assurgentiflora, assurgentiflorum

With flowers in ascending clusters, as in *Lavatera assurgentiflora*

assyriacus

ass-see-re-AH-kus

assyriaca, assyriacum

Connected with Assyria, as in *Fritillaria assyriaca*

Astelia

ah-STEE-lee-uh

From Greek *a*, meaning "without," and *steli*, meaning "column," referencing the lack of a trunk (*Asteliaceae*)

Aster

A-stur

From Greek *aster*, meaning "star," because the capitulae are starlike (*Asteraceae*)

Asteranthera

a-stur-ANTH-er-uh

From Greek *aster*, meaning "star," and *anthera*, meaning "anther," because the anthers are fused together forming a star shape (*Gesneriaceae*)

asteroides

ass-ter-OY-deez

Resembling *Aster*, as in *Amellus asteroides*

Astilbe

ah-STIL-bee

From Greek *a*, meaning "without," and *stilbe*, meaning "shine" or "gloss," because some species have dull, mat leaves (*Saxifragaceae*)

Astilboides

ah-stil-BOY-deez

Resembling the related genus *Astilbe* (*Saxifragaceae*)

Astragalus

ah-STRA-ga-loos

From Greek *astragalos*, an ankle or vertebra, perhaps alluding to the knotted roots that are said to resemble bones (*Fabaceae*)

Astrantia [2]

ah-STRAN-tee-uh

From Greek *aster*, meaning "star," because the flower heads are starlike, or from Latin *magister*, or "master," a common name being masterwort (*Apiaceae*)

Astrophytum

as-tro-FIE-toom

From Greek *aster*, meaning "star," and *phyton*, meaning "plant," because some of these cacti have a star-shaped body in outline (*Cactaceae*)

asturiensis

ass-tur-ee-EN-sis

asturiensis, asturiense

From the province of Asturias, Spain, as in *Narcissus asturiensis*

Asyneuma

a-sin-OOM-ah

From Greek *a*, meaning "without," and *syn*, meaning "together," referencing the divided petals (*Campanulaceae*)

Atherosperma

ah-ther-oh-SPER-muh

From Greek *ather*, meaning "barb," and *spermum*, meaning "seed," because the seed is furnished with awns (*Atherospermataceae*)

Athrotaxis

ah-thro-TAX-iss

From Greek *athros*, meaning "crowded," and *taxis*, meaning "arrangement," because the scalelike leaves are densely arranged (*Cupressaceae*)

Athyrium

ah-THI-ree-um

From Greek *a*, meaning "without," and *thyra*, meaning "door," because the sporangia do not appear to open (*Athyriaceae*)

atkinsianus

at-kin-see-AH-nus

atkinsiana, atkinsianum

—

atkinsii

at-KIN-see-eye

Named after James Atkins (1802–84), British nurseryman, as in *Petunia* × *atkinsiana*

atlanticus

at-LAN-tih-kus

atlantica, atlanticum

Connected with the Atlantic shoreline, or from the Atlas Mountains, as in *Cedrus atlantica*

Atriplex

A-tree-plex

The classical Latin name for *A. hortensis* (*Amaranthaceae*)

atriplicifolius

at-ry-pliss-ih-FOH-lee-us

atriplicifolia, atriplicifolium

With leaves like orache or saltbush (*Atriplex*), as in *Perovskia atriplicifolia*

atro-

Used in compound words to denote dark

atrocarpus

at-ro-KAR-pus

atrocarpa, atrocarpum

With black or especially dark fruit, as in *Berberis atrocarpa*

Atropa [1]
AH-tro-pah
Named after Atropos, one of the three Greek goddesses of destiny. Atropos ended life by cutting the mortal thread; deadly nightshade is, of course, toxic (*Solanaceae*)

atropurpureus
at-ro-pur-PURR-ee-us
atropurpurea, atropurpureum
Dark purple, as in *Scabiosa atropurpurea*

atrorubens
at-roh-ROO-benz
Dark red, as in *Helleborus atrorubens*

atrosanguineus
at-ro-san-GWIN-ee-us
atrosanguinea, atrosanguineum
Dark blood red, as in *Rhodochiton atrosanguineus*

atroviolaceus
at-roh-vy-oh-LAH-see-us
atroviolacea, atroviolaceum
Dark violet, as in *Dendrobium atroviolaceum*

atrovirens
at-ro-VY-renz
Dark green, as in *Chamaedorea atrovirens*

attenuatus
at-ten-yoo-AH-tus
attenuata, attenuatum
With a narrow point, as in *Haworthia attenuata*

atticus
AT-tih-kus
attica, atticum
Connected with Attica, Greece, as in *Ornithogalum atticum*

Aubrieta
awe-BREE-shuh
Named for Claude Aubriet (1665–1742), French botanical artist (*Brassicaceae*)

aubrietioides
au-bre-teh-OY-deez
aubrietiodes
Resembling *Aubrieta*, as in *Arabis aubrietiodes*

aucheri
aw-CHER-ee
Named after Pierre Martin Rémi Aucher-Éloy (1792–1838), French pharmacist and botanist, as in *Iris aucheri*

Aucuba
awe-KOO-bah
From Japanese *aokiba*, the vernacular name (*Garryaceae*)

aucuparius
awk-yoo-PAH-ree-us
aucuparia, aucuparium
Of bird catching, as in *Sorbus aucuparia*

augustinii
aw-gus-TIN-ee-eye
augustinei
Named after the Irish plantsman and botanist Dr. Augustine Henry (1857–1930), as in *Rhododendron augustinii*

augustissimus
aw-gus-TIS-sih-mus
augustissima, augustissimum
—

augustus
aw-GUS-tus
augusta, augustum
Majestic, noteworthy, as in *Abroma augusta*

aurantiacus
aw-ran-ti-AH-kus
aurantiaca, aurantiacum
—

aurantius
aw-RAN-tee-us
aurantia, aurantium
Orange, as in *Pilosella aurantiaca*

aurantiifolius
aw-ran-tee-FOH-lee-us
aurantiifolia, aurantiifolium
With leaves like an orange tree (*Citrus aurantium*), as in *Citrus aurantiifolia*

auratus
aw-RAH-tus
aurata, auratum
With golden rays, as in *Lilium auratum*

aureo-
Used in compound words to denote golden

aureosulcatus
aw-ree-oh-sul-KAH-tus
aureosulcata, aureosulcatum
With yellow furrows, as in *Phyllostachys aureosulcata*

aureus
AW-re-us
aurea, aureum
Golden yellow, as in *Phyllostachys aurea*

auricomus
aw-RIK-oh-mus
auricoma, auricomum
With golden hair, as in *Ranunculus auricomus*

auriculatus
aw-rik-yoo-LAH-tus
auriculata, auriculatum
—

auriculus
aw-RIK-yoo-lus
auricula, auriculum
—

auritus
aw-RY-tus
aurita, auritum
With ears or ear-shaped appendages, as in *Plumbago auriculata*

Aurinia
awe-RIN-ee-uh
From Latin *aurum*, meaning "gold," and *inia*, meaning "color," because most species have yellow flowers (*Brassicaceae*)

australiensis
aw-stra-li-EN-sis
australiensis, australiense
From Australia, as in *Idiospermum autraliense*

australis
aw-STRAH-lis
australis, australe
Southern, as in *Cordyline australis*

austriacus
oss-tree-AH-kus
austriaca, austriacum
Connected with Austria, as in *Doronicum austriacum*

austrinus
oss-TREE-nus
austrina, austrinum
Southern, as in *Rhododendron austrinum*

Austrocedrus
os-troh-SEED-rus
From Latin *australis*, meaning "southern" and Greek *kedros*, meaning "cedar," because this conifer occurs in only the southern hemisphere (*Cupressaceae*)

Austrocylindropuntia
os-troh-sil-in-droh-PUN-tee-uh
From Latin *australis*, meaning "southern," and *Cylindropuntia*, a related genus (*Cactaceae*)

1

Atropa belladonna

2

Anchusa azurea

Austroderia

os-troh-DEER-ee-uh

From Latin *australis*, meaning "southern," and *Cortaderia*, a related genus (*Poaceae*)

autumnalis

aw-tum-NAH-lis

autumnalis, autumnale

Relating to fall (from "autumn"), as in *Colchicum autumnale*

avellanus

av-el-AH-nus

avellana, avellanum

Connected with Avella, Italy, as in *Corylus avellana*

Avena

ah-VEE-nuh

The classical Latin name for oats (*A. sativa*) (*Poaceae*)

avenaceus

a-vee-NAY-see-us

avenacea, avenaceum

Like *Avena* (oats), as in *Agrostis avenacea*

Averrhoa

ah-vuh-ROH-ah

Named after Ibn Rushd (1126–98), Andalusian Muslim philosopher, whose name was Latinized as Averroes (*Oxalidaceae*)

avium

AY-ve-um

Relating to birds, as in *Prunus avium*

axillaris

ax-ILL-ah-ris

axillaris, axillare

Growing in the axil, as in *Petunia axillaris*

Azara

a-ZAR-uh

Named after Félix Manuel de Azara (1742–1821), Spanish military officer and naturalist, or perhaps his brother José Nicolás de Azara (1730–1804), Spanish diplomat and patron of science (*Salicaceae*)

azedarach

a-ZED-ur-ack

From the Persian for noble tree, as in *Melia azedarach*

Azolla

ah-ZOLL-uh

From Greek *azo*, meaning "to dry," and *ollyo*, meaning "to kill," because these aquatic ferns are intolerant of drought (*Salviniaceae*)

Azorella

ah-zaw-RELL-uh

From the Azores, plus Latin diminutive *ella*, indicating small stature (*Apiaceae*)

azoricus

a-ZOR-ih-kus

azorica, azoricum

Connected with the Azores Islands, as in *Jasminum azoricum*

Azorina

ah-zaw-REE-nuh

From the Azores (*Campanulaceae*)

Aztekium

azz-TEE-kee-um

Named after the Aztec people of present-day Mexico (*Cactaceae*)

azureus [2]

a-ZOOR-ee-us

azurea, azureum

Azure; sky blue, as in *Muscari azureum*

Babiana
ba-bee-AH-nah
From Dutch *baviaan*, meaning "baboon," monkeys that consume the corms (*Iridaceae*)

babylonicus
bab-il-LON-ih-kus
babylonica, babylonicum
Connected with Babylonia, Mesopotamia (Iraq), as in *Salix babylonica*, which Linnaeus mistakenly believed to be from southwest Asia

baccans
BAK-kanz
—

bacciferus
bak-IH-fer-us
baccifera, bacciferum
With berries, as in *Erica baccans*

baccatus
BAK-ah-tus
baccata, baccatum
With fleshy berries, as in *Malus baccata*

Baccharis
BAK-uh-riss
Named after Bacchus, Roman god of wine, for reasons unknown (*Asteraceae*)

bacillaris
bak-ILL-ah-ris
bacillaris, bacillare
Like a stick, as in *Cotoneaster bacillaris*

backhousia
bak-how-zee-AH
backhouseanus, backhouseana, backhouseanum
—

backhousianus
bak-how-zee-AH-nus
backhousiana, backhousianum
—

backhousei
bak-HOW-zee-eye
Named after James Backhouse (1794–1869), British nurseryman, as in *Correa backhouseana*

badius
bad-ee-AH-nus
badia, badium
Chestnut brown, as in *Trifolium badium*

Baeckea
BEK-ee-uh
Named after Abraham Bäck (1713–95), Swedish physician and naturalist (*Myrtaceae*)

baicalensis
by-kol-EN-sis
baicalensis, baicalense
From Lake Baikal, eastern Siberia, as in *Anemone baicalensis*

baileyi
BAY-lee-eye
—

baileyanus
bay-lee-AH-nus
baileyana, baileyanum
Named after one of the following: Frederick Manson Bailey (1827–1915), Australian botanist; Lt. Colonel Frederick Marshman Bailey (1882–1967), Indian Army soldier who collected plants on the Tibetan borders from 1913; Major Vernon Bailey (1864–1942), American Army soldier who collected cacti from 1900; Liberty Hyde Bailey (1858–1954), author and professor of horticulture at Cornell University, New York, as in *Rhododendron baileyi* (named after Lt. Colonel Frederick Marshman Bailey)

bakeri
BAY-ker-eye
—

bakerianus
bay-ker-ee-AH-nus
bakeriana, bakerianum
Usually honoring John Gilbert Baker (1834–1920) of Kew, London, as in *Aloe bakeri*, but also George Percival Baker (1856–1951), British plant collector

baldensis [1]
bald-EN-sis
baldensis, baldense

baldianus
bald-ee-AN-ee-us
baldiana, baldianum
From or of Monte Baldo, Italy, as in *Hieracium baldense*

baldschuanicus
bald-SHWAN-ih-kus
baldschuanica, baldschuanicum
Connected with Baljuan, Turkistan, as in *Fallopia baldschuanica*

balearicus
bal-AIR-ih-kus
balearica, balearicum
Connected with the Balearic Islands, Spain, as in *Buxus balearica*

1

Anemone baldensis

Bambusa bambos

Banksia squarrosa

Ballota
bah-LOT-uh
The classical Greek name for *B. nigra* (*Lamiaceae*)

balsameus
bal-SAM-ee-us
balsamea, balsameum
Like balsam, as in *Abies balsamea*

balsamiferus
bal-sam-IH-fer-us
balsamifera, balsamiferum
Producing balsam, as in *Aeonium balsamiferum*

Balsamorhiza
bal-SAM-oh-rye-zuh
From Greek *balsamon*, a fragrant gum, and *rhiza*, meaning "root," because the crushed roots are resinous (*Asteraceae*)

balticus
BOL-tih-kus
baltica, balticum
Connected with the Baltic Sea region, as in *Cotoneaster balticus*

Bambusa [1]
bam-BOO-suh
Original source unknown, but could be Malay *bambu* or Kannada (India) *bambu*; also source of English word "bamboo" (*Poaceae*)

bambusoides
bam-BOO-soy-deez
Resembling bamboo (*Bambusa*), as in *Phyllostachys bambusoides*

banaticus
ba-NAT-ih-kus
banatica, banaticum
Connected with the Banat region of Central Europe, as in *Crocus banaticus*

Banksia [2]
BANK-see-uh
Named after Sir Joseph Banks (1743–1820), British botanist and founding member of the Royal Horticultural Society (*Proteaceae*)

banksianus
banks-ee-AH-nus
banksiana, banksianum
—

banksii
BANK-see-eye
Named after Sir Joseph Banks (1743–1820), British botanist and plant collector, as in *Cordyline banksii*; *banksiae* commemorates his wife, Lady Dorothea Banks (1758–1828)

bannaticus
ban-AT-ih-kus
bannatica, bannaticum
Connected with Banat, Central Europe, as in *Echinops bannaticus*

Baptisia tinctoria

Barleria recurva

Baptisia [3]
bap-TIZ-ee-uh
From Greek *bapto*, meaning "to dye," because plants were used as a substitute for *Indigofera*, the source of indigo dye (*Fabaceae*)

Barbarea
bar-BUH-ree-uh
Named after Saint Barbara, perhaps because this plant is one of the few edibles available on her feast day, December 4 (*Brassicaceae*)

barbarus
BAR-bar-rus
barbara, barbarum
Foreign, as in *Lycium barbarum*

barbatulus
bar-BAT-yoo-lus
barbatula, barbatulum

barbatus
bar-BAH-tus
barbata, barbatum
Bearded; with long, weak hairs, as in *Hypericum barbatum*

barbigerus
bar-BEE-ger-us
barbigera, barbigerum
With beards or barbs, as in *Bulbophyllum barbigerum*

barbinervis
bar-bih-NER-vis
barbinervis, barbinerve
With bearded or barbed veins, as in *Clethra barbinervis*

barbinodis
bar-bin-OH-dis
barbinodis, barbinode
With beards at the nodes or joints, as in *Bothriochloa barbinodis*

barbulatus
bar-bul-AH-tus
barbulata, barbulatum
With a short or less significant beard, as in *Anemone barbulata*

barcinonensis
bar-sin-oh-NEN-sis
barcinonensis, barcinonense
From Barcelona, Spain, as in *Galium* × *barcinonense*

Barleria [4]
bar-LEER-ee-uh
Named after Jacques Barrelier (1606–73), French botanist and monk (*Acanthaceae*)

baselloides
bar-sell-OY-deez
Resembling *Basella*, as in *Boussangaultia baselloides*

Begonia diadema

basilaris

bas-il-LAH-ris

basilaris, basilare

Relating to the base or bottom, as in *Opuntia basilaris*

basilicus

bass-IL-ih-kus

basilica, basilicum

With princely or royal properties, as in *Ocimum basilicum*

Bassia

BAH-see-uh

Named after Ferdinando Bassi (1710–74), Italian naturalist (*Amaranthaceae*)

Bauera

BOW-er-ah

Named after brothers Ferdinand (1760–1826) and Franz Bauer (1758–1840), Austrian botanical illustrators (*Cunoniaceae*)

baueri

baw-WARE-eye

—

bauerianus

baw-ware-ee-AH-nus

baueriana, bauerianum

Named after Ferdinand Bauer (1760–1826), Austrian botanical artist to Flinders' Australian expedition, as in *Eucalyptus baueriana*

Bauhinia

bau-EEN-ee-uh

Named after brothers Gaspard (1560–1624) and Jean Bauhin (1541–1613), Swiss physicians and botanists (*Fabaceae*)

baurii

BOUR-ee-eye

Named after Dr. Georg Herman Carl Ludwig Baur (1859–98), German plant collector, as in *Rhodohypoxis baurii*

Beaucarnea

bow-CAR-nee-uh

Named after Jean-Baptiste Beaucarne (dates unknown), Belgian plant collector (*Asparagaceae*)

Beaufortia

bow-FOR-tee-uh

Named for Mary Somerset, Duchess of Beaufort (1630–1715), British gardener and botanist (*Myrtaceae*)

Beaumontia

bow-MON-tee-uh

Named after Lady Diana Beaumont (1765–1831), British patron of horticulture (*Apocynaceae*)

Beesia

BEE-zee-uh

Named after Bees Nursery (est. 1903/4), now Ness Botanic Gardens in England, which sponsored plant-hunting trips by George Forrest, Reginald Farrer, and Frank Kingdon-Ward (*Ranunculaceae*)

beesianus

bee-zee-AH-nus

beesiana, beesianum

Named after Bees Nursery, Chester, England, as in *Allium beesianum*

Begonia

beh-GO-nee-uh

Named after Michel Bégon (1638–1710), French Intendant of Saint-Domingue (now Haiti) and patron of botany (*Begoniaceae*)

belladonna

bel-uh-DON-nuh

Beautiful lady, as in *Amaryllis belladonna*

Bellevalia

bel-VAL-ee-uh

Named after Pierre Richer de Belleval (1564–1632), French physician and botanist (*Asparagaceae*)

bellidifolius

bel-lid-ee-FOH-lee-us

bellidifolia, bellidifolium

With leaves like a daisy (*Bellis*), as in *Ageratina bellidifolia*

bellidiformis

bel-id-EE-for-mis

bellidiformis, bellidiforme

Like a daisy (*Bellis*), as in *Dorotheanthus bellidiformis*

bellidioides

bell-id-ee-OY-deez

Resembling *Bellium*, as in *Silene bellidioides*

Bellis

BEH-liss

From Latin *bellus*, meaning "pretty" (*Asteraceae*)

bellus

BELL-us

bella, bellum

Beautiful; handsome, as in *Graptopetalum bellum*

benedictus

ben-uh-DICK-tus

benedicta, benedictum

A blessed plant; spoken of favorably, as in *Centaurea benedicta*

benghalensis

ben-gal-EN-sis

benghalensis, benghalense

Also *bengalensis*; from Bengal, India, as in *Ficus benghalensis*

Berberidopsis

bur-beh-ri-DOP-sis

Resembling the genus *Berberis* (*Berberidopsidaceae*)

Berberis

BUR-beh-riss

From Latin *barbaris*, itself derived from the Arabic word for North Africa, the Barbary Coast inhabited by the Berber people (*Berberidaceae*)

Berchemia

bur-SHEE-mee-uh

Named after Jacob Pierre Berthoud van Berchem (1763–1832), Dutch mineralogist and naturalist (*Rhamnaceae*)

Bergenia

bur-JEE-nee-uh

Named after Karl August von Bergen (1704–1759), German anatomist and botanist (*Saxifragaceae*)

Berkheya

bur-KAY-uh

Named after Johannes le Francq van Berkhey (1729–1812), Dutch painter, poet, and scientist (*Asteraceae*)

bermudianus

ber-myoo-dee-AH-nus

bermudiana, bermudianum

Connected with Bermuda, as in *Juniperus bermudiana*

berolinensis

ber-oh-lin-EN-sis

berolinensis, berolinense

From Berlin, Germany, as in *Populus × berolinensis*

berthelotii

berth-eh-LOT-ee-eye

Named after Sabin Berthelot (1794–1880), French naturalist, as in *Lotus berthelotii*

Bertolonia

bur-tah-LOW-nee-uh

Named after Antonio Bertoloni (1775–1869), Italian physician and botanist (*Melastomataceae*)

Beschorneria

besh-or-NEAR-ee-uh

Named after Friedrich Beschorner (1806–1873), German botanist (*Asparagaceae*)

Bessera

BES-uh-ruh

Named after Wilibald von Besser (1784–1842), Austrian botanist (*Asparagaceae*)

Beta [1]

BEE-tuh

From Latin *beta*, meaning "beet," as in the vegetable beet (*Amaranthaceae*)

betaceus

bet-uh-KEE-us

betacea, betaceum

Like a beet (*Beta*), as in *Solanum betaceum*

betonicifolius

bet-on-ih-see-FOH-lee-us

betonicifolia, betonicifolium

Like betony (*Stachys*), as in *Meconopsis betonicifolia*

Betula [2]

BET-you-luh

From classical Latin name for birch; Pliny used *betula* as a word for pitch, which is extracted from birch bark (*Betulaceae*)

betulifolius

bet-yoo-lee-FOH-lee-us

betulifolia, betulifolium

With leaves like a birch (*Betula*), as in *Pyrus betulifolia*

betulinus

bet-yoo-LEE-nus

betulina, betulinum

—

betuloides

bet-yoo-LOY-deez

Resembling or like a birch (*Betula*), as in *Carpinus betulinus*

Biarum

BYE-air-oom

From Latin *bis*, meaning "two," plus *Arum*, a related genus, although the exact reason for this name is unclear (*Araceae*)

bicolor

BY-kul-ur

With two colors, as in *Caladium bicolor*

bicornis

BY-korn-is

bicornis, bicorne

—

bicornutus

by-kor-NOO-tus

bicornuta, bicornutum

With two horns or hornlike spurs, as in *Passiflora bicornis*

Bidens

BYE-denz

From Latis *bis*, meaning "two," and *dens*, meaning "tooth," because some species have fruit tipped with two teeth (*Asteraceae*)

bidentatus

by-den-TAH-tus

bidentata, bidentatum

With two teeth, as in *Allium bidentatum*

biennis

by-EN-is

biennis, bienne

Biennial, as in *Oenothera biennis*

bifidus

BIF-id-us

bifida, bifidum

Cleft in two parts, as in *Rhodophila bifida*

biflorus

BY-flo-rus

biflora, biflorum

With twin flowers, as in *Geranium biflorum*

1

Beta vulgaris

2

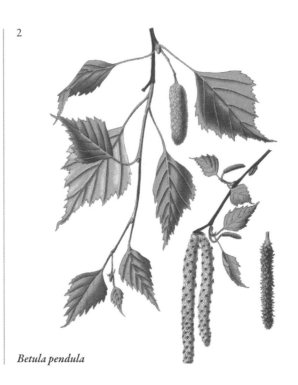

Betula pendula

Bismarckia

The Bismarck palm (*Bismarckia nobilis*) is widely grown in tropical countries, primarily for the beauty of its large, silvery leaves. Native to Madagascar, it was first described by German botanists Johann Hildebrandt (1847–81) and Hermann Wendland (1825–1903), honoring the first German Chancellor, Otto von Bismarck. It is unusual for plants to be named after politicians and perhaps this choice of name did not sit well with France, who colonized Madagascar. Two attempts were made to transfer the palm to a related (and less political) genus, such as *Medemia nobilis*, but current taxonomic opinion is that *Bismarckia* and *Medemia* are separate and distinct.

Bismarckia nobilis

bifolius
by-FOH-lee-us
bifolia, bifolium
With twin leaves, as in *Scilla bifolia*

bifurcatus
by-fur-KAH-tus
bifurcata, bifurcatum
Divided into equal stems or branches, as in *Platycerium bifurcatum*

Bignonia
big-NOH-nee-uh
Named after Jean-Paul Bignon (1662–1743), French preacher, statesman, and librarian (*Bignoniaceae*)

bignonioides
big-non-YOY-deez
Resembling crossvine (*Bignonia*), as in *Catalpa bignonioides*

bijugus
bih-JOO-gus
bijuga, bijugum
Two pairs joined together, as in *Pelargonium bijugum*

Billardiera
bil-ar-dee-AIR-uh
Named after Jacques-Julien Houtou de Labillardière (1755-1834), French botanist (*Pittosporaceae*)

Billbergia
bil-BUR-gee-uh
Named after Gustaf Johann Billberg (1772–1844), Swedish botanist (*Bromeliaceae*)

bilobatus
by-low-BAH-tus
bilobata, bilobatum
—
bilobus
by-LOW-bus
biloba, bilobum
With two lobes, as in *Ginkgo biloba*

Biophytum
bye-oh-FY-toom
From Greek *bios*, meaning "life," and *phyton*, meaning "plant," probably an allusion to the leaves, which can rapidly move and give the plant the appearance of springing back to life (*Oxalidaceae*)

bipinnatus
by-pin-NAH-tus
bipinnata, bipinnatum
A leaf that is doubly pinnate, as in *Cosmos bipinnatus*

biserratus
by-ser-AH-tus
biserrata, biserratum
A leaf that is double-toothed, as in *Nephrolepis biserrata*

Bismarckia
biz-MAR-kee-uh
Named after Otto von Bismarck (1815–98), Prussian/German politician (*Arecaceae*)

biternatus
by-ter-NAH-tus
biternata, biternatum
A leaf that is twice ternate, as in *Actaea biternata*

bituminosus
by-tu-min-OH-sus
bituminosa, bituminosum
Like bitumen, sticky, as in *Bituminaria bituminosa*

bivalvis
by-VAL-vis
bivalvis, bivalve
With two valves, as in *Ipheion bivalve*

Bixa
BICKS-uh
From indigenous Carib name *biché* or *bija* (*Bixaceae*)

Blandfordia
bland-FOR-dee-uh
Named after George Spencer-Churchill, Marquess of Blandford (1766–1840), British politician and collector (*Blandfordiaceae*)

blandus

BLAN-dus

blanda, blandum

Mild or charming, as in *Anemone blanda*

Blechnum

BLEK-noom

From Greek *blechnon*, a generic name for ferns (*Blechnaceae*)

blepharophyllus

blef-ar-oh-FIL-us

blepharophylla, blepharophyllum

With leaves that are fringed like eyelashes, as in *Arabis blepharophylla*

Bletilla

bleh-TILL-uh

From Latin *Bletia*, a related genus, plus diminutive *ella*, implying a resemblance; *Bletia* is named after Luis Blet (dates unknown), Spanish apothecary (*Orchidaceae*)

LANGUAGE BIAS

The system of taxonomy used to classify plants has its origins in Europe, and a bias toward Europe is noticeable when surveying plant names. Genera named after people most often venerate Europeans, although this has begun to change and many newly described genera honor botanists and collectors from around the globe. A handful of generic names are based on native names for the plant instead of using a purely Latin moniker. *Catalpa* derives from Muscogee, a Native American language, while *Luma* originates in the Mapuche language of Chile. *Manihot*, *Ananas*, *Luffa*, *Puya*, and *Nelumbo* are all based on local languages.

Blossfeldia

blow-FEL-tee-uh

Named after Harry Blossfeld (1913–86), German botanist (*Cactaceae*)

bodinieri

boh-din-ee-ER-ee

Named after Émile-Marie Bodinier (1842–1901), French missionary who collected plants in China, as in *Callicarpa bodinieri*

bodnantense

bod-nan-TEN-see

Named after Bodnant Gardens, Wales, as in *Viburnum* × *bodnantense*

Boehmeria

bow-MEER-ee-uh

Named after Georg Rudolf Böhmer (1723–1803), German botanist (*Urticaceae*)

Boenninghausenia

bur-ning-how-ZEN-ee-uh

Named after Clemens Maria Franz von Bönninghausen (1785–1864), Dutch lawyer and botanist (*Rutaceae*)

Bolax

BOH-lax

From Greek *bolax*, meaning "lump," possibly in reference to its mounded habit or the shape of the flower clusters (*Apiaceae*)

Boltonia

bowl-TOH-nee-uh

Named after James Bolton (1735–99), British naturalist and illustrator (*Asteraceae*)

Bolusanthus

boh-lus-ANTH-us

Named for Harry Bolus (1834–1911), South African botanist and philanthropist (*Fabaceae*)

Bomarea

boh-MA-ree-uh

Named for Jacques-Christophe Valmont de Bomare (1731–1807), French botanist and naturalist (*Alstroemeriaceae*)

Bombax [1]

BOM-backs

From Greek *bombyx*, meaning "silkworm," because the seed is covered in silky hairs (*Malvaceae*)

bonariensis

bon-ar-ee-EN-sis

bonariensis, bonariense

From Buenos Aires, as in *Verbena bonariensis*

Bongardia

bon-GAR-dee-uh

Named after Gustav Heinrich von Bongard (1786–1839), German botanist (*Berberidaceae*)

1

Bombax ceiba

Borago officinalis

Bowenia spectabilis

bonus
BOW-nus
bona, bonum
In compound words, good, as in *Chenopodium bonus-henricus*, good (King) Henry

Borago [2]
buh-RAY-go
From Latin *burra*, meaning a "shaggy garment," and *ago*, meaning "like," referring to the hairy foliage (*Boraginaceae*)

borbonicus
bor-BON-ih-kus
borbonica, borbonicum
Connected with Réunion Island in the Indian Ocean, formerly known as Île Bourbon. Also refers to the French Bourbon kings, as in *Watsonia borbonica*

borealis
bor-ee-AH-lis
borealis, boreale
Northern, as in *Erigeron borealis*

Borinda
buh-RIN-duh
Named after Norman Loftus Bor (1893–1972), Irish taxonomist (*Poaceae*)

borinquenus
bor-in-KAH-nus
borinquena, borinquenum
From Borinquen, local name for Puerto Rico, as in *Roystonea borinquena*

borneensis
bor-nee-EN-sis
borneensis, borneense
From Borneo, as in *Gaultheria borneensis*

Boronia
buh-ROW-nee-uh
Named after Francesco Borone (1769–94), Italian naturalist (*Rutaceae*)

Bothriochloa
BOTH-ree-oh-klo-ah
From Greek *bothros*, meaning "pit," and *chloa*, "a blade of grass"; some species have glumes (floral bracts) that are pitted (*Poaceae*)

botryoides
bot-ROY-deez
Resembling a bunch of grapes, as in *Muscari botryoides*

Bougainvillea
boo-gun-VILL-ee-uh
Named after Louis-Antoine, Comte de Bougainville (1729–1811), French admiral and explorer (*Nyctaginaceae*)

Bouteloua
boo-tuh-LOO-ah
Named after brothers Claudio (1774–1842) and Esteban Boutelou (1776–1813), Spanish botanists (*Poaceae*)

Bouvardia
boo-VAR-dee-uh
Named after Charles Bouvard (1572–1658), French chemist and physician (*Rubiaceae*)

bowdenii
bow-DEN-ee-eye
Named after plantsman Athelstan Cornish-Bowden (1871–1942), as in *Nerine bowdenii*

Bowenia [3]
boh-WEE-nee-uh
Named for George Ferguson Bowen (1821–99), British colonial administrator and first governor of Queensland (*Zamiaceae*)

Bowiea
BOH-wee-uh
Named after James Bowie (1789–1869), British botanist (*Asparagaceae*)

Boykinia
boy-KIN-ee-uh
Named after Samuel Boykin (1786–1848), American physician and naturalist (*Saxifragaceae*)

brachiatus

brak-ee-AH-tus

brachiata, brachiatum

With branches at right angles; like arms, as in *Clematis brachiata*

brachy-

Used in compound words to denote short

brachybotrys

brak-ee-BOT-rees

With short clusters, as in *Wisteria brachybotrys*

brachycerus

brak-ee-SER-us

brachycera, brachycerum

With short horns, as in *Gaylussacia brachycera*

Brachychiton

bra-kee-KY-tun

From Greek *brachys*, meaning "short," and *chiton*, "tunic," in reference to the coating of shorts hairs on the seed (*Malvaceae*)

Brachyglottis

bra-kee-GLOT-iss

From Greek *brachys*, meaning "short," and *glottis*, meaning "tongue," because the ray florets are short (*Asteraceae*)

brachypetalus

brak-ee-PET-uh-lus

brachypetala, brachypetalum

With short petals, as in *Cerastium brachypetalum*

brachyphyllus

brak-ee-FIL-us

brachyphylla, brachyphyllum

With short leaves, as in *Colchicum brachyphyllum*

Brachypodium

bra-kee-POH-dee-um

From Greek *brachys*, meaning "short," and *podion*, meaning "foot," because the flower stalks (pedicels) are short (*Poaceae*)

Brachyscome

bra-kee-SKO-mee

From Greek *brachys*, meaning "short," and *kome*, meaning "hair," because the pappus hairs are short (*Asteraceae*)

Brachystelma

bra-kee-STEL-muh

From Greek *brachys*, meaning "short," and *stelma*, meaning "crown," referring to the short corona (*Apocynaceae*)

bracteatus

brak-tee-AH-tus

bracteata, bracteatum

—

bracteosus

brak-tee-OO-tus

bracteosa, bracteosum

—

bractescens

brak-TES-senz

With bracts, as in *Veltheimia bracteata*

Brahea

bra-HEE-uh

Named after Tycho Brahe (1546–1601), Danish astronomer (*Arecaceae*)

brasilianus

bra-sill-ee-AHN-us

brasiliana, brasilianum

—

brasiliensis

bra-sill-ee-EN-sis

brasiliensis, brasiliense

From or of Brazil, as in *Begonia brasiliensis*

Brassavola

BRAS-ah-voh-lah

Named after Antonio Musa Brassavola (1500–55), Italian physician and botanist (*Orchidaceae*)

Brassia

BRAS-ee-uh

Named after William Brass (?–1783), British botanist (*Orchidaceae*)

Brassica

BRAS-i-kuh

The classical Latin name for cabbage (*Brassicaceae*)

brevifolius

brev-ee-FOH-lee-us

brevifolia, brevifolium

With short leaves, as in *Gladiolus brevifolius*

brevipedunculatus

brev-ee-ped-un-kew-LAH-tus

brevipedunculata, brevipedunculatum

With a short flower stalk, as in *Olearia brevipedunculata*

brevis

BREV-is

brevis, breve

Short, as in *Androsace brevis*

breviscapus

brev-ee-SKAY-pus

breviscapa, breviscapum

With a short scape, as in *Lupinus breviscapus*

Breynia

BRY-nee-uh

Named after Jacob Breyne (1637–97) and his son Johann Philipp Breyne (1680–1764), Polish botanists (*Phyllanthaceae*)

Brickellia

bri-KEH-lee-uh

Named after John Brickell (1748–1809), Irish physician and naturalist (*Asteraceae*)

Briggsia

BRIG-zee-uh

Named after Munro Briggs Scot (1887–1917), Scottish botanist (*Gesneriaceae*)

Briza

BREE-zuh

From Greek *brizo*, meaning "to nod, sleep," in reference to the drooping flower spikes (*Poaceae*)

Brodiaea

broh-dee-IE-uh

Named after James Brodie (1744–1824), Scottish politician and botanist (*Asparagaceae*)

Bromelia

bruh-MEE-lee-uh

Named after Olaf Bromel (1629–1705), Swedish botanist (*Bromeliaceae*)

bromoides

brom-OY-deez

Resembling brome grass (*Bromus*), as in *Stipa bromoides*

Bromus

BROH-muss

From Greek *bromos*, meaning fodder (*Poaceae*)

bronchialis

bron-kee-AL-lis

bronchialis, bronchiale

Used in the past as a treatment for bronchitis, as in *Saxifraga bronchialis*

Brassica oleracea

Broussonetia

broo-son-ET-ee-uh

Named after Pierre Marie Auguste Broussonet (1761–1807), French naturalist (*Moraceae*)

Browallia

bro-WAH-lee-uh

Named after Johann Browall (1707–55), Swedish botanist and cleric (*Solanaceae*)

Brownea

BROW-nee-uh

Named after Patrick Browne (1720–90), Irish physician and historian (*Fabaceae*)

Browningia

brow-NIN-gee-uh

Named after Webster Browning (1869–1942), director of the Instituto Inglés, Santiago, Chile (*Cactaceae*)

Brugmansia

broog-MAN-zee-uh

Named after Sebald Justinus Brugmans (1763–1819), Dutch physician and botanist (*Solanaceae*)

Brunfelsia

broon-FELL-zee-uh

Named after Otto Brunfels (1488–1534), German theologian and botanist (*Solanaceae*)

Brunnera

BRUN-uh-ruh

Named after Samuel Brunner (1790–1844), Swiss botanist (*Boraginaceae*)

brunneus

BROO-nee-us

brunnea, brunneum

Deep brown, as in *Coprosma brunnea*

Brunsvigia

broonz-VIG-gee-uh

Named after Karl Wilhelm Ferdinand (1713–80), Duke of Brunswick-Lunenburg, patron of arts and sciences (*Amaryllidaceae*)

bryoides

bri-ROY-deez

Resembling moss, as in *Dionysia bryoides*

Bryonia

bry-OH-nee-uh

From Greek *bruein*, meaning "to swell" or "grow," because these herbaceous climbers regrow each year from the roots (*Cucurbitaceae*)

buckleyi

BUK-lee-eye

For those named Buckley, such as William Buckley, American geologist, as in *Schlumbergera* × *buckleyi*

Buddleja [1]

BUD-lee-uh

Named after Adam Buddle (1660–1715), British botanist and cleric (*Scrophulariaceae*)

bufonius

buf-OH-nee-us

bufonia, bufonium

Relating to toads; grows in damp places, as in *Juncus bufonius*

Buglossoides

boo-glos-OY-deez

Resembling the related genus *Buglossum* (=*Anchusa*; *Boraginaceae*)

bulbiferus

bulb-IH-fer-us

bulbifera, bulbiferum

—

bulbiliferus

bulb-il-IH-fer-us

bulbilifera, bulbiliferum

With bulbs, often referring to bulbils, as in *Lachenalia bulbifera*

Bulbine

bul-BYE-nee

From Greek *bolbos*, meaning "bulb" (*Asphodelaceae*)

Bulbinella

bul-bi-NEL-uh

From *Bulbine*, a related genus, plus diminutive *ella*, implying a resemblance (*Asphodelaceae*)

bulbocodium

bulb-oh-KOD-ee-um

With a woolly bulb, as in *Narcissus bulbocodium*

1

Buddleja colvilei

Bulbophyllum [2]
bul-boh-FIL-um
From Greek *bolbos*, meaning "bulb" and *phyllon* meaning "leaf,"
because each pseudobulb has one leaf (*Orchidaceae*)

bulbosus
bul-BOH-sus
bulbosa, bulbosum
A bulbous, swollen stem that grows underground; resembling a bulb,
as in *Ranunculus bulbosus*

bulgaricus
bul-GAR-ih-kus
bulgarica, bulgaricum
Connected with Bulgaria, as in *Cerastium bulgaricum*

bullatus
bul-LAH-tus
bullata, bullatum
With blistered or puckered leaves, as in *Cotoneaster bullatus*

bulleyanus
bul-ee-YAH-nus
bulleyana, bulleyanum
—

bulleyi
bul-ee-YAH-eye
Named after Arthur Bulley (1861–1942), founder of Ness Botanic
Gardens, Cheshire, England, as in *Primula bulleyana*

bungeanus
bun-jee-AH-nus
bungeana, bungeanum
Named after Dr. Alexander von Bunge (1803–90), Russian botanist,
as in *Pinus bungeana*

Buphthalmum
boof-THAL-mum
From Greek *bous*, meaning "ox," and *opthalmos*, meaning "eye," to
which the inflorescence resembles (*Asteraceae*)

Bupleurum
boo-PLUR-um
From classical Greek name that means "ox rib" (*Apiaceae*)

Burbidgea
bur-BIJ-ee-uh
Named after Frederick William Burbidge (1847–1905), British
botanist and explorer (*Zingiberaceae*)

Burchellia
bur-CHEE-lee-uh
Named after William John Burchell (1781–1863), British explorer
(*Rubiaceae*)

burkwoodii
berk-WOOD-ee-eye
Named after brothers Arthur and Albert Burkwood, nineteenth-
century hybridizers, as in *Viburnum × burkwoodii*

2

Bulbophyllum anceps

Bursera
BER-suh-ruh
Named after Joachim Burser (1583–1639), German botanist
(*Burseraceae*)

Butia
BOO-tee-uh
From indigenous South American name *mbotiá*, meaning "to make
teeth," probably referring to the spines on the petiole (*Arecaceae*)

Butomus
boo-TOH-mus
From Greek *bous*, meaning "cow," and *temno*, meaning "to cut,"
because the sharp leaves were said to damage the mouths of grazing
cattle (*Butomaceae*)

buxifolius
buks-ih-FOH-lee-us
buxifolia, buxifolium
With leaves like box (*Buxus*), as in *Cantua buxifolia*

Buxus
BUK-sus
The classical Latin name for this plant (*Buxaceae*)

byzantinus
biz-an-TEE-nus
byzantina, byzantinum
Connected with Istanbul, Turkey, as in *Colchicum byzantinum*

Cabomba

kah-BOM-buh

From the Guyanese vernacular name for *C. aquatica* (*Cabombaceae*)

cacaliifolius

ka-KAY-see-eye-FOH-lee-us

cacaliifolia, cacaliifolium

With leaves like *Cacalia*, as in *Salvia cacaliifolia*

cachemiricus

kash-MI-rih-kus

cachemirica, cachemiricum

Connected with Kashmir, as in *Gentiana cachemirica*

cadierei

kad-ee-AIR-eye

Named after R. P. Cadière, twentieth-century plant collector in Vietnam, as in *Pilea cadierei*

cadmicus

KAD-mih-kus

cadmica, cadmicum

Metallic; like tin, as in *Ranunculus cadmicus*

caerulescens

see-roo-LES-enz

Turning blue, as in *Euphorbia caerulescens*

caeruleus

see-ROO-lee-us

caerulea, caeruleum

Dark blue, as in *Passiflora caerulea*

Caesalpinia

sees-al-PIN-ee-uh

Named after Andrea Caesalpino (1519–1603), Italian physician, philosopher, and botanist (*Fabaceae*)

caesius

KESS-ee-us

caesia, caesium

Bluish gray, as in *Allium caesium*

caespitosus

kess-pi-TOH-sus

caespitosa, caespitosum

Growing in a dense clump, as in *Eschscholzia caespitosa*

caffer

KAF-er

caffra, caffrum

—

caffrorum

kaf-ROR-um

Connected with South Africa, as in *Erica caffra*

Cakile

ka-KY-lee

From the Arabic name *qaqulleh* (*Brassicaceae*)

calabricus

ka-LA-brih-kus

calabrica, calabricum

Connected with the Calabria region of Italy, as in *Thalictrum calabricum*

Caladium

kuh-LAY-dee-um

From indigenous Indian name *kaladi*, for this or a related plant (*Araceae*)

Calamagrostis

kal-uh-muh-GROS-tis

From Greek *kalamos*, meaning "reed," and *agrostis*, meaning "grass" (*Poaceae*)

Calamintha

kah-luh-MIN-thuh

From Greek *kallos*, meaning "beautiful," and *Mentha*, a related genus (*Lamiaceae*)

Calamus

KAL-uh-mus

From Greek *kalamos*, meaning "reed," because rattan palms have long reedlike stems (*Arecaceae*)

GENUS SPOTLIGHT

Calceolaria

Slipper flower gets its name from its pouchlike lower petal. The genus name, too, has similar origins in the Greek word *calceolus*, meaning "little shoe." Slipperlike flowers occur throughout *Calceolaria*, but also in the orchid family, such as in *Cypripedium calceolus*. *Calceolaria* floral pouches contain oil-secreting hairs that attract bees, which use the oils to charm their mates. *Calceolaria uniflora* is an exception, because it does not produce oils. Instead, it develops a large white food body on the pouch, which is pecked away by hungry birds, inadvertently pollinating the flowers.

Calceolaria diffusa

A
B
C
D
E
F
G
H
I
J
K
L
M
N
O
P
Q
R
S
T
U
V
W
X
Y
Z

i

Callicarpa dichotoma   *Callistemon speciosus*

Calandrinia
kah-lan-DRI-nee-uh
Named after Jean-Louis Calandrini
(1703–58), Swiss scientist (*Portulacaceae*)

calandrinioides
ka-lan-DREEN-ee-oy-deez
Resembling *Calandrinia*, as in *Ranunculus
calandrinioides*

Calanthe
kah-LAN-thee
From Greek *kallos*, meaning "beautiful," and
anthos, meaning "flower" (*Orchidaceae*)

Calathea
kah-LA-thee-uh
From Greek *kalathos*, meaning "basket,"
because the flowers are enclosed within
bracts (*Marantaceae*)

calcaratus
kal-ka-RAH-tus
calcarata, calcaratum
With spurs, as in *Viola calcarata*

calcareus
kal-KAH-ree-us
calcarea, calcareum
Relating to lime, as in *Titanopsis calcarea*

Calceolaria
kal-see-oh-LAIR-ee-uh
From Latin *calceolus*, "a slipper," in reference
to the pouched petal (*Calceolariaceae*)

Calendula
kah-LEN-dew-luh
From Latin *calendae*, meaning "first day of
the month," an allusion to the marigold's
long flowering period (*Asteraceae*)

calendulaceus
kal-en-dew-LAY-see-us
calendulacea, calendulaceum
The color of the yellow-flowered marigold
(*Calendula officinalis*), as in *Rhododendron
calendulaceum*

Calibrachoa
kah-lee-bra-KOH-uh
Named after Antonio de la Cal y Bracho
(1766–1833), Mexican botanist and
pharmacologist (*Solanaceae*)

californicus
kal-ih-FOR-nih-kus
californica, californicum
Connected with California, as in
Zauschneria californica

Calla
KAH-luh
From Greek *kallos*, meaning "beautiful"
(*Araceae*)

calleryanus
kal-lee-ree-AH-nus
calleryana, calleryanum
Named after Joseph-Marie Callery (1810–
62), nineteenth-century French missionary
who was a plant hunter in France, as in *Pyrus
calleryana*

Calliandra
ka-lee-AN-druh
From Greek *kallos*, meaning "beautiful," and
andros, meaning "stamen" (*Fabaceae*)

callianthus
kal-lee-AN-thus
calliantha, callianthum
With beautiful flowers, as in *Berberis
calliantha*

Callicarpa [1]
ka-lee-KAR-puh
From Greek *kallos*, meaning "beautiful," and
karpos, "fruit" (*Lamiaceae*)

callicarpus
kal-ee-KAR-pus
callicarpa, callicarpum
With beautiful fruit, as in *Sambucus
callicarpa*

Calochortus clavatus

Calystegia sepium

Callirhoe
ka-lee-ROW-ee
From Greek *kallos*, meaning "beautiful," and *rhoias*, a corn poppy, or after the daughter of the Greek river god Achelous (*Malvaceae*)

Callisia
ka-LIZ-ee-uh
From Greek *kallos*, meaning "beautiful" (*Commelinaceae*)

Callistemon [2]
ka-lis-STEM-on
From Greek *kallos*, meaning "beautiful," and *stemon*, meaning "stamen" (*Myrtaceae*)

Callistephus
ka-lee-STEF-us
From Greek *kallos*, meaning "beautiful," and *stephos*, meaning "crown," a reference to the shape of the inflorescence (*Asteraceae*)

Callitriche
ka-li-TREE-kee
From Greek *kallos*, meaning "beautiful," and *thrix*, meaning "hair," perhaps an allusion to the hairlike leaves (*Plantaginaceae*)

Callitris
ka-LEE-triss
From Greek *kallos*, meaning "beautiful," referring to the tree (*Cupressaceae*)

callizonus
kal-ih-ZOH-nus
callizona, callizonum
With beautiful bands or zones, as in *Dianthus callizonus*

callosus
kal-OH-sus
callosa, callosum
With thick skin; with calluses, as in *Saxifraga callosa*

Calluna
ka-LOO-nuh
From Greek *kallyno*, meaning "to sweep," because their stems were traditionally used as brooms (*Ericaceae*)

Calocedrus
kal-oh-SEED-rus
From Greek *kallos*, meaning "beautiful," and *Cedrus*, another conifer genus (*Cupressaceae*)

Calochortus [3]
kal-oh-KOR-tus
From Greek *kallos*, meaning "beautiful," and *chortus*, meaning "grass" (*Liliaceae*)

calophyllus
kal-ee-FIL-us
calophylla, calophyllum
With beautiful leaves, as in *Dracocephalum calophyllum*

Calothamnus
cal-oh-THAM-nus
From Greek *kallos*, meaning "beautiful," and *thamnos*, meaning "shrub" (*Myrtaceae*)

Caltha
KAL-thuh
From Greek *kalathos*, meaning "goblet," in reference to the cup-shaped flowers; in Latin, the name for a marigold (*Ranunculaceae*)

calvus
KAL-vus
calva, calvum
Without hair; naked, as in *Viburnum calvum*

Calycanthus
ca-lee-CAN-thus
From Greek *kalyx* ("calyx') and *anthos* ("flower'), referring to the blooms in which the sepals and petals are similar (*Calycanthaceae*)

calycinus
ka-lih-KEE-nus
calycina, calycinum
Like a calyx, as in *Halimium calycinum*

Calypso
ka-LIP-soh
From Greek mythology, Calypso was a nymph and the daughter of Atlas, who entertained Odysseus (*Orchidaceae*)

calyptratus
kal-lip-TRA-tus
calyptrata, calyptratum
With a calyptra, a caplike covering of a flower or fruit, as in *Podalyria calyptrata*

Calystegia [4]
ka-lee-STEE-jee-uh
From Greek *kalux*, meaning "cup," and *stegos*, meaning "covering," perhaps referring to the bracts that surround the flower (*Convolvulaceae*)

Calytrix
ka-LEE-tricks
From Greek *kalyx*, meaning "calyx," and *thrix*, meaning "hair," because the sepals have elongated tips (*Myrtaceae*)

Camassia
ka-MAS-see-uh
From the Shoshone name *camas* or *quamash* (*Asparagaceae*)

cambricus

KAM-brih-kus

cambrica, cambricum

Connected with Wales, as in *Papaver cambricum*

Camellia [1]

kuh-MEE-lee-uh

Named after Georg Josef Kamel (1661–1706), Moravian/Czech missionary and botanist (*Theaceae*)

Campanula (also campanula)

kam-PAN-yu-luh

From Latin *campana*, meaning "bell," plus diminutive *ella*, because the small flowers are bell-shaped (*Campanulaceae*)

campanularius

kam-pan-yoo-LAH-ri-us

campanularia, campanularium

With bell-shaped flowers, as in *Phacelia campanularia*

campanulatus

kam-pan-yoo-LAH-tus

campanulata, campanulatum

In the shape of a bell, as in *Enkianthus campanulatus*

campbellii

kam-BEL-ee-eye

Named after Dr Archibald Campbell (1805–74), Superintendent of Darjeeling, who accompanied Hooker to the Himalayas, as in *Magnolia campbellii*

campestris

kam-PES-tris

campestris, campestre

Of fields or open plains, as in *Acer campestre*

camphoratus

kam-for-AH-tus

camphorata, camphoratum

—

camphora

kam-for-AH

Like camphor, as in *Thymus camphoratus*

Campsis

KAMP-siss

From Greek *kampe*, meaning "bent," because the stamens curve (*Bignoniaceae*)

campylocarpus

kam-plo-KAR-pus

campylocarpa, campylocarpum

With curved fruit, as in *Rhododendron campylocarpum*

camtschatcensis

kam-shat-KEN-sis

camtschatcensis, camtschatcense

—

camtschaticus

kam-SHAY-tih-kus

camtschatica, camtschaticum

From or of the Kamchatka Peninsula, Russia, as in *Lysichiton camtschatcensis*

canadensis

ka-na-DEN-sis

canadensis, canadense

From Canada, although once also applied to northeastern parts of the United States, as in *Cornus canadensis*

canaliculatus

kan-uh-lik-yoo-LAH-tus

canaliculata, canaliculatum

With channels or grooves, as in *Erica canaliculata*

Cananga

kuh-NAN-guh

From Tagalog *ilang-ilang*, the name for *C. odorata*; the vernacular name "ylang-ylang" has the same origin (*Annonaceae*)

canariensis

kuh-nair-ee-EN-sis

canariensis, canariense

From the Canary Islands, Spain, as in *Phoenix canariensis*

Canarina

cah-nair-EE-nuh

Named for the Canary Islands, where one species (*C. canariensis*) is native (*Campanulaceae*)

canbyi

KAN-bee-eye

Named after William Marriott Canby (1831–1904), American botanist, as in *Quercus canbyi*

cancellatus

kan-sell-AH-tus

cancellata, cancellatum

With cross bars, as in *Phlomis cancellata*

1

Camellia japonica

Rosa canina

candelabrum
kan-del-AH-brum
Branched like a candelabra, as in *Salvia candelabrum*

candicans
KAN-dee-kanz
—

candidus
KAN-dee-dus
candida, candidum
Shining white, as in *Echium candicans*

canescens
kan-ESS-kenz
With off-white or gray hairs, as in *Populus × canescens*

caninus [2]
kay-NEE-nus
canina, caninum
Relating to dogs, often meaning inferior, as in *Rosa canina*

Canistrum
kah-NISS-trum
From Greek *kanistron*, a type of basket, perhaps alluding to the rosette shape of the foliage or the inflorescence (*Bromeliaceae*)

Canna
KA-nuh
From Greek *kanna*, a reedlike plant (*Cannaceae*)

Cannabis
KA-nuh-biss
From Greek *kannabis*, meaning "hemp" (*Cannabaceae*)

cannabinus
kan-na-BEE-nus
cannabina, cannabinum
Like hemp (*Cannabis*), as in *Eupatorium cannabinum*

Cannomois
KAH-no-moys
From Greek *kanna*, meaning "reed," and *omoios*, meaning "similar" (*Restionaceae*)

cantabricus
kan-TAB-rih-kus
cantabrica, cantabricum
Connected with the Cantabria region of Spain, as in *Narcissus cantabricus*

Cantua
KAN-too-uh
From the indigenous Peruvian name for the plant (*Polemoniaceae*)

canus
kan-nus
cana, canum
Off-white; an ash color, as in *Calceolaria cana*

capensis
ka-PEN-sis
capensis, capense
From the Cape of Good Hope, South Africa, as in *Phygelius capensis*

capillaris
kap-ill-AH-ris
capillaris, capillare
Particularly slender, like fine hair, as in *Tillandsia capillaris*

capillatus
kap-ill-AH-tus
capillata, capillatum
With fine hairs, as in *Stipa capillata*

capillifolius
kap-ill-ih-FOH-lee-us
capillifolia, capillifolium
With hairy leaves, as in *Eupatorium capillifolium*

capilliformis
kap-il-ih-FOR-mis
capilliformis, capilliforme
Like hair, as in *Carex capilliformis*

capillipes
cap-ILL-ih-peez
With slender feet, as in *Acer capillipes*

capillus-veneris
KAP-il-is VEN-er-is
Venus's hair, as in *Adiantum capillus-veneris*

capitatus
kap-ih-TAH-tus
capitata, capitatum
Flowers, fruit, or whole plant growing in a dense head, as in *Cornus capitata*

capitellatus
kap-ih-tel-AH-tus
capitellata, capitellatum
—

capitellus
kap-ih-TELL-us
capitella, capitellum
—

capitulatus
kap-ih-tu-LAH-tus
capitulata, capitulatum
With a small head, as in *Primula capitellata*

cappadocicus
kap-puh-doh-SIH-kus
cappadocica, cappadocicum
Connected with the ancient province of
Cappadocia, Asia Minor, as in *Omphalodes
cappadocica*

Capparis
KAH-par-iss
From Arabic *kabar*, meaning "caper,"
C. spinosa (*Capparidaceae*)

capreolatus
kap-ree-oh-LAH-tus
capreolata, capreolatum
With tendrils, as in *Bignonia capreolata*

capreus
KAP-ray-us
caprea, capreum
Relating to goats, as in *Salix caprea*

capricornis
kap-ree-KOR-nis
capricornis, capricorne
Of or below the Tropic of Capricorn in the
southern hemisphere; shaped like a goat's
horn, as in *Astrophytum capricorne*

caprifolius
kap-rih-FOH-lee-us
caprifolia, caprifolium
With leaves having some characteristic of
goats, as in *Lonicera caprifolium*

Capsicum
KAP-sih-kum
From Greek *kapto*, meaning "to bite,"
referring to the pepper's bitter taste
(*Solanaceae*)

capsularis
kap-SYOO-lah-ris
capsularis, capsulare
With capsules, as in *Corchorus capsularis*

caracasanus
kar-ah-ka-SAH-nus
caracasana, caracasanum
Connected with Caracas, Venezuela, as in
Serjania caracasana

Caragana
ka-ruh-GAH-nuh
From Mongolian *caragan*, the name for
C. arborescens (*Fabaceae*)

Caralluma
ka-ruh-LOO-muh
Possibly from Telugu (India) name *car-allam*
(*Apocynaceae*)

Cardamine
kar-DA-mi-nee
From Greek *kardamon*, meaning "cress"
(*Brassicaceae*)

cardinalis
kar-dih-NAH-lis
cardinalis, cardinale
Bright scarlet; cardinal red, as in *Lobelia
cardinalis*

Cardiocrinum
kar-dee-oh-KRY-num
From Greek *kardia*, meaning "heart," and
krinon, meaning "lily," because this lily
relative has heart-shaped leaves (*Liliaceae*)

cardiopetalus
kar-dee-oh-PET-uh-lus
cardiopetala, cardiopetalum
With heart-shaped petals, as in *Silene
cardiopetala*

Cardiospermum
kar-dee-oh-SPER-mum
From Greek *kardia*, meaning "heart," and
spermum, meaning "seed," because the black
seed each has a white, heart-shaped mark
(*Sapindaceae*)

carduaceus
kard-yoo-AY-see-us
carduacea, carduaceum
Like a thistle, as in *Salvia carduacea*

cardunculus
kar-DUNK-yoo-lus
carduncula, cardunculum
Like a small thistle, as in *Cynara cardunculus*

Carduus
KAR-doo-us
From classical Greek name for thistle
(*Asteraceae*)

Carex
KAIR-ex
From classical Latin name for a sedge
(*Cyperaceae*)

caribaeus
kuh-RIB-ee-us
caribaea, caribaeum
Connected with the Caribbean, as in
Pinus caribaea

Carica
KA-ri-kuh
From Latin for dried fig (*Ficus carica*), to
which the fruit (pawpaw) may resemble, or
from their similar leaves, or possibly due to
the mistaken belief that these plants hailed
from ancient Caria in modern-day Turkey
(*Caricaceae*)

caricinus
kar-ih-KEE-nus
caricina, caricinum
—

caricosus
kar-ee-KOH-sus
caricosa, caricosum
Like sedge (*Carex*), as in *Dichanthium
caricosum*

carinatus
kar-IN-uh-tus
carinata, carinatum
—

cariniferus
kar-in-IH-fer-us
carinifera, cariniferum
With a keel, as in *Allium carinatum*

carinthiacus
kar-in-thee-AH-kus
carinthiaca, carinthiacum
Connected with the Carinthia region of
Austria, as in *Wulfenia carinthiaca*

Carissa
kuh-RIS-uh
Possibly from Sanskrit *krishnapakphula*,
the name for *C. carandas* (*Apocynaceae*)

Carduus crispa

carlesii

KARLS-ee-eye

Named after William Richard Carles
(1848–1929) of the British consular service
in China, who collected plants in Korea, as
in *Viburnum carlesii*

Carlina

kar-LIE-nuh

Named after Charlemagne (742–814), Holy
Roman Emperor, after he supposedly cured
a plague among his soldiers using this plant
(*Asteraceae*)

Carludovica

car-loo-doh-VEE-kuh

Named after Charles IV of Spain
(1748–1819) and his wife Maria Luisa
(1751–1819) (*Cyclanthaceae*)

Carmichaelia

kar-my-KEE-lee-uh

Named after Dugald Carmichael
(1772–1827), Scottish surgeon and
naturalist (*Fabaceae*)

carminatus

kar-MIN-uh-tus

carminata, carminatum

—

carmineus

kar-MIN-ee-us

carminea, carmineum

Carmine; bright crimson, as in *Metrosideros
carminea*

Carnegiea

kar-NEE-gee-uh

Named after Andrew Carnegie
(1835–1919), Scottish industrialist and
philanthropist (*Cactaceae*)

carneus

KAR-nee-us

carnea, carneum

Flesh color; deep pink, as in *Androsace
carnea*

carnicus

KAR-nih-kus

carnica, carnicum

Like flesh, as in *Campanula carnica*

carniolicus

kar-nee-OH-lih-kus

carniolica, carniolicum

Connected with the historical region of
Carniola, now in Slovenia, as in *Centaurea
carniolica*

Hoya carnosa

carnosulus

karn-OH-syoo-lus

carnosula, carnosulum

Somewhat fleshy, as in *Hebe carnosula*

carnosus [1]

kar-NOH-sus

carnosa, carnosum

Fleshy, as in *Hoya carnosa*

carolinianus

kair-oh-lin-ee-AH-nus

caroliniana, carolinianum

—

carolinensis

kair-oh-lin-ee-EN-sis

carolinensis, carolinense

—

carolinus

kar-oh-LEE-nus

carolina, carolinum

From or of North Carolina or South
Carolina, as in *Halesia carolina*

carota

kar-OH-tuh

Carrot, as in *Daucus carota*

carpaticus

kar-PAT-ih-kus

carpatica, carpaticum

Connected with the Carpathian Mountains,
as in *Campanula carpatica*

Carpinus betulus

Carpenteria

kar-pen-TEE-ree-uh

Named after William Marbury Carpenter
(1811–48), American physician
(*Hydrangeaceae*)

carpinifolius

kar-pine-ih-FOH-lee-us

carpinifolia, carpinifolium

With leaves like hornbeam (*Carpinus*), as in
Zelkova carpinifolia

Carpinus [2]

kar-PIE-noos

From classical Latin name for hornbeam,
C. betulus (*Betulaceae*)

Carpobrotus

kar-poh-BRO-tus

From Greek *karpos*, meaning "fruit," and
brota, meaning "edible" (*Aizoaceae*)

Carrierea

kah-ree-AIR-ee-uh

Named after Élie-Abel Carrière (1818–96),
French botanist (*Salicaceae*)

Carthamus

KAR-thu-mus

From Arabic *qartam*, the name for safflower
(*C. tinctorius*), deriving from the verb "to
paint," because the flowers yield a brilliant
dye (*Asteraceae*)

carthusianorum

kar-thoo-see-an-OR-um

Of Grande Chartreuse, Carthusian monastery near Grenoble, France, as in *Dianthus carthusianorum*

cartilagineus

kart-ill-uh-GIN-ee-us

cartilaginea, cartilagineum

Like cartilage, as in *Blechnum cartilagineum*

cartwrightianus

kart-RITE-ee-AH-nus

Named after John Cartwright, nineteenth-century British consul to Constantinople (now Istanbul, Turkey) , as in *Crocus cartwrightianus*

Carum

KA-room

Both Latin and common name (caraway) originate with the Arabic *karawya*, the vernacular name, although Pliny suggested the name derives from Caria (now part of Turkey), its native country (*Apiaceae*)

Carya

KA-ree-uh

From Greek *karyon*, meaning "nut," these trees produce pecans (*Juglandaceae*)

caryophyllus [3]

kar-ee-oh-FIL-us

caryophylla, caryophyllum

Walnut-leaved (from Greek *karya*); likened to clove for their smell, and thence to clove pink, as in *Dianthus caryophyllus*

caryopteridifolius

kar-ee-op-ter-id-ih-FOH-lee-us

caryopteridifolia, caryopteridifolium

With leaves like *Caryopteris*, as in *Buddleja caryopteridifolia*

Caryopteris

ka-ree-OP-tuh-ris

From Greek *karyon* meaning "nut," and *pteron*, meaning "wing," because the fruit is winged (*Lamiaceae*)

Caryota

ka-ree-OH-tah

From Greek *karyon*, meaning "nut," because the fruit is vaguely nutlike (*Arecaceae*)

caryotideus

kar-ee-oh-TID-ee-us

caryotidea, caryotideum

Like fishtail palms (*Caryota*), as in *Cyrtomium caryotideum*

cashmerianus

kash-meer-ee-AH-nus

cashmeriana, cashmerianum

—

cashmirianus

kash-meer-ee-AH-nus

cashmiriana, cashmirianum

—

cashmiriensis

kash-meer-ee-EN-sis

cashmiriensis, cashmiriense

From or of Kashmir, as in *Cupressus cashmeriana*

caspicus

KAS-pih-kus

caspica, caspicum

—

caspius

KAS-pee-us

caspia, caspium

Connected with the Caspian Sea, as in *Ferula caspica*

Cassia

KA-see-uh

From classical Greek name for the plant that produces medicinal senna pods (*Fabaceae*)

Cassinia

ka-SIN-ee-uh

Named after Alexandre Henri Gabriel de Cassini (1781–1832), French botanist (*Asteraceae*)

Cassiope

ka-SEE-oh-pee

From Greek mythology, the wife of Cepheus, King of Aethiopia, and mother to Andromeda (*Ericaceae*)

3

Dianthus caryophyllus

Catalpa bignonioides

Catasetum purum

Castanea

kas-TAN-ee-uh

From Greek *kastanaion karuon*, or "nut from Castania," probably referring either to Kastanaia in Pontus, Turkey, or Castana in Thessaly, Greece (*Fagaceae*)

Castanopsis

kas-tan-OP-sis

Resembling the related genus *Castanea* (*Fagaceae*)

Castanospermum

kas-tan-oh-SPER-mum

A combination of Latin *castanea*, meaning "chestnut," and Greek *spermum*, meaning "seed," in reference to their large seed (*Fabaceae*)

Castilleja

kah-stee-LAY-uh

Named after Domingo Castillejo (eighteenth century, dates uncertain), Spanish botanist (*Orobanchaceae*)

Casuarina

kah-shoo-ah-REE-nuh

Named after the Australian cassowary (*Casuarius*) because the thin, pendulous stems resemble its plumage (*Casuarinaceae*)

Catalpa [1]

kah-TAL-puh

From the indigenous North American name (*Bignoniaceae*)

catalpifolius

ka-tal-pih-FOH-lee-us

catalpifolia, catalpifolium

With leaves like *Catalpa*, as in *Paulownia catalpifolia*

Catananche

kat-uh-NAN-kee

From Greek *katananke*, meaning "strong force," because it was a common ingredient in love potions (*Asteraceae*)

cataria

kat-AR-ee-uh

Relating to cats, as in *Nepeta cataria*

catarractae

kat-uh-RAK-tay

Of waterfalls, as in *Parahebe catarractae*

Catasetum [2]

kat-uh-SEE-toom

From Greek *kata*, meaning "downward," and *seta*, meaning "bristle," because the male flowers have hanging bristles that act as triggers, launching the pollen when touched (*Orchidaceae*)

catawbiensis

ka-taw-bee-EN-sis

catawbiensis, catawbiense

From the Catawba River, North Carolina, as in *Rhododendron catawbiense*

catesbyi

KAYTS-bee-eye

Named after Mark Catesby (1682–1749), British naturalist, as in *Sarracenia × catesbyi*

Catharanthus

kath-uh-RAN-thus

From Greek *katharos*, meaning "pure," and *anthos*, meaning "flower" (*Apocynaceae*)

catharticus

kat-AR-tih-kus

carthartica, catharticum

Cathartic; purgative, as in *Rhamnus cathartica*

cathayanus

kat-ay-YAH-nus

cathayana, cathayanum

—

cathayensis

kat-ay-YEN-sis

cathayensis, cathayense

From or of China, as in *Cardiocrinum cathayana*

Cattleya

KAT-lay-uh

Named after William Cattley (1788–1835), British merchant and horticulturist (*Orchidaceae*)

caucasicus

kaw-KAS-ih-kus

caucasica, caucasicum

Connected with the Caucasus, as in *Symphytum caucasicum*

caudatus

kaw-DAH-tus

caudata, caudatum

With a tail, as in *Asarum caudatum*

caulescens

kawl-ESS-kenz

With a stem, as in *Kniphofia caulescens*

cauliflorus

kaw-lih-FLOR-us

cauliflora, cauliflorum

With flowers on the stem or trunk, as in *Saraca cauliflora*

Caulophyllum

kaw-low-FIL-um

From Greek *kaulon*, meaning "stem," and *phyllon*, meaning "leaf," because these plants typically have a single stem with one divided leaf on top (*Berberidaceae*)

causticus
KAWS-tih-kus
caustica, causticum
With a caustic or burning taste, as in *Lithraea caustica*

cauticola
kaw-TIH-koh-luh
Growing on cliffs, as in *Sedum cauticola*

Cautleya
KAWT-ley-uh
Named after Sir Proby Thomas Cautley (1802–71), British engineer and palaeontologist (*Zingiberaceae*)

cautleyoides
kawt-ley-OY-deez
Resembling *Cautleya*, as in *Roscoea cautleyoides*

Cavendishia
kav-un-DISH-ee-uh
Named after William George Spencer Cavendish (1790–1858), 6th Duke of Devonshire, politician, and horticulturist (*Ericaceae*)

cavus
KA-vus
cava, cavum
Hollow, as in *Corydalis cava*

Ceanothus
see-uh-NOH-thoos
From Greek *keanothus*, the classical name for a spiny plant, but not this one (*Rhamnaceae*)

cebennensis
kae-ben-EN-sis
cebennensis, cebennense
From Cévennes, France, as in *Saxifraga cebennensis*

Cedronella
see-dro-NEL-luh
From the genus *Cedrus* (cedar), plus Latin diminutive *ella*, indicating small stature; fragrant herbaceous perennials (*Lamiaceae*)

Cedrus (also cedrus)
SEE-druss
The classical Latin name for cedar (*Pinaceae*)

Ceiba
SEE-buh
From indigenous South American name (*Malvaceae*)

celastrinus
seh-lass-TREE-nus
celastrina, celastrinum
Like bittersweet (*Celastrus*), as in *Azara celastrina*

Celastrus
sel-AS-trus
From Greek *kelastros*, meaning "evergreen tree," possibly holly (*Celastraceae*)

Celosia argentea

Celmisia
sel-MIZ-ee-uh
Named after Celmisios, son of Alciope in Greek mythology; this genus was so named because a related genus in South Africa already bore the name *Alciope* (now *Capelio*), (*Asteraceae*)

Celosia [3]
sel-OH-see-uh
From Greek *kelos*, meaning "burned," a reference to flower color (*Amaranthaceae*)

Celtis
KEL-tis
The classical Greek name for a tree with sweet fruit (*Cannabaceae*)

Centaurea
sen-TAW-ree-uh
After the centaurs of Greek mythology, who are said to have discovered the medicinal properties of this plant (*Asteraceae*)

Centaurium
sen-TAW-ree-um
For the centaur Chiron, an expert in medicinal herbs (*Gentianaceae*)

centifolius
sen-tih-FOH-lee-us
centifolia, centifolium
With many leaves; with a hundred leaves, as in *Rosa × centifolia*

Centradenia
sen-tra-DEE-nee-uh
From Greek *kentron*, meaning "spur," and *aden*, meaning "gland," because the anthers have a spurlike gland (*Melastomataceae*)

centralis
sen-tr-AH-lis
centralis, centrale
Central (for example, in distribution), as in *Diplocaulobium centrale*

centranthifolius
sen-tran-thih-FOH-lee-us
centranthifolia, centranthifolium
With leaves like valerian (*Centranthus*), as in *Penstemon centranthifolius*

Centranthus
sen-TRAN-thus
From Greek *kentron*, meaning "spur," and *anthos*, meaning "flower," because the flowers are spurred (*Caprifoliaceae*)

Ceratostigma willmottianum

cepa

KEP-uh

The Roman name for an onion, as in *Allium cepa*

Cephalanthus

kef-uh-LAN-thus

From Greek *kephale*, meaning "head," and *anthos*, meaning "flower," because the flowers are arranged in a globe-shaped inflorescence (*Rubiaceae*)

Cephalaria

kef-uh-LAIR-ee-uh

From Greek *kephale*, meaning "head," because the flowers are in circular clusters (*Caprifoliaceae*)

Cephalocereus

kef-al-oh-SER-ee-us

From Greek *kephale*, meaning "head," and *Cereus*, a related genus, because this cactus forms a woolly head when flowering (*Cactaceae*)

cephalonicus

kef-al-OH-nih-kus

cephalonica, cephalonicum

Connected with Cephalonia, Greece, as in *Abies cephalonica*

Cephalotaxus

kef-al-oh-TAX-us

From Greek *kephale*, meaning "head," and *Taxus*, a related genus, because the pollen cones are spherical (*Taxaceae*)

cephalotes

sef-ah-LOH-tees

Like a small head, as in *Gypsophila cephalotes*

ceraceus

ke-ra-KEE-us

ceracea, ceraceum

With a waxy texture, as in *Wahlenbergia ceracea*

ceramicus

ke-RA-mih-kus

ceramica, ceramicum

Like pottery, as in *Rhopaloblaste ceramica*

cerasiferus

ke-ra-SIH-fer-us

cerasifera, cerasiferum

A plant that bears cherries or cherrylike fruit, as in *Prunus cerasifera*

cerasiformis

see-ras-if-FOR-mis

cerasiformis, cerasiforme

Shaped like a cherry, as in *Oemleria cerasiformis*

cerasinus

ker-ras-EE-nus

cerasina, cerasinum

Cherry red, as in *Rhododendron cerasinum*

cerastiodes

ker-ras-tee-OY-deez

cerastioides

Resembling mouse-ear chickweed (*Cerastium*), as in *Arenaria cerastioides*

Cerastium

suh-RAS-tee-um

From Greek *keras*, meaning "horn," a reference to the shape of the fruit (*Caryophyllaceae*)

cerasus

KER-uh-sus

Latin for cherry, as in *Prunus cerasus*

Ceratopetalum

suh-rah-toh-PET-al-um

From Greek *keras*, meaning "horn," and *petalon*, meaning "petal," because one species has antlerlike petals (*Cunoniaceae*)

Ceratophyllum

suh-rah-toh-FIL-um

From Greek *keras*, meaning "horn," and *phyllon*, meaning "leaf," because the leaves resemble antlers (*Ceratophyllaceae*)

Ceratopteris

suh-rah-TOP-ter-us

From Greek *keras*, meaning "horn," and *Pteris*, a related genus (*Pteridaceae*)

Ceratostigma

suh-rah-to-STIG-muh

From Greek *keras*, meaning "horn," because the stigma has a hornlike protrusion (*Plumbaginaceae*)

Ceratozamia

suh-rah-to-ZAY-mee-uh

From Greek *keras*, meaning "horn," and *Zamia*, a related genus, because the cones have horned scales (*Zamiaceae*)

cercidifolius

ser-uh-sid-ih-FOH-lee-us

cercidifolia, cercidifolium

With leaves like redbud tree (*Cercis*), as in *Disanthus cercidifolius*

Cercidiphyllum

sur-sid-i-FIL-um

From the genus *Cercis*, plus Greek *phyllon*, meaning "leaf," because the two genera have similarly shaped foliage (*Cercidiphyllaceae*)

Cercis

SUR-sis

From Greek *kerkis*, meaning "weaver's shuttle," to which the pods resemble (*Fabaceae*)

cerealis

ser-ee-AH-lis

cerealis, cereale

Relating to agriculture, derived from Ceres, the goddess of farming, as in *Secale cereale*

CLERICAL CLASSIFICATION

From Copernicus to Gregor Mendel, clergy have long held an important role in the development of science. Historically, they were some of the only members of society to receive a university education, and while their clerical studies enabled them to care for the spiritual welfare of their respective flocks, scientific curiosity was far from uncommon. The ubiquitous genus *Buddleja* honors such a man, the Reverend Adam Buddle, vicar to a parish in Essex, England. He was an authority on bryophytes (mosses and liverworts) and wrote an unpublished Flora of England.

cerefolius
ker-ee-FOH-lee-us
cerefolia, cerefolium
With waxy leaves, as in *Anthriscus cerefolium*

Cereus
seh-REE-us
From Latin *cereus*, meaning "wax candle," a reference to the shape of this columnar cactus (*Cactaceae*)

Cerinthe
suh-RIN-thee
From Greek *keros*, meaning "wax," and *anthos*, meaning "flower," possibly due to the waxy texture of the flowers, or the mistaken belief that bees collected wax from the blooms (*Boraginaceae*)

ceriferus
ker-IH-fer-us
cerifera, ceriferum
Producing wax, as in *Morella cerifera*

cerinthoides
ser-in-THOY-deez
Resembling honeywort (*Cerinthe*), as in *Tradescantia cerinthoides*

cerinus
ker-REE-nus
cerina, cerinum
Waxy, as in *Narcissus cerinus*

cernuus
SER-new-us
cernua, cernuum
Drooping or nodding, as in *Enkianthus cernuus*

Ceropegia
seer-oh-PEE-gee-uh
From Greek *keros*, meaning "wax," and *pege*, meaning "fountain," because the flowers are waxy (*Apocynaceae*)

Cestrum
SES-trum
From Greek *kestron*, the name of a plant, or *kestrum*, a tool used for engraving, which the anthers supposedly resemble (*Solanaceae*)

ceterach
KET-er-ak
Derived from an Arabic word applied to spleenworts (*Asplenium*), as in *Asplenium ceterach*

Chaenomeles [2]
kah-NOM-uh-leez
From Greek *chaino*, meaning "to split," and *melon*, meaning "apple," because the fruit was thought to split open (*Rosaceae*)

Chaenostoma
kee-noh-STOW-muh
From Greek *chaino*, meaning "to split," and *stoma*, meaning "mouth," because these snapdragon-like flowers gape (*Scrophulariaceae*)

chaixii
kay-IKX-ee-eye
Named after Dominique Chaix (1730–99), French botanist, as in *Verbascum chaixii*

chalcedonicus
kalk-ee-DON-ih-kus
chalcedonica, chalcedonicum
Connected with Chalcedon, the ancient name for a district of Istanbul, Turkey, as in *Lychnis chalcedonica*

chamaebuxus
kam-ay-BUKS-us
Dwarf boxwood, as in *Polygala chamaebuxus*

Chamaecyparis [1]
kam-ee-SIP-uh-ris
From Greek *chamai*, meaning "on the ground" or "low growing," and *cyparissos*, meaning "cypress" (*Cupressaceae*)

chamaecyparissus
kam-ee-ky-pah-RIS-us
Like *Chamaecyparis*, as in *Santolina chamaecyparissus*

Chamaecytisus
kam-ee-sy-TIS-us
From Greek *chamai*, meaning "on the ground" or "low growing," and *Cytisus*, a related genus (*Fabaceae*)

Chamaedaphne
kam-ee-DAF-nee
From Greek *chamai*, meaning "on the ground" or "low growing," and *Daphne*, another genus, inappropriate for this upright plant (*Ericaceae*)

Chamaedorea
kam-ee-DORE-ee-uh
From Greek *chamai*, meaning "on the ground" or "low growing," and *dorea*, "a gift," referencing the small stature of many of these pretty palms (*Arecaceae*)

chamaedrifolius
kam-ee-drih-FOH-lee-us
chamaedrifolia, chamaedrifolium, chamaedryfolius, chamaedryfolia, chamaedryfolium
With leaves like *Chamaedrys*, as in *Aloysia chamaedrifolia* (Note: *Chamaedrys* is now listed under *Teucrium*).

Chamaelirium
kam-ee-LI-ree-um
From Greek *chamai*, meaning "on the ground" or "low growing," and *lirion*, meaning "lily" (*Melanthiaceae*)

Chamaemelum
kam-ee-MEL-um
From Greek *chamai*, meaning "on the ground" or "low growing," and *melon*, meaning "apple," because the foliage has an apple scent (*Asteraceae*)

Chamaenerion
kam-ee-NEER-ee-on
From Greek *chamai*, meaning "on the ground" or "low growing," and *Nerium*, the oleander genus, on account of their similar leaves (*Onagraceae*)

Chamaerops
KAM-air-opps
From Greek *chamai*, meaning "on the ground" or "low growing," and *rhops*, "a bush" (*Arecaceae*)

Chamelaucium
kam-ee-LAW-see-um
From Greek *chamai*, meaning "on the ground" or "low growing," and *leucos*, meaning "white," perhaps referencing flower color (*Myrtaceae*)

chantrieri
shon-tree-ER-ee
Named after the French nursery Chantrier Frères, as in *Tacca chantrieri*

charianthus
kar-ee-AN-thus
chariantha, charianthum
With elegant flowers, as in *Ceratostema charianthum*

Chasmanthe
kas-MAN-thee
From Greek *chasme*, meaning "gaping," and *anthos*, meaning "flower" (*Iridaceae*)

Chamaecyparis obtusa

Chaenomeles japonica

Chasmanthium
kas-MAN-thee-um
Same derivation as *Chasmanthe* (*Poaceae*)

chathamicus
chath-AM-ih-kus
chathamica, chathamicum
Connected with the Chatham Islands, in the South Pacific, as in *Astelia chathamica*

Cheilanthes
ky-LAN-theez
From Greek *cheilos*, meaning "lip," and *anthos*, meaning "flower," because the sporangia are tucked under the leaf edges (*Pteridaceae*)

cheilanthus
kay-LAN-thus
cheilantha, cheilanthum
With flowers that have a lip, as in *Delphinium cheilanthum*

cheiri
kye-EE-ee
Perhaps from the Greek word *cheir*, "a hand," as in *Erysimum cheiri*

chelidonioides
kye-li-don-OY-deez
Resembling greater celandine (*Chelidonium*), as in *Calceolaria chelidonioides*

Chelidonium
kel-i-DOW-nee-um
From Greek *chelidon*, meaning "swallow," because plants bloom when the swallows arrive in spring; a more fanciful derivation from Greek mythology is that swallows bathe their chick's eyes in saliva and this plant is used medicinally to treat ailments of the eye (*Papaveraceae*)

Chelone
kee-LOH-nee
From Greek *chelone*, meaning "tortoise," which the flowers resemble (*Plantaginaceae*)

Chelonopsis
kee-loh-NOP-sis
Resembling the genus *Chelone* (*Lamiaceae*)

Chenopodium
kee-noh-POD-ee-um
From Greek *chen*, meaning "goose," and *podion*, meaning "foot," which the leaves resemble (*Amaranthaceae*)

chilensis
chil-ee-EN-sis
chilensis, chilense
From Chile, as in *Blechnum chilense*

chiloensis
kye-loh-EN-sis
chiloensis, chiloense
From the island of Chiloe, Chile, as in *Fragaria chiloensis*

Chimaphila

ky-muh-FIL-uh

From Greek *cheima*, meaning "winter," and *philos*, meaning "loving," because these herbs are evergreen (*Ericaceae*)

Chimonanthus

kim-oh-NAN-thus

From Greek *cheima*, meaning "winter," and *anthos*, meaning "flower," referencing the bloom period (*Calycanthaceae*)

Chimonobambusa

kim-oh-no-bam-BOO-suh

From Greek *cheima*, meaning "winter," and *Bambusa*, a related genus, because some species shoot in winter (*Poaceae*)

chinensis

CHI-nen-sis

chinensis, chinense

From China, as in *Stachyurus chinensis*

Chionanthus

kee-oh-NAN-thus

From Greek *chion*, meaning "snow," and *anthos*, meaning "flower," because the blooms are white (*Oleaceae*)

Chionochloa

kee-on-oh-KLO-uh

From Greek *chion*, meaning "snow," and *chloa*, "a blade of grass," because some species grow above the snow line (*Poaceae*)

Chionohebe

kee-on-oh-HE-bee

From Greek *chion*, meaning "snow," and *Hebe*, a related genus (*Plantaginaceae*)

Chirita

chi-REE-tuh

From vernacular name in India for a species of gentian (*Gesneriaceae*)

Chlidanthus

kli-DAN-thus

From Greek *chlide*, meaning "luxury," and *anthos*, meaning "flower" (*Amaryllidaceae*)

Chloranthus

klor-AN-thus

From Greek *chloros*, meaning "green," and *anthos*, meaning "flower" (*Chloranthaceae*)

Chloris

KLOR-iss

From Greek mythology, Khloris was the goddess of flowers (*Poaceae*)

chlorochilon

klor-oh-KY-lon

With a green lip, as in *Cycnoches chlorochilon*

Chlorogalum

klor-oh-GA-lum

From Greek *chloros*, meaning "green," and *gala*, meaning "milk," because crushed bulbs produce a green sap (*Asparagaceae*)

chloropetalus

klo-ro-PET-al-lus

chloropetala, chloropetalum

With green petals, as in *Trillium chloropetalum*

Chlorophytum

klor-oh-FY-tum

From Greek *chloros*, meaning "green," and *phyton*, meaning "plant" (*Asparagaceae*)

Choisya [1]

CHOI-see-uh

Named after Jacques Denis Choisy (1799–1859), Swiss cleric and botanist (*Rutaceae*)

Chorizema

kor-is-ZEE-muh

From Greek *chora*, meaning "place," and *zema*, meaning "drink," because the plant was discovered growing over a usable water source; alternatively, could derive from Greek *chorizo*, meaning "to separate," and *nema*, meaning "thread," for reasons unknown (*Fabaceae*)

Chrysanthemum

kri-SAN-thuh-mum

From Greek *khrysos*, meaning "gold," and *anthos*, meaning "flower" (*Asteraceae*)

chrysanthus

kris-AN-thus

chrysantha, chrysanthum

With golden flowers, as in *Crocus chrysanthus*

chryseus

KRIS-ee-us

chrysea, chryseum

Golden, as in *Dendrobium chryseum*

1

Choisya ternata

Chrysosplenium oppositifolium

Cichorium intybus

Erodium cicutarium

chrysocarpus

kris-oh-KAR-pus

chrysocarpa, chrysocarpum

With golden fruit, as in *Crataegus chrysocarpa*

chrysocomus

kris-oh-KOH-mus

chrysocoma, chrysocomum

With golden hairs, as in *Clematis chrysocoma*

Chrysogonum

kry-soh-GOH-num

From Greek *khrysos*, meaning "gold," and *gonia*, meaning "knee," because the inflorescences develop at stem joints (*Asteraceae*)

chrysographes

kris-oh-GRAF-ees

With gold markings, as in *Iris chrysographes*

Chrysolepis

kry-soh-LEE-pis

From Greek *khrysos*, meaning "gold," and *lepis*, meaning "scale," because many parts of the plant have golden glands (*Fagaceae*)

chrysoleucus

kris-roh-LEW-kus

chrysoleuca, chrysoleucum

Gold and white, as in *Hedychium chrysoleucum*

chrysophyllus

kris-oh-FIL-us

chrysophylla, chrysophyllum

With golden leaves, as in *Phlomis chrysophylla*

Chrysosplenium [2]

kry-soh-SPLEE-nee-um

From Greek *khrysos*, meaning "gold," and *splen*, meaning "spleen," because this golden-flowered plant has alleged medicinal value (*Saxifragaceae*)

chrysostoma

kris-oh-STO-muh

With a golden mouth, as in *Lasthenia chrysostoma*

Chusquea

CHUS-kwee-uh

From indigenous South America name *chusque* (*Poaceae*)

Cibotium

si-BOH-tee-um

From Greek *kibotion*, meaning "box," because the sporangia are enclosed within the indusium (*Cibotiaceae*)

Cicer

SY-sur

The classical Latin name for chickpea (*Fabaceae*)

Cicerbita

sy-SUR-bi-tuh

From Italian name for related plant, either sowthistle (*Sonchus*) or perhaps chicory (*Cichorium*), (*Asteraceae*)

Cichorium [3]

ki-CHOR-ee-um

From Arabic *chikouryeh*, the vernacular name for chicory (*Asteraceae*)

cicutarius [4]

kik-u-tah-ree-us

cicutaria, cicutarium

Like poison hemlock (*Conium maculatum*, formerly *Cicuta*), as in *Erodium cicutarium*

ciliaris

sil-ee-AH-ris

ciliaris, ciliare

—

ciliatus

sil-ee-ATE-us

ciliata, ciliatum

With leaves and petals that are fringed with hairs, as in *Tropaeolum ciliatum*

cilicicus

kil-LEE-kih-kus

cilicica, cilicicum

Connected with Lesser Armenia (formerly Cilicia), as in *Colchicum cilicicum*

ciliicalyx

kil-LEE-kal-ux

With a fringed calyx, as in *Menziesia ciliicalyx*

ciliosus

sil-ee-OH-sus

ciliosa, ciliosum

With a small fringe, as in *Sempervivum ciliosum*

cinctus

SINK-tus

cincta, cinctum

With a girdle, as in *Angelica cincta*

Cineraria
sin-uh-RARE-ree-uh
From Latin *cinerea*, meaning "an ash color," due to the silvery leaves (*Asteraceae*)

cinerariifolius
sin-uh-rar-ee-ay-FOH-lee-us
cinerariifolia, cinerariifolium
With leaves like *Cineraria*, as in *Tanacetum cinerariifolium*

cinerarius
sin-uh-RAH-ree-us
cineraria, cinerarium
Ash-gray, as in *Centaurea cineraria*

cinerascens
sin-er-ASS-enz
Turning to ash-gray, as in *Senecio cinerascens*

cinereus
sin-EER-ee-us
cinerea, cinereum
The color of ash, as in *Veronica cinerea*

cinnabarinus
sin-uh-bar-EE-nus
cinnabarina, cinnabarinum
Vermilion, as in *Echinopsis cinnabarina*

cinnamomeus
sin-uh-MOH-mee-us
cinnamomea, cinnamomeum
Cinnamon brown, as in *Osmunda cinnamomea*

cinnamomifolius
sin-uh-mom-ih-FOH-lee-us
cinnamomifolia, cinnamomifolium
With leaves like cinnamon (*Cinnamomum*), as in *Viburnum cinnamomifolium*

Cinnamomum
sin-uh-MOH-mum
From Greek *kinnamomon*, the vernacular name for cinnamon (*Lauraceae*)

circinalis
kir-KIN-ah-lis
circinalis, circinale
Coiled in form, as in *Cycas circinalis*

circum-
Used in compound words to denote around

cirratus
sir-RAH-tus
cirrata, cirratum
—

cirrhosus
sir-ROH-sus
cirrhosa, cirrhosum
With tendrils, as in *Clematis cirrhosa*

Cirsium
SUR-see-um
From Greek *kirsion*, a type of thistle (*Asteraceae*)

cissifolius
kiss-ih-FOH-lee-us
cissifolia, cissifolium
With leaves like ivy (from the Greek *kissos*), as in *Acer cissifolium*

Cissus
SIS-oos
From Greek *kissos*, meaning "ivy," because this is a climbing plant (*Vitaceae*)

cistena
sis-TEE-nuh
Of dwarf habit, from the Sioux word for baby, as in *Prunus* × *cistena*

Cistus [1]
SIS-toos
From Greek *kistos*, the name for a red-flowered shrub (*Cistaceae*)

citratus
sit-TRAH-tus
citrata, citratum
Like *Citrus*, as in *Mentha citrata*

citrinus
sit-REE-nus
citrina, citrinum
Lemon yellow or like *Citrus*, as in *Callistemon citrinus*

citriodorus
sit-ree-oh-DOR-us
citriodora, citriodorum
With the scent of lemons, as in *Thymus citriodorus*

citrodora
sit-roh-DOR-uh
With a lemon scent, as in *Aloysia citrodora*

Citrullus
si-TROO-lus
From genus *Citrus*, plus diminutive *ella*, suggesting a resemblance between a watermelon (*C. lanatus*) and citrus fruit, or their color (*Cucurbitaceae*)

Citrus [2]
SI-troos
From classical Latin name for *C. medica*, although possibly derived from Greek *kedros* ("cedar"), because they share a similar fragrance (*Rutaceae*)

Cladanthus
kla-DAN-thus
From Greek *klados*, meaning "branch," and *anthos*, meaning "flower," because the inflorescences are at the branch tips (*Asteraceae*)

cladocalyx
kla-do-KAL-iks
From the Greek *klados*, "a branch," referring to flowers borne on leafless branches, as in *Eucalyptus cladocalyx*

Cladrastis
kla-DRAS-tis
From Greek *klados*, meaning "branch," and *thraustos*, meaning "fragile," because this tree's limbs are notoriously brittle (*Fabaceae*)

clandestinus
klan-des-TEE-nus
clandestina, clandestinum
Hidden; concealed, as in *Lathraea clandestina*

1

Cistus × *purpureus*

clandonensis

klan-don-EN-sis

From Clandon, England, as in *Caryopteris* × *clandonensis*

clarkei

KLAR-kee-eye

Commemorates various noteworthy people with the surname Clarke, including Charles Baron Clarke (1832–1906), superintendent of the Calcutta Botanic Gardens and former president of the Linnean Society, as in *Geranium clarkei*

Clarkia

KLAR-kee-uh

Named after William Clark (1770–1838), American explorer (*Onagraceae*)

clausus

KLAW-sus

clausa, clausum

Closed; shut, as in *Pinus clausa*

clavatus

KLAV-ah-tus

clavata, clavatum

Shaped like a club, as in *Calochortus clavatus*

Claytonia

klay-TOW-nee-uh

Named after John Clayton (1694–1773), British-born American botanist (*Montiaceae*)

claytonianus

klay-ton-ee-AH-nus

claytoniana, claytonianum

Named after John Clayton (1694–1773), plant collector in Virginia, as in *Osmunda claytoniana*

Cleistocactus

KLY-stoh-kak-tus

From Greek *cleistos*, meaning "closed," because the flowers of this cactus barely open (*Cactaceae*)

clematideus

klem-AH-tee-dus

clematidea, clematideum

Like *Clematis*, as in *Agdestis clematidea*

Clematis [3]

KLEM-uh-tis

From classical Greek name for a climbing plant (*Ranunculaceae*)

Cleome

KLEE-oh-mee

Derivation unknown, although perhaps from Greek *kleos*, meaning "glory," or after Kleo, a Greek muse (*Cleomaceae*)

Clerodendrum

kleh-roh-DEN-drum

From Greek *kleros*, meaning "chance," and *dendron*, meaning "tree," probably referencing medicinal properties (*Lamiaceae*)

2

Citrus trifoliata

3

Clematis 'Jackmanii'

1

Clivia miniata

2

Cocos nucifera

Clethra
KLETH-ruh
From Greek *klethra*, a name for alder (*Alnus*), which has similar leaves (*Clethraceae*)

clethroides
klee-THROY-deez
Resembling white alder (*Clethra*), as in *Lysimachia clethroides*

clevelandii
kleev-LAN-dee-eye
Named after Daniel Cleveland, nineteenth-century American collector and fern expert, as in *Bloomeria clevelandii*

Cleyera
klay-AIR-uh
Named after Andreas Cleyer (1634–98), German physician and botanist (*Pentaphylacaceae*)

Clianthus
klee-ANTH-us
From Greek *kleios*, meaning "glory," and *anthos*, meaning "flower" (*Fabaceae*)

Clintonia
klin-TOW-nee-uh
Named after De Witt Clinton (1769–1828), American politician and naturalist, sixth Governor of New York State (*Liliaceae*)

Clitoria
kly-TOR-ee-uh
From Latin *clitoris*, because the flowers resemble this part of the female anatomy (*Fabaceae*)

Clivia [1]
KLI-vee-uh
Named after Charlotte Florentia Clive (1787–1866), Duchess of Northumberland (*Amaryllidaceae*)

Clusia
KLOO-see-uh
Named after Charles de l'Écluse (1526–1609), Flemish botanist (*Clusiaceae*)

clusianus
kloo-zee-AH-nus
clusiana, clusianum
Named after Charles de l'Écluse (1526–1609), Flemish botanist, as in *Tulipa clusiana*

clypeatus
klye-pee-AH-tus
clypeata, clypeatum
Like the round Roman shield, as in *Fibigia clypeata*

clypeolatus
klye-pee-OH-la-tus
clypeolata, clypeolatum
Shaped like a shield, as in *Achillea clypeolata*

Clytostoma
kly-toh-STO-muh
From Greek *klytos*, meaning "beautiful," and *stoma*, meaning "mouth," in reference to the tubular flowers (*Bignoniaceae*)

cneorum
suh-NOR-um
From the Greek for a small shrublike olive, possibly a kind of *Daphne*; as in *Convolvulus cneorum*

coarctatus
koh-ARK-tah-tus
coarctata, coarctatum
Pressed or crowded together, as in *Achillea coarctata*

Cobaea
KOH-bee-uh
Named after Bernabé Cobo (1582–1657),
Spanish missionary (*Polemoniaceae*)

cocciferus
koh-KIH-fer-us
coccifera, cocciferum
—

coccigerus
koh-KEE-ger-us
coccigera, coccigerum
Producing berries, as in *Eucalyptus coccifera*

coccineus
kok-SIN-ee-us
coccinea, coccineum
Scarlet, as in *Musa coccinea*

Coccoloba
ko-koh-LOW-buh
From Greek *kokkos*, meaning "berry," and
lobos, meaning "pod," because the dry fruit
has a fleshy covering (*Polygonaceae*)

Coccothrinax
ko-koh-THRI-nax
From Greek *kokkos*, meaning "berry," and
Thrinax, a related genus (*Arecaceae*)

Cochlearia
kok-LEER-ee-uh
From Greek *kokhliarion*, meaning "spoon,"
the shape of the leaves in some species
(*Brassicaceae*)

cochlearis
kok-lee-AH-ris
cochlearis, cochleare
Shaped like a spoon, as in *Saxifraga cochlearis*

cochleatus
kok-lee-AH-tus
cochleata, cochleatum
Shaped like a spiral, as in *Lycaste cochleata*

cockburnianus
kok-burn-ee-AH-nus
cockburniana, cockburnianum
Named after the family Cockburn, residents
of China, as in *Rubus cockburnianus*

Cocos [2]
KOH-koss
From Portuguese *coco*, meaning "ape,"
because coconut shells have a facelike
appearance (*Arecaceae*)

Codiaeum
koh-die-EE-um
From Malay *kodiho*, the vernacular name
(*Euphorbiaceae*)

Codonanthe
koh-duh-NAN-thee
From Greek *codon*, meaning "bell," and
anthe, meaning "flower" (*Gesneriaceae*)

Codonopsis
koh-duh-NOP-sis
From Greek *codon*, meaning "bell," and *opsis*,
meaning "like" (*Campanulaceae*)

coelestinus
koh-el-es-TEE-nus
coelestina, coelestinum
—

coelestis
koh-el-ES-tis
coelestis, coeleste
Sky blue, as in *Phalocallis coelestis*

Coelia
SEE-lee-uh
From Greek *koilos*, meaning "hollow," due to
the mistaken belief that the pollen masses
(pollinia) are hollow (*Orchidaceae*)

Coelogyne
see-LOJ-uh-nee
From Greek *koilos*, meaning "hollow," and
gyne, meaning "female," because the stigma
is depressed (*Orchidaceae*)

coeruleus
ko-er-OO-lee-us
coerulea, coeruleum
Blue, as in *Satureja coerulea*

Coffea [3]
KOFF-ee-uh
From Dutch *koffie*, itself derived originally
from Arabic *qah-wah*, the vernacular name
for the drink (*Rubiaceae*)

3

Coffea arabica

cognatus

kog-NAH-tus

cognata, cognatum

Closely related to, as in *Acacia cognata*

Coix

KOH-ix

The Greek name for the sub-Saharan palm *Hyphaene thebaica*, although it is unknown why this was applied to a grass (*Poaceae*)

Cola

KOH-luh

From indigenous African name *kola* (*Malvaceae*)

Colchicum

KOL-chi-kum

Named after Colchis, a state on the Black Sea in modern-day Georgia, where autumn crocus is said to be plentiful (*Colchicaceae*)

colchicus

KOHL-chih-kus

colchica, colchicum

Connected with the coastal region of the Black Sea, Georgia, as in *Hedera colchica*

colensoi

co-len-SO-ee

Named after Revd William Colenso (1811–99), New Zealand plant collector, as in *Pittosporum colensoi*

Coleonema

koh-lee-oh-NEE-muh

From Greek *koleos*, meaning "sheath," and *nema*, meaning "thread," because some of the stamen filaments are enclosed within petal folds (*Rutaceae*)

Coleus

KOH-lee-us

From Greek *koleos*, meaning "sheath," because the stamen filaments fuse together, forming a tube around the style (*Lamiaceae*)

Colletia

ko-LEET-ee-uh

Named after Philibert Collet (1643–1718), French botanist (*Rhamnaceae*)

Collinsia

ko-LIN-see-uh

Named after Zaccheus Collins (1764–1831), American botanist (*Plantaginaceae*)

Collinsonia

ko-lin-SOH-nee-uh

Named after Peter Collinson (1694–1768), British merchant and horticulturist (*Lamiaceae*)

collinus

kol-EE-nus

collina, collinum

Relating to hills, as in *Geranium collinum*

Collomia

ko-LOH-mee-uh

From Greek *kolla*, meaning "glue," because the seed become mucilaginous when wet (*Polemoniaceae*)

Colocasia

ko-loh-KAY-see-uh

From Greek *kolokasia*, the classical name for an edible root, possibly of *Nelumbo nucifera* (*Araceae*)

colorans

kol-LOR-anz

—

coloratus

kol-or-AH-tus

colorata, coloratum

Color, as in *Silene colorata*

Colquhounia

koh-HOON-ee-uh

Named after Robert David Colquhoun (1786–1838), British soldier and plant collector (*Lamiaceae*)

colubrinus

kol-oo-BREE-nus

colubrina, colubrinum

Like a snake, as in *Opuntia colubrina*

columbarius

kol-um-BAH-ree-us

columbaria, columbarium

Like a dove, as in *Scabiosa columbaria*

columbianus

kol-um-bee-AH-nus

columbiana, columbianum

Connected with British Columbia, Canada, as in *Aconitum columbianum*

columellaris

kol-um-EL-ah-ris

columellaris, columellare

Relating to a small pillar or pedestal, as in *Callitris columellaris*

columnaris

kol-um-nah-ris

columnaris, columnare

In the shape of a column, as in *Eryngium columnare*

Columnea

ko-LUM-nee-uh

Named after Fabio Colonna (1567–1640), Italian botanist (*Gesneriaceae*)

Colutea

ko-LOO-tee-uh

From Greek *kolutea*, the classical name for these plants (*Fabaceae*)

Coluteocarpus

ko-loo-tee-oh-KAR-pus

From Greek *karpos*, meaning "fruit," because this plant's capsules resemble those of *Colutea* (*Brassicaceae*)

Colvillea

kol-VILL-ee-uh

Named after Charles Colville (1770–1843), British soldier and third Governor of Mauritius (*Fabaceae*)

colvillei

koh-VIL-ee-eye

Named after either Sir James William Colville (1801–80), Scottish lawyer and judge in Calcutta, or James Colville, nineteenth-century nurseryman, as in *Gladiolus* × *colvillei* (after the latter)

comans

KO-manz

—

comatus

kom-MAH-tus

comata, comatum

Tufted, as in *Carex comans*

Comarum

koh-MAR-um

From Greek *komaros*, the name for the strawberry tree (*Arbutus unedo*), because the fruit is said to appear similar (*Rosaceae*)

Combretum

kom-BREE-tum

From Latin for climbing plant, as used by Pliny, although he was referring to a different plant (*Combretaceae*)

Colchicum autumnale

Pyrus communis

Lycopodium complanatum

Commelina
ko-muh-LEE-nuh
Named after brothers Jan (1629–92) and Caspar Commelijn (1668–1731), Dutch botanists, and a third unnamed brother who died young. The flowers have two large, showy petals and one that is small and inconspicuous (*Commelinaceae*)

commixtus
kom-MIKS-tus
commixta, commixtum
Mixed; mingled together, as in *Sorbus commixta*

communis [1]
KOM-yoo-nis
communis, commune
Growing in groups; common, as in *Myrtus communis*

commutatus
kom-yoo-TAH-tus
commutata, commutatum
Changed, for example, when formerly included in another species, as in *Papaver commutatum*

comosus
kom-OH-sus
comosa, comosum
With tufts, as in *Eucomis comosa*

compactus
kom-PAK-tus
compacta, compactum
Compact; dense, as in *Pleiospilos compactus*

complanatus [2]
kom-plan-NAH-tus
complanata, complanatum
Flat; level, as in *Lycopodium complanatum*

complexus
kom-PLEKS-us
complexa, complexum
Complex; encircled; as in *Muehlenbeckia complexa*

complicatus
kom-plih-KAH-tus
complicata, complicatum
Complicated; complex, as in *Adenocarpus complicatus*

compressus
kom-PRESS-us
compressa, compressum
Compressed, flattened, as in *Conophytum compressum*

Comptonia [3]
komp-TOW-nee-uh
Named after Henry Compton (1632–1713), Bishop of London and patron of botany (*Myricaceae*)

comptoniana
komp-toh-nee-AH-nuh
For various people with the surname Compton, as in *Hardenbergia comptoniana*

concavus
kon-KAV-us
concava, concavum
Hollowed out, as in *Conophytum concavum*

conchifolius

con-chee-FOH-lee-us

conchifolia, conchifolium

With leaves shaped like seashells, as in *Begonia conchifolia*

concinnus

KON-kin-us

concinna, concinnum

With a neat or elegant form, as in *Parodia concinna*

concolor

KON-kol-or

All the same color, as in *Abies concolor*

condensatus

kon-den-SAH-tus

condensata, condensatum

—

condensus

kon-DEN-sus

condensa, condensum

Crowded together, as in *Alyssum condensatum*

confertiflorus

kon-fer-tih-FLOR-us

confertiflora, confertiflorum

With flowers crowded together, as in *Salvia confertiflora*

confertus

KON-fer-tus

conferta, confertum

Crowded together, as in *Polemonium confertum*

confusus

kon-FEW-sus

confusa, confusum

Confused or uncertain, as in *Sarcococca confusa*

congestus

kon-JES-tus

congesta, congestum

Congested or crowded together, as in *Aciphylla congesta*

conglomeratus

kon-glom-er-AH-tus

conglomerata, conglomeratum

Crowded together, as in *Cyperus conglomeratus*

conicus

KON-ih-kus

conica, conicum

In the shape of a cone, as in *Carex conica*

coniferus

koh-NIH-fer-us

conifera, coniferum

With cones, as in *Magnolia conifera*

Coniogramme

koh-nee-oh-GRA-mee

From Greek *konion*, meaning "dust," and *gramme*, meaning "line," because the sporangia are arranged along the veins (*Pteridaceae*)

Conium [4]

KOH-nee-um

From Greek *koneion*, meaning "hemlock," perhaps originating in *konas*, meaning "to whorl," because dizziness is a common symptom of hemlock poisoning (*Apiaceae*)

3

Comptonia peregrina

4

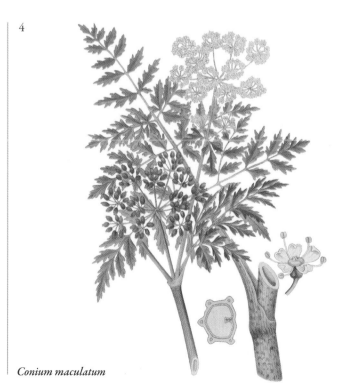

Conium maculatum

conjunctus

kon-JUNK-tus

conjuncta, conjunctum

Joined, as in *Alchemilla conjuncta*

connatus

kon-NAH-tus

connata, connatum

United; twin; opposite leaves joined together at the base, as in *Bidens connata*

conoideus

ko-NOY-dee-us

conoidea, conoideum

Resembling a cone, as in *Silene conoidea*

Conophytum

koh-no-FY-tum

From Greek *konos*, meaning "cone," and *phyton*, meaning "plant," because this succulent plant has conical leaves (*Aizoaceae*)

conopseus

kon-OP-see-us

conopsea, conopseum

Gnatlike, from Greek *konops*, as in *Gymnadenia conopsea*

consanguineus

kon-san-GWIN-ee-us

consanguinea, consanguineum

Related, as in *Vaccinium consanguineum*

Consolida

kon-SOL-i-duh

From Latin *consolidatus*, meaning "to become firm," from its reputed medicinal value (*Ranunculaceae*)

conspersus

kon-SPER-sus

conspersa, conspersum

Scattered, as in *Primula conspersa*

conspicuus

kon-SPIK-yoo-us

conspicua, conspicuum

Conspicuous, as in *Sinningia conspicua*

constrictus

kon-STRIK-tus

constricta, constrictum

Constricted, as in *Yucca constricta*

contaminatus

kon-tam-in-AH-tus

contaminata, contaminatum

Contaminated; defiled, as in *Lachenalia contaminata*

continentalis

kon-tin-en-TAH-lis

continentalis, continentale

Continental, as in *Aralia continentalis*

contortus

kon-TOR-tus

contorta, contortum

Twisted; contorted, as in *Pinus contorta*

contra-

Used in compound words to denote against

contractus

kon-TRAK-tus

contracta, contractum

Contracted; drawn together, as in *Fargesia contracta*

controversus

kon-troh-VER-sus

controversa, controversum

Controversial; doubtful, as in *Cornus contraversa*

Convallaria [1]

kon-val-AIR-ee-uh

From Latin *convallis*, meaning "valley," appropriate for lily-of-the-valley, *C. majalis* (*Asparagaceae*)

convallarioides

kon-va-lar-ee-OY-deez

Resembling lily-of-the-valley (*Convallaria*), as in *Speirantha convallarioides*

convolvulaceus

kon-vol-vu-la-SEE-us

convolvulacea, convolvulaceum

Somewhat like *Convolvulus*, as in *Codonopsis convolvulacea*

Convolvulus

kon-VOL-voo-lus

From Latin *convolvo*, meaning "to twine around," in reference to the manner of climbing employed by bindweeds (*Convolvulaceae*)

1

Convallaria majalis

Cordia myxa

Coreopsis verticillata

Conyza

kuh-NY-zuh

The classical name for fleabane, perhaps from Greek *konops*, meaning "flea," or *konis*, meaning "dust," because the powdered plant is used to repel insects (*Asteraceae*)

conyzoides

kon-ny-ZOY-deez

Resembling *Conyza*, as in *Ageratum conyzoides*

copallinus

kop-al-EE-nus

copallina, copallinum

With gum or resin, as in *Rhus copallinum*

Copernicia

kop-ur-NIS-ee-uh

Named after Nicolaus Copernicus (1473–1543), Polish astronomer (*Arecaceae*)

Copiapoa

koh-pee-uh-POH-uh

Named for the city of Copiapó in the Atacama Desert of Chile (*Cactaceae*)

Coprosma

kuh-PROZ-muh

From Greek *kopros*, meaning "dung," and *osme*, meaning "fragrance," relating to the fetid odor of the crushed foliage (*Rubiaceae*)

Coptis

KOP-tis

From Greek *kopto*, meaning "to cut," because the foliage is finely dissected (*Ranunculaceae*)

coralliflorus

kaw-lih-FLOR-us

coralliflora, coralliflorum

With coral-red flowers, as in *Lampranthus coralliflorus*

corallinus

kor-al-LEE-nus

corallina, corallinum

Coral red, as in *Ilex corallina*

coralloides

kor-al-OY-deez

Resembling coral, as in *Ozothamnus coralloides*

cordatus

kor-DAH-tus

cordata, cordatum

In the shape of a heart, as in *Pontederia cordata*

Cordia [2]

KOR-dee-uh

Named after Enricus Cordus (1486–1535) and his son Valerius Cordus (1515–1544), German physicians and botanists (*Boraginaceae*)

cordifolius

kor-di-FOH-lee-us

cordifolia, cordifolium

With heart-shaped leaves, as in *Crambe cordifolia*

cordiformis

kord-ih-FOR-mis

cordiformis, cordiforme

Heart-shaped, as in *Carya cordiformis*

Cordyline

kor-di-LIE-nee

From Greek *cordyle*, meaning "club," in reference to the shape of the trunk (*Asparagaceae*)

coreanus

kor-ee-AH-nus

coreana, coreanum

Connected with Korea, as in *Hemerocallis coreana*

Coreopsis [3]

kor-ee-OP-sis

From Greek *korios*, meaning "bedbug," because the seed resembles a tick (*Asteraceae*)

coriaceus

kor-ee-uh-KEE-us

coriacea, coriaceum

Thick, tough, and leathery, as in *Paeonia coriacea*

Coriandrum

kor-ee-AN-drum

From Greek *korios,* meaning "bedbug," to which the fragrance of the leaves and unripe fruit is supposed to resemble (*Apiaceae*)

Coriaria

kor-ee-AIR-ee-uh

From Latin *corium*, meaning "leather," because some species are used in the tanning process (*Coriariaceae*)

coriarius

kor-i-AH-ree-us

coriaria, coriarium

Like leather, as in *Caesalpinia coriaria*

coridifolius

kor-id-ee-FOH-lee-us

coridifolia, coridifolium

coriophyllus

kor-ee-uh-FIL-us

coriophylla, coriophyllum

With leaves like *Coris*, as in *Erica corifolia*

corifolius

kor-ee-FOH-lee-us

corifolia, corifolium

—

coriifolius

kor-ee-eye-FOH-lee-us

coriifolia, coriifolium

With leathery leaves, as in *Erica corifolia*

corniculatus

korn-ee-ku-LAH-tus

corniculata, corniculatum

With small horns, as in *Lotus corniculatus*

corniferus

korn-IH-fer-us

conifera, coniferum

—

corniger

korn-ee-ger

cornigera, cornigerum

With horns, as in *Coryphantha cornifera*

cornucopiae

korn-oo-KOP-ee-ay

Of a cornucopia or horn of plenty, as in *Fedia cornucopiae*

Cornus [1]

KOR-nus

From Latin *cornu*, meaning "hard," a reference to the wood; also, the classical name for cornelian cherry (*C. mas*), (*Cornaceae*)

cornutus

kor-NOO-tus

cornuta, cornutum

With horns or shaped like a horn, as in *Viola cornuta*

Corokia

kuh-ROH-kee-uh

From Maori (New Zealand) *korokio*, the vernacular name (*Argophyllaceae*)

corollatus

kor-uh-LAH-tus

corollata, corollatum

Like a corolla, as in *Fuchsia corollata*

coronans

kor-OH-nanz

—

coronatus

kor-oh-NAH-tus

coronata, coronatum

Crowned, as in *Lychnis coronata*

coronarius

kor-oh-NAH-ree-us

coronaria, coronarium

Used for garlands, as in *Anemone coronaria*

Coronilla

ko-roh-NIL-uh

From Latin *corona*, meaning "crown," plus diminutive *illa*, because the flowers are arranged in a small ring (*Fabaceae*)

coronopifolius

koh-ron-oh-pih-FOH-lee-us

coronopifolia, coronopifolium

With leaves like *Coronopus*, as in *Lobelia coronopifolia*

Correa

KOR-ee-uh

Named after José Francisco Correia da Serra (1750–1823), Portuguese abbot, politician, and scientist (*Rutaceae*)

corrugatus

kor-yoo-GAH-tus

corrugata, corrugatum

Corrugated; wrinkled, as in *Salvia corrugata*

corsicus

KOR-sih-kus

corsica, corsicum

Connected with Corsica, France, as in *Crocus corsicus*

Cortaderia

kor-tuh-DEER-ee-uh

From Argentinian name for pampas grass, *cortadera*, derived from Spanish *cortar*, meaning "to cut," because the leaves have sharp edges (*Poaceae*)

Cortusa

kor-TOO-suh

Named after Jacobi Antonii Cortusi (1513–93), Italian botanist (*Primulaceae*)

cortusoides

kor-too-SOY-deez

Resembling *Cortusa*, as in *Primula cortusoides*

Corydalis

ko-RI-dah-lis

From Greek *korydallis*, meaning "crested lark," because the flower spurs resemble the bird's crest (*Papaveraceae*)

1

Cornus canadensis

Corylus avellana

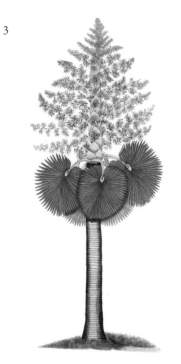

Corypha taliera

Cotoneaster moupinensis

corylifolius
kor-ee-lee-FOH-lee-us
corylifolia, corylifolium
With leaves like hazelnut (*Corylus*), as in
Betula corylifolia

Corylopsis
ko-ri-LOP-sis
Resembling the genus *Corylus*
(*Hamamelidaceae*)

Corylus [2]
KO-ri-lus
From Greek *korus*, meaning "helmet,"
a reference to the shape and strength of
the nuts (*Betulaceae*)

Corymbia
kuh-RIM-bee-uh
From Latin *corymbium*, meaning "cluster,"
because the flowers are arranged in corymbs
(*Myrtaceae*)

corymbiferus
kor-im-BIH-fer-us
corymbifera, corymbiferum
With a corymb, as in *Linum corymbiferum*

corymbiflorus
kor-im-BEE-flor-us
corymbiflora, corymbiflorum
With flowers produced in a corymb, as in
Solanum corymbiflorum

corymbosus
kor-rim-BOH-sus
corymbosa, corymbosum
With corymbs, as in *Vaccinium corymbosum*

Corynocarpus
kuh-rye-noh-KAR-pus
From Greek *koryne*, meaning "club," and
karpos, meaning "fruit" (*Corynocarpaceae*)

Corypha [3]
kuh-RYE-fuh
From Greek *koryphe*, meaning "tip" or
"summit," because this palm produces a
massive inflorescence that terminates the
trunk (*Arecaceae*)

Coryphantha
ko-ree-FAN-thuh
From Greek *koryphe*, meaning "tip," and
anthos, meaning "flower," because the
flowers appear at the stem tip, unlike in the
closely-related *Mammillaria*, where they
develop in a ring around the stem
(*Cactaceae*)

cosmophyllus
kor-mo-FIL-us
cosmophylla, cosmophyllum
With leaves like *Cosmos*, as in *Eucalyptus
cosmophylla*

Cosmos
KOZ-moss
From Greek *kosmo*, meaning "ornamental"
(*Asteraceae*)

costatus
kos-TAH-tus
costata, costatum
With ribs, as in *Aglaonema costatum*

Costus (also costus)
KOS-toos
From Greek *kostos*, meaning "aromatic
plant," in reference to the fragrant root,
although also applied to *Saussurea costus*
(*Costaceae*)

cotinifolius
kot-in-ih-FOH-lee-us
cotinifolia, cotinifolium
With leaves like smoke tree (*Cotinus*), as in
Euphorbia cotinifolia

Cotinus
KOT-i-nus
From Greek *kotinus*, meaning "olive"
(*Anacardiaceae*)

Cotoneaster [4]
(also cotoneaster)
kot-oh-nee-AS-ter
From Latin *cotoneum*, meaning "quince," and
aster, meaning "similar" (*Rosaceae*)

Cotula

KOT-ew-luh

From Greek *kotule*, meaning "small cup," alluding to the small capitulae (*Asteraceae*)

Cotyledon

kot-i-LEE-don

From Greek *kotyle*, meaning "cup-shaped hollow," because some species have concave leaves (*Crassulaceae*)

coulteri

kol-TER-ee-eye

Named after Dr. Thomas Coulter (1793–1843), Irish botanist, as in *Romneya coulteri*

coum

KOO-um

Connected with Kos, Greece, as in *Cyclamen coum*

Crambe

KRAM-bee

From Greek *krambe*, meaning "cabbage" (*Brassicaceae*)

Craspedia

kras-PEE-dee-uh

From Greek *kraspedon*, meaning "hem," because some species have a woolly fringe on the leaves (*Asteraceae*)

crassicaulis

krass-ih-KAW-lis

crassicaulis, crassicaule

With a thick stem, as in *Begonia crassicaulis*

crassifolius

krass-ih-FOH-lee-us

crassifolia, crassifolium

With thick leaves, as in *Pittosporum crassifolium*

crassipes

KRASS-ih-peez

With thick feet or thick stems, as in *Quercus crassipes*

crassiusculus

krass-ih-US-kyoo-lus

crassiuscula, crassiusculum

Quite thick, as in *Acacia crassiuscula*

Crassula [1]

KRAS-ew-luh

From Greek *krassus*, meaning "thick," because many species have succulent leaves (*Crassulaceae*)

Crassula coccinea

Crocus speciosus

crassus

KRASS-us

crassa, crassum

Thick; fleshy, as in *Asarum crassum*

crataegifolius

krah-tee-gi-FOH-lee-us

crataegifolia, crataegifolium

With leaves like hawthorn (*Crataegus*), as in *Acer crataegifolium*

Crataegus

kra-TEE-gus

From Greek *kratos*, meaning "strength," and *akis*, meaning "sharp tip," in reference to the thorns (*Rosaceae*)

crenatiflorus

kren-at-ih-FLOR-us

crenatiflora, crenatiflorum

With flowers cut into rounded scallops, as in *Calceolaria crenatiflora*

crenatus

kre-NAH-tus

crenata, crenatum

Scalloped, crenate, as in *Ilex crenata*

crenulatus

kren-yoo-LAH-tus

crenulata, crenulatum

Scalloped, as in *Boronia crenulata*

crepidatus

krep-id-AH-tus

crepidata, crepidatum

Shaped like a sandal or slipper, as in *Dendrobium crepidatum*

Crepis

KRE-pis

From Greek *krepis*, meaning "slipper," perhaps referring to the shape of the fruit (*Asteraceae*)

crepitans

KREP-ih-tanz

Rustling; crackling, as in *Hura crepitans*

cretaceus

kret-AY-see-us

cretacea, cretaceum

Relating to chalk, as in *Dianthus cretaceus*

creticus

KRET-ih-kus

cretica, creticum

Connected with Crete, Greece, as in *Pteris cretica*

crinitus

krin-EE-tus

crinita, crinitum

With long, weak hairs, as in *Acanthophoenix crinita*

Crinodendron

kry-noh-DEN-dron

From Greek *krinon*, meaning "lily," and *dendron*, meaning "tree" (*Elaeocarpaceae*)

Crinum

KRY-num

From Greek *krinon*, meaning "lily" (*Amaryllidaceae*)

crispatus

kriss-PAH-tus

crispata, crispatum

—

crispus

KRISP-us

crispa, crispum

Closely curled, as in *Mentha crispa*

cristatus

kris-TAH-tus

cristata, cristatum

With tassel-like tips, as in *Iris cristata*

crithmifolius

krith-mih-FOH-lee-us

crithmifolia, crithmifolium

With leaves like *Crithmum*, as in *Achillea crithmifolia*

—

Crithmum

KRITH-moom

From Greek *krithe*, meaning "barley," because the ribbed seed resembles this grain

crocatus

kroh-KAH-tus

crocata, crocatum

—

croceus

KRO-kee-us

crocea, croceum

Saffron yellow, as in *Tritonia crocata*

Crocosmia

kro-KOZ-mee-uh

From Greek *krokos*, meaning "saffron," and *osme*, meaning "fragrance," because this *Crocus* relative has scented flowers (*Iridaceae*)

crocosmiiflorus

kroh-koz-mee-eye-FLOR-us

crocosmiiflora, crocosmiiflorum

With flowers like *Crocosmia*. The genus of *Crocosmia × crocosmiiflora* was originally *Montbretia*; the name therefore meant crocosmia-flowered montbretia

Crocus [2]

KRO-kus

From Greek *krokos*, meaning "saffron," extracted from *C. sativus* (*Iridaceae*)

Crossandra

kros-AND-ruh

From Greek *krossos*, meaning "fringe," and *andros*, meaning "male," because the anthers are fringed (*Acanthaceae*)

Crotalaria [3]

kro-tuh-LAIR-ee-uh

From Greek *krotalon*, meaning a "rattle," because the seed rattles within the pods (*Fabaceae*)

3

Crotalaria juncea

99

Croton
KRO-ton
From Greek *kroton*, meaning a "tick," because the seed resembles this parasite (*Euphorbiaceae*)

Crowea
KROW-ee-uh
Named after James Crowe (1750–1807), British surgeon and botanist (*Fabaceae*)

cruciatus
kruks-ee-AH-tus
cruciata, crusiatum
In the shape of a cross, as in *Gentiana cruciata*

cruentus
kroo-EN-tus
cruenta, cruentum
Bloody, as in *Lycaste cruenta*

crus-galli
krus GAL-ee
Cockspur, as in *Crataegus crus-galli*

crustatus
krus-TAH-tus
crustata, crustatum
Encrusted, as in *Saxifraga crustata*

Cryptanthus
krip-TAN-thus
From Greek *krypte*, meaning "hidden," and *anthos*, meaning "flower," because the lower part of the flowers is concealed among bracts (*Bromeliaceae*)

Cryptocoryne
krip-toh-kor-AY-nee
From Greek *krypte*, meaning "hidden," and *koryne*, meaning "club," because the spadix is concealed within a tubular spathe (*Araceae*)

Cryptogramma
krip-toh-GRAM-uh
From Greek *krypte*, meaning "hidden," and *gramme*, meaning "line," because the line of sporangia are secreted within the curled leaf margin (*Pteridaceae*)

Cryptomeria [1]
krip-toh-MEER-ee-uh
From Greek *krypte*, meaning "hidden," and *meris*, meaning "part," because the fertile structures are hidden in cones (*Cupressaceae*)

crystallinus
kris-tal-EE-nus
crystallina, crystallinum
Crystalline, as in *Anthurium crystallinum*

Ctenanthe
ten-AN-thee
From Greek *ktenion*, meaning "comb," and *anthos*, meaning "flower," because the blooms emerge from parallel comblike bracts (*Marantaceae*)

cucullatus
kuk-yoo-LAH-tus
cucullata, cucullatum
Like a hood, as in *Viola cucullata*

cucumerifolius
ku-ku-mer-ee-FOH-lee-us
cucumerifolia, cucumerifolium
With leaves like a cucumber, as in *Cissus cucumerifolia*

cucumerinus
ku-ku-mer-EE-nus
cucumerina, cucumerinum
Like a cucumber, as in *Trichosanthes cucumerina*

Cucumis [2]
koo-KEW-mis
From Latin *cucumis*, meaning "cucumber" (*Cucurbitaceae*)

Cucurbita
koo-KUR-bit-uh
From Latin *cucurbita*, meaning "gourd" (*Cucurbitaceae*)

cultorum
kult-OR-um
Relating to gardens, as in *Trollius × cultorum*

cultratus
kul-TRAH-tus
cultrata, cultratum
—

cultriformis
kul-tre-FOR-mis
cultriformis, cultriforme
Shaped like a knife, as in *Angraecum cultriforme*

Cuminum
koo-MINE-um
From Greek *kuminon*, the classical name for cumin (*Apiaceae*)

1

Cryptomeria japonica

2

Cucumis melo

Helianthemum cupreum

cuneatus
kew-nee-AH-tus
cuneata, cuneatum
In the shape of a wedge, as in *Prostanthera cuneata*

cuneifolius
kew-nee-FOH-lee-us
cuneifolia, cuneifolium
With leaves shaped like a wedge, as in *Primula cuneifolia*

cuneiformis
kew-nee-FOR-mis
cuneiformis, cuneiforme
In the form of a wedge, as in *Hibbertia cuneiformis*

Cunninghamia
kuh-ning-HAY-mee-uh
Named after James Cunningham (dates unknown), British surgeon with East India Company (*Cupressaceae*)

cunninghamianus
kun-ing-ham-ee-AH-nus
cunninghamiana, cunninghamianum
—

cunninghamii
kuh-ning-HAY-mee-eye
May commemorate various people called Cunningham, including Alan Cunningham (1791–1839), British plant collector and botanist, as in *Archontophoenix cunninghamiana*

Cunonia
koo-NOH-nee-uh
Named after Johann Christian Cuno (1708–83), German poet, merchant, and botanist (*Cunoniaceae*)

Cuphea
KOO-fee-uh
From Greek *kyphos*, meaning "curved" or "humped," in reference to the shape of the fruit or the protuberance on the calyx (*Lythraceae*)

cupreatus
kew-pree-AH-tus
cupreata, cupreatum
—

cupreus [3]
kew-pree-US
cuprea, cupreum
The color of copper, as in *Alocasia cuprea*

cupressinus
koo-pres-EE-nus
cupressina, cupressinum
—

cupressoides
koo-press-OY-deez
Resembling cypress, as in *Fitzroya cupressoides*

Cupressus
koo-PRES-us
From Greek *kuparissos*, the classical name for a cypress tree, probably *C. sempervirens* (*Cupressaceae*)

curassavicus
ku-ra-SAV-ih-kus
curassavica, curassavicum
From Curaçao, Lesser Antilles, as in *Asclepias curassavica*

Curcuma
kur-KOO-muh
From Arabic *kurkum*, meaning "turmeric," extracted from *C. longa* (*Zingiberaceae*)

curtus
KUR-tus
curta, curtum
Shortened, as in *Ixia curta*

curvatus
KUR-va-tus
curvata, curvatum
Curved, as in *Adiantum curvatum*

curvifolius
kur-vi-FOH-lee-us
curvifolia, curvifolium
With curved leaves, as in *Ascocentrum curvifolium*

cuspidatus
kus-pi-DAH-tus
cuspidata, cuspidatum
With a stiff point, as in *Taxus cuspidata*

cuspidifolius
kus-pi-di-FOH-lee-us
cuspidifolia, cuspidifolium
With leaves with a stiff point, as in *Passiflora cuspidifolia*

Cyananthus
sy-uh-NAN-thus
From Greek *cyanos*, meaning "blue," and *anthos*, meaning "flower" (*Campanulaceae*)

cyaneus
sy-AN-ee-us
cyanea, cyaneum
—

cyanus
sy-AH-nus
Blue, as in *Allium cyaneum*

cyanocarpus
sy-an-o-KAR-pus
cyanocarpa, cyanocarpum
Bearing blue fruit, as in *Rhododendron cyanocarpum*

101

Cyanotis

sy-uh-NO-tis

From Greek *cyanos*, meaning "blue," and *otos*, meaning "ear," to which the petals resemble (*Commelinaceae*)

Cyathea

sy-ATH-ee-uh

From Greek *kyathios*, meaning "little cup," because the membrane covering the sporangia is cuplike in some species (*Cyatheaceae*)

cyatheoides

sigh-ath-ee-OY-deez

Resembling *Cyathea*, as in *Sadleria cyatheoides*

Cycas

SY-cass

From Greek *kykas*, probably a spelling error from *koikas*, meaning "palm tree," which these gymnosperms resemble (*Cycadaceae*)

Cyclamen [1]

SIK-luh-min

From Greek *kyklos*, meaning "circle," in reference to the circular tubers, the coiling stem holding the fruit, or perhaps the use of leaves in garlands (*Primulaceae*)

cyclamineus

SIGH-kluh-min-ee-us

cyclaminea, cyclamineum

Like *Cyclamen*, as in *Narcissus cyclamineus*

cyclocarpus

sigh-klo-KAR-pus

cyclocarpa, cyclocarpum

With fruit arranged in a circle, as in *Enterolobium cyclocarpum*

Cycnoches

sik-NOH-kees

From Greek *kykneios*, meaning "swan," referring to the gracefully arching column in male flowers (*Orchidaceae*)

Cydonia

sy-DOH-nee-uh

Named for the city of Kydonia (now Chania) in Crete, Greece (*Rosaceae*)

cylindraceus

sil-in-DRA-see-us

cylindracea, cylindraceum

cylindricus

sil-IN-drih-kus

cylindrica, cylindricum

Long and cylindrical, as in *Vaccinium cylindraceum*

Cylindropuntia

sil-in-droh-PUN-tee-uh

From Latin *cylindrus*, meaning "cylindrical," and *Opuntia*, a related genus; unlike *Opuntia*, which has flat, padlike stems, these cacti have cylindrical stems (*Cactaceae*)

cylindrostachyus

sil-in-dro-STAK-ee-us

cylindrostachya, cylindrostachyum

With a cylindrical spike, as in *Betula cylindrostachya*

Cymbalaria

sim-buh-LAIR-ee-uh

From Greek *kymbalon*, meaning "cymbal," an allusion to leaf shape in some species (*Plantaginaceae*)

Cymbidium

sim-BID-ee-um

From Greek *kymbe*, meaning "boat," because the lip of the flower is concave (*Orchidaceae*)

cymbiformis

sim-BIH-for-mis

cymbiformis, cymbiforme

Shaped like a boat, as in *Haworthia cymbiformis*

GENUS SPOTLIGHT

Cyperus

The genus *Cyperus* is found on all continents but Antarctica, and these grasslike plants are especially common in wet areas, so not surprisingly the genus name derives from a Greek word for marsh plant. The name does not appear connected with Cyprus (the island), which is thought to derive from the Greek word for cypress (*Cupressus sempervirens*) or an older word for copper. *Cyperus* plants are harvested for their edible tubers (for example, *C. esculentus*), can be troublesome weeds (for example, *C. rotundus*), or are grown as houseplants (for example, *C. alternifolius*). Papyrus paper was manufactured using the stems of *Cyperus papyrus*.

Cyperus alternifolius

Cymbopogon

sim-boh-POH-gon

From Greek *kymbe*, meaning "boat," and *pogon*, meaning "beard," because the flower spikelets have numerous bristlelike awns and boat-shaped glumes (*Poaceae*)

cymosus

sy-MOH-sus

cymosa, cymosum

With flower clusters that flower from the center outward, as in *Rosa cymosa*

Cynanchum

sy-NAN-kum

From Greek *kynos*, meaning "dog," and *anchein*, meaning "to choke," in reference to the many poisonous species (*Apocynaceae*)

Cynara

sy-NAR-uh

From Greek *kynara*, meaning "artichoke" (harvested from *C. cardunculus*), perhaps derived from *kynos*, meaning "dog," because the stiff involucral bracts are toothlike (*Asteraceae*)

cynaroides

sin-nar-OY-deez

Resembling *Cynara*, as in *Protea cynaroides*

Cynodon

SY-noh-don

From Greek *kynos*, meaning "dog," and *odontos*, meaning "tooth," but the object of this description is uncertain (*Poaceae*)

Cynoglossum

sy-noh-GLOS-um

From Greek *kynos*, meaning "dog," and *glossa*, meaning "tongue," because the coarsely textured leaves are tonguelike (*Boraginaceae*)

Cynosurus

sy-noh-SOOR-us

From Greek *kynos*, meaning "dog," and *oura*, meaning "tail," because the inflorescence is so shaped (*Poaceae*)

cyparissias

sy-pah-RIS-ee-as

Latin name for a kind of spurge, as in *Euphorbia cyparissias*

Cypella

sy-PEL-uh

From Greek *kyphella*, meaning the "hollow of the ear," because the innermost petals are cupped (*Iridaceae*)

1

Cyclamen hederifolium

Cyperus

SY-per-us

From Greek *kypeiros*, the classical name for an aromatic marsh plant, probably *C. longus* (*Cyperaceae*)

Cyphostemma

sy-foh-STEM-uh

From Greek *kyphos*, meaning "hump," and *stemma*, meaning "wreath," because foliage sits on a swollen succulent stem (*Vitaceae*)

Cypripedium

sip-ri-PEED-ee-um

From Greek *Kypris* (=Aphrodite) and *pedilon*, meaning "slipper" (*Orchidaceae*)

cyprius

SIP-ree-us

cypria, cyprium

Connected with the island of Cyprus, as in *Cistus* × *cyprius*

Cyrilla

si-RIL-uh

Named after Domenico Maria Leone Cirillo (1739–99), Italian physician and botanist (*Cyrillaceae*)

Cyrtanthus

sir-TAN-thus

From Greek *kyrtos*. meaning "curved," and *anthos* meaning "flower" (*Amaryllidaceae*)

Cyrtomium

sir-TOH-mee-um

From Greek *kyrtos*, meaning "curved," referring to the arching leaf veins or fronds (*Dryopteridaceae*)

Cyrtostachys

sir-toh-STA-kis

From Greek *kurtos*, meaning "curved," and *stachys*, meaning "ear of grain," a reference to the curved inflorescence stalks (*Arecaceae*)

Cystopteris

sis-TOP-ter-us

From Greek *kystos*, meaning "bladder," and *pteris*, meaning "fern," because the membrane protecting the sporangia is inflated when young (*Woodsiaceae*)

cytisoides

sit-iss-OY-deez

Resembling broom (*Cytisus*), as in *Lotus cytisoides*

Cytisus

sy-TIS-us

From Greek *kytisos*, the classical name for several leguminous shrubs (*Fabaceae*)

Daboecia
da-BEE-see-uh
Named after Saint Dabheog (dates unknown), Irish founder of a monastery in Lough Derg (*Ericaceae*)

Dacrydium
dah-CRID-ee-um
From Greek *dakrydion*, meaning "small tear," because the wood weeps resin (*Podocarpaceae*)

Dactylicapnos
dak-til-ee-KAP-nos
From Greek *daktylos*, meaning "finger," and *kapnos*, meaning "smoke"; the fruit is fingerlike, while the soft, flimsy foliage appears smoky (*Papaveraceae*)

dactyliferus
dak-ty-LIH-fer-us
dactylifera, dactyliferum
With fingers; finger-like, as in *Phoenix dactylifera*

Dactylis
DAK-til-iss
From Greek *daktylos*, meaning "finger" or probably "toe," the inflorescence resembling a bird foot (*Poaceae*)

dactyloides
dak-ty-LOY-deez
Resembling fingers, as in *Hakea dactyloides*

Dactylorhiza
dak-til-oh-RYE-zuh
From Greek *daktylos*, meaning "finger," and *rhiza*, meaning "root" (*Orchidaceae*)

Dahlia
DAH-lee-uh
Named after Anders Dahl (1751–89), Swedish botanist (*Asteraceae*)

dahuricus
da-HYUR-ih-kus
dahurica, dahuricum
Connected with Dahuria, a region incorporating parts of Siberia and Mongolia, as in *Codonopsis dahurica*

Dais
DYE-us
From Greek *dais*, meaning "torch," because the flower clusters sit atop long, straight stems (*Thymelaeaceae*)

Dalbergia
dal-BURG-ee-uh
Named after Nils Ericsson Dalberg (1736–1820), Swedish botanist (*Fabaceae*)

Dalea
DAY-lee-uh
Named after Samuel Dale (1659–1739), British apothecary and botanist (*Fabaceae*)

Dalechampia
dayle-CHAM-pee-uh
Named after Jacques Daléchamps (1513–88), French surgeon and botanist (*Euphorbiaceae*)

dalhousiae
dal-HOO-zee-ay
dalhousieae
Named after Susan Georgiana Ramsay, Marchioness of Dalhousie (1817–53), as in *Rhododendron dalhousiae*

dalmaticus
dal-MAT-ih-kus
dalmatica, dalmaticum
Connected with Dalmatia, Croatia, as in *Geranium dalmaticum*

damascenus
dam-ASK-ee-nus
damascena, damascenum
Connected with Damascus, Syria, as in *Nigella damascena*

dammeri
DAM-mer-ee
Named after Carl Lebrecht Udo Dammer (1860–1920), German botanist, as in *Cotoneaster dammeri*

Dampiera
dam-pee-AIR-uh
Named after William Dampier (1651–1715), British explorer and pirate (*Goodeniaceae*)

Danae
DAN-ay-ee
Named after Danaë, mother of Perseus in Greek mythology (*Asparagaceae*)

danfordiae
dan-FORD-ee-ay
Named after Mrs. C. G. Danford, nineteenth-century traveler, as in *Iris danfordiae*

danicus
DAN-ih-kus
danica, danicum
From Denmark, as in *Erodium danicum*

Danthonia
dan-THO-nee-uh
Named after Étienne Danthoine (1739–94), French botanist (*Poaceae*)

Daphne
DAF-nee
The classical Greek name for laurel and a nymph of the same name (*Thymelaeaceae*)

Daphniphyllum
daf-ni-FIL-um
Leaves resembling *Daphne*, from Greek *phyllon*, meaning "leaf" (*Daphniphyllaceae*)

daphnoides
daf-NOY-deez
Resembling *Daphne*, as in *Salix daphnoides*

darleyensis
dar-lee-EN-sis
Of Darley Dale nursery (James Smith & Sons), Derbyshire, England, as in *Erica × darleyensis*

Darlingtonia
dar-ling-TOH-nee-uh
Named after William Darlington (1782–1863), American politician and botanist (*Sarraceniaceae*)

Darmera
DAR-muh-ruh
Named after Karl Darmer (1843–1918), German botanist and horticulturist (*Saxifragaceae*)

Darwinia
dar-WIN-ee-uh
Named after Erasmus Darwin (1731–1802), British physician and botanist, grandfather to Charles Darwin (*Myrtaceae*)

darwinii
dar-WIN-ee-eye
Named after Charles Darwin (1809–82), British naturalist, as in *Berberis darwinii*

Dasiphora
da-zee-FOR-uh
From Greek *dasys*, meaning "hairy," and *phoros*, meaning "bearing," and may refer to the hairy bark, seed, or hypanthium (*Rosaceae*)

dasyacanthus

day-see-uh-KAN-thus

dasyacantha, dasyacanthum

With thick spines, as in *Escobaria dasyacantha*

dasyanthus

day-see-AN-thus

dasyantha, dasyanthum

With shaggy flowers, as in *Spiraea dasyantha*

dasycarpus

day-see-KAR-pus

dasycarpa, dasycarpum

With hairy fruit, as in *Angraecum dasycarpum*

Dasylirion

da-zee-LI-ree-on

From Greek *dasy*, meaning "dense," and *lirion*, meaning "lily," because this lily relation has an inflorescence with densely packed flowers (*Asparagaceae*)

dasyphyllus

das-ee-FIL-us

dasyphylla, dasyphyllum

With shaggy leaves, as in *Sedum dasyphyllum*

dasystemon

day-see-STEE-mon

With hairy stamens, as in *Tulipa dasystemon*

Datisca

dah-TIS-kuh

From Greek *datiska*, the vernacular name for *Catananche caerulea* (*Asteraceae*), probably from *datessai*, meaning "to heal," because the latter plant has medicinal properties; Linnaeus applied the name to the unrelated *Datisca* (*Datiscaceae*)

Datura [1]

dah-TEW-ruh

From Hindi *dhatura*, the vernacular name (*Solanaceae*)

daucifolius

daw-ke-FOH-lee-us

daucifolia, daucifolium

With leaves like carrot (*Daucus*), as in *Asplenium daucifolium*

daucoides

do-KOY-deez

Resembling carrot (*Daucus*), as in *Erodium daucoides*

Daucus [2]

DOW-kus

The Greek vernacular name for carrot, *D. carota* (*Apiaceae*)

dauricus

DOR-ih-kus

daurica, dauricum

Connected with Siberia, as in *Lilium dauricum*

Davallia

duh-VAY-lee-uh

Named after Edmund Davall (1762–98), Swiss botanist (*Davalliaceae*)

Davidia [3]

duh-VID-ee-uh

Named after Père Armand David (1826–1900), French missionary, botanist, and zoologist (*Nyssaceae*)

davidianus

duh-vid-ee-AH-nus

davidiana, davidianum

—

davidii

davidii duh-vid-ee-eye

Named after Père Armand David (1826–1900), French naturalist and missionary, as in *Buddleja davidii*

davuricus

dav-YUR-ih-kus

davurica, davuricum

Connected with Siberia, as in *Juniperus davurica*

dawsonianus

daw-son-ee-AH-nus

dawsoniana, dawsonianum

Named after Jackson T. Dawson (1841–1916), the first Superintendent of the Arnold Arboretum, Boston, Massachusetts, as in *Malus × dawsoniana*

Datura metel

Daucus carota

Davidia involucrata

Decaisnea fargesii

Rhododendron decorum

Decumaria barbara

dealbatus
day-al-BAH-tus
dealbata, dealbatum
Covered with an opaque white powder, as in *Acacia dealbata*

debilis
deb-IL-is
debilis, debile
Weak and frail, as in *Asarum debile*

Decaisnea [4]
duh-KAYZ-nee-uh
Named after Joseph Decaisne (1807–82), French botanist (*Lardizabalaceae*)

decaisneanus
de-kane-ee-AY-us
decaisneana, decaisneanum
—
decaisnei
de-KANE-ee-eye
Named after Joseph Decaisne (1807–82), French botanist, as in *Aralia decaisneana*

decandrus
dek-AN-drus
decandra, decandrum
With ten stamens, as in *Combretum decandrum*

decapetalus
dek-uh-PET-uh-lus
decapetala, decapetalum
With ten petals, as in *Caesalpinia decapetala*

Decarya
deh-CAH-ree-uh
Named after Raymond Decary (1891–1973), French botanist and anthropologist (*Didiereaceae*)

deciduus
dee-SID-yu-us
decidua, deciduum
Deciduous, as in *Larix decidua*

decipiens
de-SIP-ee-enz
Deceptive; not obvious, as in *Sorbus decipiens*

declinatus
dek-lin-AH-tus
declinata, declinatum
Bending downward, as in *Cotoneaster declinatus*

decompositus
de-kom-POZ-ee-tus
decomposita, decompostitum
Divided several times, as in *Paeonia decomposita*

decoratus
dek-kor-RAH-tus
decorata, decoratum
—
decorus [5]
dek-kor-RUS
decora, decorum
Decorative, as in *Rhododendron decorum*

decumanus
dek-yoo-MAH-nus
decumana, decumanum
Very large, as in *Phlebodium decumanum*

Decumaria [6]
dek-ew-MAH-ree-uh
From Latin *decem*, meaning "ten," because the flower parts are in tens (*Hydrangeaceae*)

decumbens
de-KUM-benz
Trailing with upright tips, as in *Correa decumbens*

decurrens
de-KUR-enz
Running down the stem, as in *Calocedrus decurrens*

decussatus
de-KUSS-ah-tus
decussata, decussatum
With leaves that are borne in pairs at right angles to each other, as in *Microbiota decussata*

deflexus
de-FLEKS-us
deflexa, deflexum
Bending downward, as in *Enkianthus deflexus*

deformis
de-FOR-mis
deformis, deforme
Deformed; misshapen, as in *Haemanthus deformis*

degronianum

de-gron-ee-AH-num

Named after Henri Joseph Degron, director of the French Post Office in Yokohama, 1865–80, as in *Rhododendron degronianum*

Deinanthe

dye-NAN-thee

From Greek *deinos*, meaning "wondrous," and *anthos*, meaning "flower" (*Hydrangeaceae*)

dejectus

dee-JEK-tus

dejecta, dejectum

Debased, as in *Opuntia dejecta*

delavayi

del-uh-VAY-ee

Named after Père Jean Marie Delavay (1834–95), French missionary, explorer, and botanist, as in *Magnolia delavayi*

delicatus

del-ih-KAH-tus

delicata, delicatum

Delicate, as in *Dendrobium* × *delicatum*

deliciosus

de-lis-ee-OH-sus

deliciosa, deliciosum

Delicious, as in *Monstera deliciosa*

Delonix [1]

deh-LON-iks

From Greek *delos*, meaning "visible," and *onux*, meaning "claw," because the petals have a narrow limb or claw (*Fabaceae*)

Delosperma

del-oh-SPER-muh

From Greek *delos*, meaning "visible," and *sperma*, meaning "seed," because the fruit opens to reveal the seed (*Aizoaceae*)

delphiniifolius

del-fin-uh-FOH-lee-us

delphiniifolia, delphiniifolium

With leaves like *Delphinium*, as in *Aconitum delphiniifolium*

Delphinium

del-FIN-ee-um

From Greek *delphin*, meaning "dolphin," because the blue flowers are said to resemble these aquatic mammals (*Ranunculaceae*)

deltoides

del-TOY-deez

—

deltoideus

el-TOY-dee-us

deltoidea, deltoideum

Triangular, as in *Dianthus deltoides*

demersus

DEM-er-sus

demersa, demersum

Living under water, as in *Ceratophyllum demersum*

deminutus

dee-MIN-yoo-tus

deminuta, deminutum

Small, diminished, as in *Rebutia deminuta*

demissus

dee-MISS-us

demissa, demissum

Hanging downward; weak, as in *Cytisus demissus*

Dendrobium

den-DRO-bee-um

From Greek *dendron*, meaning "tree," and *bios*, meaning "life," because most of these orchids are epiphytyic, living perched on tree branches (*Orchidaceae*)

Dendrocalamus

den-dro-KAL-uh-mus

From Greek *dendron*, meaning "tree," and *kalamos*, meaning "reed," because this is a genus of large, woody bamboos (*Poaceae*)

Dendrochilum

den-dro-KY-lum

From Greek *dendron*, meaning "tree," and *cheilos*, meaning "lip," because the floral lip is large, or possibly *chilos*, meaning "green food," because these orchids grow as epiphytes (*Orchidaceae*)

dendroides

den-DROY-deez

—

dendroideus

den-DROY-dee-us

dendroidea, dendroideum

Resembling a tree, as in *Sedum dendroideum*

Dendromecon

den-dro-ME-kon

From Greek *dendron*, meaning "tree," and *mekon*, meaning "poppy" (*Papaveraceae*)

dendrophilus

den-dro-FIL-us

dendrophila, dendrophilum

Tree loving, as in *Tecomanthe dendrophila*

1

Delonix regia

2

Fuchsia denticulata

Denmoza

den-MOH-zuh

An anagram of Mendoza, a province in western Argentina (*Cactaceae*)

Dennstaedtia

den-STAY-tee-uh

Named after August Wilhelm Dennstädt (1776–1826), German physician and botanist (*Dennstaedtiaceae*)

dens-canis

denz-KAN-is

Term for dog's tooth, as in *Erythronium dens-canis*

densatus

den-SA-tus

densata, densatum

—

densus

den-SUS

densa, densum

Compact; dense, as in *Trichodiadema densum*

densiflorus

den-see-FLOR-us

densiflora, densiflorum

Densely flowered, as in *Verbascum densiflorum*

densifolius

den-see-FOH-lee-us

densifolia, densifolium

Densely leaved, as in *Gladiolus densifolius*

dentatus

den-TAH-tus

dentata, dentatum

With teeth, as in *Ligularia dentata*

denticulatus [2]

den-tik-yoo-LAH-tus

denticulata, denticulatum

Slightly toothed, as in *Primula denticulata*

denudatus

dee-noo-DAH-tus

denudata, denudatum

Bare; naked, as in *Magnolia denudata*

deodara

dee-oh-DAR-uh

From the Indian name for the deodar, as in *Cedrus deodara*

depauperatus

de-por-per-AH-tus

depauperata, depauperatum

Not properly developed; dwarfed, as in *Carex depauperata*

dependens

de-PEN-denz

Hanging down, as in *Celastrus dependens*

deppeanus

dep-ee-AH-nus

deppeana, deppeanum

Named after Ferdinand Deppe (1794–1861), German botanist, as in *Juniperus deppeana*

depressus

de-PRESS-us

depressa, depressum

Flattened or pressed down, as in *Gentiana depressa*

Deschampsia

deh-SHAMP-see-uh

Named after Louis Auguste Deschamps (1765–1842), French naturalist and surgeon (*Poaceae*)

deserti

DES-er-tee

Connected with the desert, as in *Agave deserti*

desertorum

de-zert-OR-um

Of deserts, as in *Alyssum desertorum*

Desfontainia

dez-fon-TAYN-ee-uh

Named after René Louiche Desfontaines (1750–1833), French botanist (*Columelliaceae*)

Desmodium

dez-MOH-dee-um

From Greek *desmos*, meaning "chain," because the seed pods break into sections (*Fabaceae*)

detonsus

de-TON-sus

detonsa, detonsum

Bare; shorn, as in *Gentianopsis detonsa*

A
B
C
D
E
F
G
H
I
J
K
L
M
N
O
P
Q
R
S
T
U
V
W
X
Y
Z

i

Dianthus caryophyllus

Deuterocohnia

dew-ter-oh-COH-nee-uh

Named after Ferdinand Julius Cohn (1828–98), German biologist; the genus *Cohnia* (=*Cordyline*) already existed, so the Greek prefix *deuter* (meaning "second") was added (*Bromeliaceae*)

deustus

dee-US-tus

deusta, deustum

Burned, as in *Tritonia deusta*

Deutzia

DOYT-zee-uh

Named after Johan van der Deutz (1743–88), Dutch lawyer and botanist (*Hydrangeaceae*)

diabolicus

dy-oh-BOL-ih-kus

diabolica, diabolicum

Devilish, as in *Acer diabolicum*

diacanthus

dy-ah-KAN-thus

diacantha, diacanthum

With two spines, as in *Ribes diacanthum*

diadema

dy-uh-DEE-ma

Crown; diadem, as in *Begonia diadema*

Dianella

dy-ah-NEL-uh

From Diana, Roman goddess of the hunt, plus Latin diminutive *ella*, perhaps alluding to a woodland habitat (*Asphodelaceae*)

diandrus

dy-AN-drus

diandra, diandrum

With two stamens, as in *Bromus diandrus*

dianthiflorus

die-AN-thuh-flor-us

dianthiflora, dianthiflorum

With flowers like pinks (*Dianthus*), as in *Episcia dianthiflora*

Dianthus

dy-AN-thus

From Greek *dios*, meaning "god," and *anthos*, meaning "flower," because pinks smell divine (*Caryophyllaceae*)

diaphanus

dy-AF-a-nus

diaphana, diaphanum

Transparent, as in *Berberis diaphana*

Diascia

dy-AS-ee-uh

From Greek *dis*, meaning "two," and *askos*, meaning "sack," because in the type species *D. bergiana*, each flower has two nectar-filled spurs (*Scrophulariaceae*)

Dicentra

dy-SEN-truh

From Greek *dis*, meaning "two," and *kentron*, meaning "sharp point," because each flower has twin spurs (*Papaveraceae*)

Dichelostemma

dy-kel-oh-STEM-uh

From Greek *dichelos*, meaning "cloven-hooved," and *stemma*, meaning "wreath," because flowers contain a ring of two-lobed appendages (*Asparagaceae*)

Dichondra

dy-KON-druh

From Greek *dis*, meaning "two," and *chondros*, meaning "grain," referring to the twin fruit (*Convolvulaceae*)

Dichorisandra

dy-kor-is-AN-druh

From Greek *dis* ("two"), *chorizo* ("to separate"), and *andros* ("male"), because each flower has stamens with two different postures (*Commelinaceae*)

dichotomus

dy-KAW-toh-mus

dichotoma, dichotomum

In forked pairs, as in *Iris dichotoma*

Dichroa

dy-KROW-uh

From Greek *dis*, meaning "two," and *chroa*, meaning "color," because the blooms can change color depending on soil pH, as in some hydrangeas (*Hydrangeaceae*)

dichroanthus

dy-kroh-AN-thus

dichroantha, dichroanthum

With flowers of two different colors, as in *Rhododendron dichroanthum*

dichromus

dy-KROH-mus

dichroma, dichromum

—

dichrous

dy-KRUS

dichroa, dichroum

With two distinct colors, as in *Gladiolus dichrous*

Dicksonia

dik-SOH-nee-uh

Named after James Dickson (1738–1822), Scottish horticulturist and botanist, founding member of the Royal Horticultural Society (*Dicksoniaceae*)

Dicliptera

dy-KLIP-ter-uh

From Greek *diklis*, meaning "twice folded," and *pteron*, meaning "wing," because the fruiting capsule has two winglike divisions (*Acanthaceae*)

Dictamnus (also dictamnus)

dik-TAM-nus

From the Greek name for *Origanum dictamnus* (*Lamiaceae*), itself derived from Mount Dikti in Crete (*Rutaceae*)

dictyophyllus

dik-tee-oh-FIL-us

dictyophylla, dictyophyllum

With leaves that have a net pattern, as in *Berberis dictyophylla*

Didierea

di-DEER-ee-uh

Named after Alfred Grandidier (1836–1921), French naturalist and explorer (*Didiereaceae*)

Didymochlaena

di-dee-mok-LAY-nuh

From Greek *didymos*, meaning "twin," and *chlaina*, meaning "cloak," because the membrane covering the sporangia opens on two sides (*Dryopteridaceae*)

didymus

DID-ih-mus

didyma, didymum

In pairs, twin, as in *Monarda didyma*

Dieffenbachia

dee-fen-BAH-kee-uh

Named after Joseph Dieffenbach (1796–1863), Austrian horticulturist (*Araceae*)

Dierama

dy-uh-RAH-muh

From Greek *dierama*, meaning "funnel," the shape of the flowers (*Iridaceae*)

Diervilla

dee-er-VIL-uh

Named after N. Dièreville (dates unknown), French surgeon (*Caprifoliaceae*)

Dietes

dy-EE-teez

From Greek *dis*, meaning "two," and *etes*, meaning "relation," an allusion to the evolutionary closeness of this genus to both *Iris* and *Moraea* (*Iridaceae*)

difformis

dif-FOR-mis

difformis, difforme

With an unusual form, unlike the rest of the genus, as in *Vinca difformis*

diffusus

dy-FEW-sus

diffusa, diffusum

With a spreading habit, as in *Cyperus diffusus*

Digitalis

di-ji-TAH-lis

From Latin *digitus*, meaning "finger," probably from the German *fingerhut*, meaning thimble (= finger hat), which the flowers resemble (*Plantaginaceae*)

digitalis

dij-ee-TAH-lis

digitalis, digitale

Like a finger, as in *Penstemon digitalis*

Digitaria

di-ji-TAIR-ee-uh

From Latin *digitus*, meaning "finger," because the inflorescence resembles a hand with spreading fingers (*Poaceae*)

dilatatus

di-la-TAH-tus

dilatata, dilatatum

Spread out, as in *Dryopteris dilatata*

Dillenia

dy-LEE-nee-uh

Named after Johann Jacob Dillen (1684–1747), German botanist (*Dilleniaceae*)

dilutus

di-LOO-tus

diluta, dilutum

Diluted (for example, pale) as in *Alstroemeria diluta*

dimidiatus

dim-id-ee-AH-tus

dimidiata, dimidiatum

Divided into two different or unequal parts, as in *Asarum dimidiatum*

Dimorphotheca

dy-morf-oh-THEE-kuh

From Greek *dis* ("two"), *morphe* ("shape"), and *theka* ("fruit"), because the ray and disk florets each produce a different-shaped achene (*Asteraceae*)

dimorphus

dy-MOR-fus

dimorpha, dimorphum

With two different forms of leaf, flower, or fruit, as in *Ceropegia dimorpha*

dioicus

dy-OY-kus

dioica, dioicum

With the male reproductive organs on one plant and the female on another, as in *Arunus dioicus*

GENUS SPOTLIGHT

Digitalis

Most gardeners are familiar with the hardy foxglove (*D. purpurea*), a biennial producing towers of pink or white flowers in its second year. The Canary Islands and Madeira, however, grow woody, shrublike foxgloves that were once separated into their own genus *Isoplexis*. It is thought that the nonwoody species colonized the islands from Europe and, in the absence of other shrubs, they evolved a woody habit. Despite their charms, woody foxgloves are not hardy, but plant breeders have crossed them with common foxgloves, creating *Digitalis* x *valinii*. This hybrid is neither woody nor biennial, but reliably perennial with flowers in a range of apricot shades.

Digitalis purpurea

Dionaea

dy-OH-nee-uh

Named after Dione, the mother of Aphrodite in Greek mythology; in Roman mythology, Aphrodite was known as Venus and Venus's flytrap gets its name from the pure white flowers and elegant leaves (*Droseraceae*)

Dionysia

dy-oh-NISS-ee-uh

Named after the Greek god Dionysus (*Primulaceae*)

Dioon

dy-OON

From Greek *dis*, meaning "two," and *oon*, meaning "egg," because the seed are in pairs (*Zamiaceae*)

Dioscorea

dy-os-KOR-ee-uh

Named after Pedanius Dioscorides (ca. AD 40–90), Greek physician and herbalist (*Dioscoreaceae*)

Diosma

dy-OZ-muh

From Greek *dios*, meaning "divine," and *osme*, meaning "fragrance," because the foliage is strongly scented (*Rutaceae*)

Diospyros

dy-os-PIE-rus

From Greek *dios*, meaning "divine," and *pyros*, meaning "wheat," referring to the delicious fruit (*Ebenaceae*)

Dipcadi

dip-CAH-dee

From the Turkish name for *Muscari*, a related genus (*Asparagaceae*)

Dipelta

dy-PEL-tuh

From Greek *dis*, meaning "two," and *pelte*, meaning "shield," because two large bracts surround each fruit (*Caprifoliaceae*)

dipetalus

dy-PET-uh-lus

dipetala, dipetalum

With two petals, as in *Begonia dipetala*

Diphylleia

dy-fil-AY-uh

From Greek *dis*, meaning "two," and *phyllon*, meaning "leaf," because each leaf is divided into two lobes (*Berberidaceae*)

diphyllus

dy-FIL-us

diphylla, diphyllum

With two leaves, as in *Bulbine diphylla*

Diplarrhena

dip-luh-REE-nuh

From Greek *dis*, meaning "two," and *arrhen*, meaning "male," because each flower has only two functional stamens (*Iridaceae*)

Diplazium

dy-PLAY-zee-um

From Greek *dis*, meaning "two," and *plasion*, meaning "oblong," because the membrane protecting the sporangia covers both sides of the leaf vein (*Athyriaceae*)

dipsaceus

dip-SAK-ee-us

dipsacea, dipsaceum

Like teasel (*Dipsacus*), as in *Carex dipsacea*

Dipsacus

DIP-sak-us

From Greek *dipsao*, meaning "thirst," because water gathers at the base of the paired leaves (*Caprifoliaceae*)

dipterocarpus

dip-ter-oh-KAR-pus

dipterocarpa, dipterocarpum

With two-winged fruit, as in *Thalictrum dipterocarpum*

Dipteronia

dip-tuh-ROH-nee-uh

From Greek *dis*, meaning "two," and *pteron*, meaning "wing"; a confusing name, because *Dipteronia* fruit each has a circular wing that surrounds the seed, while the related *Acer* fruit has two separate wings (*Sapindaceae*)

dipterus

DIP-ter-us

diptera, dipterum

With two wings, as in *Halesia diptera*

DOCTOR OF MEDICINE

Pedanius Dioscorides, after whom the yam genus *Dioscorea* is named, was a Greek physician who worked as a medic in the Roman army. He compiled a book (a pharmacopeia) listing medicines and the plants that they came from, now widely known as *De Materia Medica*. For centuries, its five volumes were the go-to reference for physicians, only being supplanted by Renaissance herbals. The father of plant taxonomy, Carl Linnaeus, adopted many of Dioscorides' plant names in his binomial classification system, although he did not always apply them to the plant that Dioscorides had perhaps intended.

---

dipyrenus

dy-pie-REE-nus

dipyrena, dipyrenum

With two seeds or kernels, as in *Ilex dipyrena*

Dirca

DUR-kuh

From Greek mythology, Dirce was the wife of Lyceus and she was transformed into a fountain by Dionysus; her connection with this shrub is uncertain (*Thymelaeaceae*)

dis-

Used in compound words to denote apart

Disa

DY-suh

Named after Disa, a heroine of Swedish mythology; challenged to appear before the god-king Freyr neither dressed or undressed, she wrapped herself in a fishing net; some *Disa* species have netlike patterning on their petals (*Orchidaceae*)

Disanthus [1]

dy-SAN-thus

From Greek *dis*, meaning "two," and *anthos*, meaning "flower," because the blooms are paired (*Hamamelidaceae*)

Discaria

dis-KAIR-ee-uh

From Greek *diskos*, meaning "disk," referring to the prominent disk within each flower (*Rhamnaceae*)

Dischidia

di-SHI-dee-uh

From Greek *dis*, meaning "two," and either *askidion*, meaning "pouch," because the paired leaves can form chambers that contain ant nests, or *schizo*, meaning "to divide," because the corona lobes are bifid (*Apocynaceae*)

disciformis

disk-ee-FOR-mis

disciformis, disciforme

Shaped like a disk, as in *Medicago disciformis*

Discocactus

DIS-koh-kak-tus

From Greek *diskos*, meaning "disk," referring to the circular shape of these cacti (*Cactaceae*)

discoideus

dis-KOY-dee-us

discoidea, discoideum

Without rays, as in *Matricaria discoidea*

discolor

DIS-kol-or

Of two completely different colors, as in *Salvia discolor*

Diselma

dy-SEL-muh

From Greek *dis*, meaning "two," and *selma*, meaning "upper deck," because each cone has two fertile scales (*Cupressaceae*)

Disocactus

DY-soh-kak-tus

From Greek *dis*, meaning "two," and *isos*, meaning "equal," because the outer and inner whorls of tepals are of equal length (*Cactaceae*)

dispar

DIS-par

Unequal; unusual for a genus, as in *Restio dispar*

dispersus

dis-PER-sus

dispersa, dispersum

Scattered, as in *Paranomus dispersus*

Disporopsis

dy-spor-OP-sis

Resembling the genus *Disporum* (*Asparagaceae*)

Disporum [2]

dy-SPOR-um

From Greek *dis*, meaning "two," and *spora*, meaning "seed," because each chamber within the ovary has two ovules (*Colchicaceae*)

1

Disanthus cercidifolius

2

Disporum cantoniense

3

Distylium racemosum

4

Dodecatheon meadia

dissectus

dy-SEK-tus

dissecta, dissectum

Deeply cut or divided, as in *Cirsium dissectum*

dissimilis

dis-SIM-il-is

dissimilis, dissimile

Differing from the norm for a particular genus, as in *Columnea dissimilis*

distachyus

dy-STAK-yus

distachya, distachyum

With two spikes, as in *Billbergia distachya*

distans

DIS-tanz

Widely apart, as in *Watsonia distans*

distichophyllus

dis-ti-koh-FIL-us

distichophylla, distichophyllum

With leaves appearing in two ranks or levels, as in *Buckleya distichophylla*

distichus

DIS-tih-kus

disticha, distichum

In two ranks or levels, as in *Taxodium distichum*

Distictis

dy-STIK-tis

From Greek *dis*, meaning "two," and *stiktos*, meaning "spotted," because the two rows of seed within each capsule will leave two rows of spots once shed (*Bignoniaceae*)

distortus

DIS-tor-tus

distorta, distortum

Misshapen, as in *Adonis distorta*

Distylium [3]

dy-STIL-ee-um

From Greek *dis*, meaning "two," and *stylos*, meaning "style," because the flowers each have two styles (*Hamamelidaceae*)

distylus

DIS-sty-lus

distyla, distylum

With two styles, as in *Acer distylum*

diurnus

dy-YUR-nus

diurna, diurnum

Flowering by day, as in *Cestrum diurnum*

divaricatus

dy-vair-ih-KAH-tus

divaricata, divaricatum

With a spreading and straggling habit, as in *Phlox divaricata*

divergens

div-VER-jenz

Spreading out a long way from the center, as in *Ceanothus divergens*

diversifolius

dy-ver-sih-FOH-lee-us

diversifolia, diversifolium

With diverse leaves, as in *Hibiscus diversifolius*

diversiformis

dy-ver-sih-FOR-mis

diversiformis, diversiforme

With diverse forms, as in *Romulea diversiformis*

divisus

div-EE-sus

divisa, divisum

Divided, as in *Pennisetum divisum*

Docynia

doh-SY-nee-uh

An anagram of *Cydonia*, a related genus (*Rosaceae*)

dodecandrus

doh-DEK-an-drus

dodecandra, dodecandrum

With twelve stamens, as in *Cordia dodecandra*

Dodecatheon [4]

doh-duh-KATH-ee-on

From Greek *dodeca*, meaning "twelve," and *theos*, meaning "gods," a name used by Pliny to refer to a primrose protected by the gods (*Primulaceae*)

Dodonaea

doh-DOH-nee-uh

Named after Rembert Dodoens (1517–85), Flemish botanist (*Sapindaceae*)

Doellingeria

doh-lin-GEER-ee-uh

Named after Ignaz Döllinger (1770–1841), German doctor and natural philosopher (*Asteraceae*)

doerfleri

DOOR-fleur-eye

Named after Ignaz Dörfler (1866–1950), German botanist, as in *Colchicum doerfleri*

dolabratus

dol-uh-BRAH-tus

dolabrata, dolabratum

—

dolabriformis

doh-la-brih-FOR-mis

dolabriformis, dolabriforme

Shaped like a hatchet, as in *Thujopsis dolabrata*

Dolichos

DOH-li-kos

From Greek *dolikhos*, meaning "long," a reference to the climbing habit (*Fabaceae*)

dolosus

do-LOH-sus

dolosa, dolosum

Deceitful; looking like another plant, as in
Cattleya × dolosa

Dombeya

dom-BAY-uh

Named after Joseph Dombey (1742–94),
French botanist (*Malvaceae*)

domesticus [1]

doh-MESS-tih-kus

domestica, domesticum

Domesticated, as in *Malus domestica*

Doodia

DOO-dee-uh

Named after Samuel Doody (1656–1706),
British botanist (*Blechnaceae*)

Doronicum

duh-RON-ik-um

From Arabic *doronigi* or *durugi*, the
vernacular name (*Asteraceae*)

Dorstenia

dor-STEE-nee-uh

Named after Theodor Dorsten (1492–
1552), German physician (*Moraceae*)

Doryanthes

dor-ee-AN-thees

From Greek *doratos*, meaning "spear," and
anthos, meaning "flower," a reference to the
shape of the inflorescence in *D. palmeri*
(*Doryanthaceae*)

Dorycnium

dor-IK-nee-um

From Greek *doratos*, meaning "spear," and
knaein, meaning "to smear," because this
plant was used to poison spear tips (*Fabaceae*)

Doryopteris

dor-ee-OP-ter-is

From Greek *doratos*, meaning "spear," and
pteris, meaning "fern," because some species
have lancelike fronds (*Pteridaceae*)

Douglasia

duh-GLA-see-uh

Named after David Douglas (1799–1834),
Scottish botanist (*Primulaceae*)

douglasianus

dug-lus-ee-AH-nus

douglasiana, douglasianum

—

1

Malus domestica

douglasii

dug-lus-ee-eye

Named after David Douglas (1799–1834),
Scottish plant hunter, as in *Limnanthes
douglasii*

drabifolius

dra-by-FOH-lee-us

drabifolia, drabifolium

With leaves like whitlow grass (*Draba*), as in
Centaurea drabifolia

Dracaena

druh-SEE-nuh

From Greek *drakaina*, meaning "female
dragon"; sap harvested from the Canary
Island native *D. draco* is initially milky,
turning blood-red when dry (*Asparagaceae*)

draco

DRAY-koh

Dragon, as in *Dracaena draco*

Dracocephalum

dray-koh-KEF-ah-lum

From Greek *drakon*, meaning "dragon," and
kephale, meaning "head," an allusion to the
shape of the corolla (*Lamiaceae*)

Dracophyllum

dray-koh-FIL-um

From Greek *drakon*, meaning "dragon," and
phyllon, meaning "leaf," because these plants
resemble the unrelated genus *Dracaena*
(*Ericaceae*)

Dracula
DRAK-ew-luh
Named after the literarary character Count Dracula, due to the somewhat gothic flowers (*Orchidaceae*)

Dracunculus (also dracunculus)
druh-KUN-koo-lus
The diminutive form of Greek *drakon*, meaning "dragon," due to the shape of the spathe (*Araceae*)

Dregea
DRAY-gee-uh
Named after Johann Franz Drège (1794–1881), German botanist and horticulturist (*Apocynaceae*)

Drimys
DRIM-is
From Greek *drimys*, meaning "acrid," referring to the taste of the bark (*Winteraceae*)

Drosanthemum
druh-SAN-thuh-mum
From Greek *drosos*, meaning "dew," and *anthos*, meaning "flower," because the leaves and some parts of the flowers are covered in glistening bladders (*Aizoaceae*)

Drosera
DRO-sur-uh
From Greek *drosos*, meaning "dew,," because the leaves of these insect-catching herbs are covered in a sticky residue (*Droseraceae*)

drummondianus
drum-mond-ee-AH-nus
drummondiana, drummondianum
—

drummondii
drum-mond-EE-eye
Named after either James Drummond (1786–1863) or Thomas Drummond (1793–1835), brothers who collected plants in Australia and North America respectively, as in *Phlox drummondii*

drupaceus
droo-PAY-see-us
drupacea, drupaceum
—

drupiferus
droo-PIH-fer-us
drupifera, drupiferum
With fleshy, hard-stone fruit that resembles a peach or cherry, as in *Hakea drupacea*

Dryas
DRY-as
Named after dryads, tree nymphs in Greek mythology, because the leaves of some species resemble those of oaks (*Rosaceae*)

Drynaria
dry-NAIR-ee-uh
From Greek *dryas*, meaning "oak leaf-shaped" (*Polypodiaceae*)

drynarioides
dri-nar-ee-OY-deez
Resembling oak-leaf fern (*Drynaria*), as in *Aglaomorpha drynarioides*

Dryopteris (also dryopteris)
dry-OP-ter-is
From Greek *dryas*, meaning "oak," and *pteris*, meaning "fern," a reference to their preferred habitat in deciduous forests (*Dryopteridaceae*)

dubius
DOO-bee-us
dubia, dubium
Doubtful, unlike the rest of the genus, as in *Ornithogalum dubium*

Duchesnea
doo-KES-nee-uh
Named after Antoine Nicolas Duchesne, (1747–1827), French botanist (*Rosaceae*)

Dudleya
DUD-lay-uh
Named after William Russel Dudley (1849–1911), American botanist (*Crassulaceae*)

dulcis
DUL-sis
dulcis, dulce
Sweet, as in *Prunus dulcis*

dumetorum
doo-met-OR-um
From hedges or bushes, as in *Fallopia dumetorum*

dumosus
doo-MOH-sus
dumosa, dumosum
Bushy; shrubby, as in *Alluaudia dumosa*

duplicatus
doo-plih-KAH-tus
duplicata, duplicatum
Double; duplicate, as in *Brachystelma duplicatum*

Duranta
duh-RAN-tuh
Named after Castore Durante (1529–90), Italian physician and botanist (*Verbenaceae*)

Durio
DEW-ree-oh
From Malay *duri*, meaning "thorn," because durian fruit (*D. zibethinus*) has a thorny skin (*Malvaceae*)

durus
DUR-us
dura, durum
Hard, as in *Blechnum durum*

Duvalia
doo-VAH-lee-uh
Named after Henri-Auguste Duval (1777–1814), French physician and botanist (*Apocynaceae*)

Dyckia
DIK-ee-uh
Named after Joseph zu Salm-Reifferscheidt-Dyck (1773–1861), German botanist (*Bromeliaceae*)

dyeri
DY-er-eye
—

dyerianus
dy-er-ee-AH-nus
dyeriana, dyerianum
Named after Sir William Turner Thiselton-Dyer (1843–1928), British botanist and Director of Kew Gardens, London, England, as in *Strobilanthes dyeriana*

Dysosma
dy-SOZ-muh
From Greek *dys*, meaning "bad," and *osme*, meaning "fragrance," because the flowers may have a putrid scent (*Berberidaceae*)

e-, ex-
Used in compound words to denote without, out of

ebeneus
eb-en-NAY-us
ebenea, ebeneum
—

ebenus
eb-en-US
ebena, ebenum
Ebony-black, as in *Carex ebenea*

ebracteatus
e-brak-tee-AH-tus
ebracteata, ebracteatum
Without bracts, as in *Eryngium ebracteatum*

eburneus
eb-URN-ee-us
eburnea, eburneum
Ivory white, as in *Angraecum eburneum*

Ecballium
ek-BAL-ee-um
From Greek *ekballein*, meaning "to cast out," because the fruit of squirting cucumber (*E. elaterium*) expels its seed in vigorous fashion (*Cucurbitaceae*)

Eccremocarpus
ek-ree-moh-KAR-pus
From Greek *ekkremes*, meaning "hanging," and *karpos*, meaning "fruit" (*Bignoniaceae*)

Echeveria
ek-uh-VEER-ee-uh
Named after Atanasio Echeverría y Godoy (ca. 1771–1803), Mexican botanical artist (*Crassulaceae*)

Echidnopsis
ek-id-NOP-sis
From Greek *echidna*, meaning "snake," and *opsis*, meaning "like," because the succulent stems are serpentine (*Apocynaceae*)

Echinacea
ek-in-AY-see-uh
From Greek *echinos*, meaning "hedgehog," because the inflorescence is prickly (*Asteraceae*)

echinatus
ek-in-AH-tus
echinata, echinatum
With prickles like a hedgehog, as in *Pelargonium echinatum*

Echinocactus
ek-i-noh-KAK-tus
From Greek *echinos*, meaning "hedgehog," plus cactus (*Cactaceae*)

Echinocereus
ek-i-noh-SER-ee-us
From Greek *echinos*, meaning "hedgehog," and *Cereus*, a related genus (*Cactaceae*)

Echinops
EK-in-ops
From Greek *echinos*, meaning "hedgehog," due to the spiny inflorescences (*Asteraceae*)

Echinopsis
ek-in-OP-sis
From Greek *echinos*, meaning "hedgehog," and *opsis*, meaning "like" (*Cactaceae*)

echinosepalus
ek-in-oh-SEP-uh-lus
echinosepala, echinosepalum
With prickly sepals, as in *Begonia echinosepala*

echioides
ek-ee-OY-deez
Resembling viper's bugloss (*Echium*), as in *Picris echioides*

Echium
EK-ee-um
From Greek *echis*, meaning "viper," because the seed resemble snake heads and the roots were used to treat snake bites (*Boraginaceae*)

ecornutus
ek-kor-NOO-tus
ecornuta, ecornutum
Without horns, as in *Stanhopea ecornuta*

Edgeworthia
ej-WURTH-ee-uh
Named after Michael Pakenham Edgeworth (1812–81), Irish botanist (*Thymelaeaceae*)
—

edgeworthianus
edj-wor-thee-AH-nus
edgeworthiana, edgeworthianum
—

edgeworthii
edj-WOR-thee-eye
Named after Michael Pakenham Edgeworth (1812–81) of the East India Company, as in *Rhododendron edgeworthii*

Edithcolea
ee-dith-KOH-lee-uh
Named after Edith Cole (1859–1940), British entomologist and botanist (*Apocynaceae*)

Edraianthus
ed-rai-ANTH-us
From Greek *hedraios*, meaning "sitting," and *anthos*, meaning "flower," because the blooms are sessile (*Campanulaceae*)

edulis
ED-yew-lis
edulis, edule
Edible, as in *Dioon edule*

effusus
eff-YOO-sus
effusa, effusum
Spreading loosely, as in *Juncus effusus*

Egeria
eh-JEER-ee-uh
From Latin *egeri*, meaning "nymph," an allusion to its aquatic habitat (*Hydrocharitaceae*)

Ehretia
eh-RET-ee-uh
Named after Georg Dionysius Ehret (1708–70), German botanist and entomologist (*Boraginaceae*)

Eichhornia
ay-KOR-nee-uh
Named after Johann Albrecht Friedrich Eichhorn (1779–1856), Prussian politician (*Pontederiaceae*)

elaeagnifolius
el-ee-ag-ne-FOH-lee-us
elaeagnifolia, elaeagnifolium
With leaves like *Elaeagnus*, as in *Brachyglottis elaeagnifolia*

Elaeagnus
ee-lee-AG-nus
From Greek *elaiagnos*, meaning "a type of willow" (*Elaeagnaceae*)

Elaeis
el-AY-is
From Greek *elaia*, meaning "olive," an allusion to the valuable oil extracted from oil palm, *E. guineensis* (*Arecaceae*)

Elaeocarpus

el-ay-oh-KAR-pus

From Greek *elaia*, meaning "olive," and *karpos*, meaning "fruit," because the fleshy fruit resembles olives (*Elaeocarpaceae*)

elasticus

ee-LASS-tih-kus

elastica, elasticum

Elastic, producing latex, as in *Ficus elastica*

Elatostema

el-ah-toh-STEM-uh

From Greek *elatos*, meaning "striking," and *stema*, meaning "stamen," referring to the explosive fashion in which the stamens eject pollen (*Urticaceae*)

elatus [1]

el-AH-tus

elata, elatum

Tall, as in *Aralia elata*

elegans [2]

el-ee-GANS

—

elegantulus

el-eh-GAN-tyoo-lus

elegantula, elegantulum

Elegant, as in *Desmodium elegans*

elegantissimus

el-ee-gan-TISS-ih-mus

elegantissima, elegantissimum

Especially elegant, as in *Schefflera elegantissima*

Elegia

uh-LEE-jee-uh

From Greek *elegeia*, meaning "song of lamentation," or elegy, perhaps an allusion to the sound of wind passing through the reedlike stems (*Restionaceae*)

elephantipes

ell-uh-fan-TY-peez

Resembling an elephant's foot, as in *Yucca elephantipes*

Elettaria

el-uh-TARE-ee-uh

From Malayalam (India) *elatarri*, the vernacular name for cardamom, *E. cardamomum* (*Zingiberaceae*)

Eleutherococcus

el-ew-thu-roh-KOK-us

From Greek *eleuthero*, meaning "free," and *kokkos*, meaning "berry," suggesting that the fruit is not fused together (*Araliaceae*)

Elliottia

el-ee-OTT-ee-uh

Named after Stephen Elliott (1771–1830), American botanist, politician, and banker (*Ericaceae*)

1

Delphinium elatum

2

Desmodium elegans

elliottianus
el-ee-ot-ee-AH-nus
elliottiana, elliottianum
Named after Captain George Henry Elliott
(1813–92), as in *Zantedeschia elliottiana*

elliottii
el-ee-ot-EE-eye
Named after Stephen Elliott (1771–1830),
American botanist, as in *Eragrostis elliottii*

ellipsoidalis
e-lip-soy-DAH-lis
ellipsoidalis, ellipsoidale
Elliptic, as in *Quercus ellipsoidalis*

ellipticus
ee-LIP-tih-kus
elliptica, ellipticum
Shaped like an ellipse, as in *Garrya elliptica*

Elodea
el-OH-dee-uh
From Greek *helos*, meaning "marsh,"
a reference to the aquatic habit
(*Hydrocharitaceae*)

elongatus
ee-long-GAH-tus
elongata, elongatum
Lengthened; elongated, as in *Mammillaria
elongata*

Elsholtzia
el-SHOLT-zee-uh
Named after Johann Sigismund Elsholtz
(1623–88), German physician and botanist
(*Lamiaceae*)

elwesii
el-WEZ-ee-eye
Named after Henry John Elwes (1846–
1922), British plant collector, one of the
inaugural recipients of the Victoria Medal
of the Royal Horticultural Society, as in
Galanthus elwesii

Elymus
EL-i-mus
From Greek *elymos*, meaning "millet"
(*Poaceae*)

emarginatus
e-mar-jin-NAH-tus
emarginata, emarginatum
Slightly notched at the margins, as in
Pinguicula emarginata

Embothrium
em-BOTH-ree-um
From Greek *en*, meaning "in," and *bothrion*,
meaning "small pit," because the anthers are
borne in pits (*Proteaceae*)

Emilia
eh-MIL-ee-uh
Named after Emilia (now Emilia-Romagna),
an Italian Region where the botanist who
named this plant worked, or possibly after
someone named Emile or Emilie, although
he did not specify (*Asteraceae*)

eminens
EM-in-enz
Eminent; prominent, as in *Sorbus eminens*

Emmenopterys
em-en-OP-ter-is
From Greek *emmeno*, meaning "abiding,"
and *pteron*, meaning "wing," because the
flowers are surrounded by persistent bracts
(*Rubiaceae*)

empetrifolius
em-pet-rih-FOH-lee-us
empetrifolia, empetrifolium
With leaves like crowberry (*Empetrum*), as
in *Berberis empetrifolia*

Empetrum
em-PET-rum
From Greek *en*, meaning "in," and *petros*,
meaning "rock," alluding to the stony habitat
(*Ericaceae*)

Encephalartos
en-kef-al-AR-tos
From Greek *en* ("in"), *kephale* ("head"), and
artos ("bread"), because the tips of the
trunks can be harvested to make a sagolike
starchy food (*Zamiaceae*)

encliandrus
en-klee-AN-drus
encliandra, encliandrum
With half the stamens inside the flower
tube, as in *Fuchsia encliandra*

Encyclia
en-SY-klee-uh
From Greek *enkyklos*, meaning "to encircle,"
because the lip petal partly encloses the
column (*Orchidaceae*)

endresii
en-DRESS-ee-eye
endressii
Named after Philip Anton Christoph
Endress (1806–31), German plant collector,
as in *Geranium endressii*

engelmannii
en-gel-MAH-nee-eye
Named after Georg Engelmann (1809–84),
German-born physician and botanist, as in
Picea engelmannii

Enkianthus
en-kee-AN-thus
From Greek *enkyos*, meaning "pregnant,"
and *anthos*, meaning "flower," because in
some species the swollen sepals give the
appearance of one flower emerging from
within another (*Ericaceae*)

enneacanthus
en-nee-uh-KAN-thus
enneacantha, enneacanthum
With nine spines, as in *Echinocereus
enneacanthus*

enneaphyllus
en-nee-a-FIL-us
enneaphylla, ennephyllum
With nine leaves or leaflets, as in *Oxalis
enneaphylla*

ensatus
en-SA-tus
ensata, ensatum
In the shape of a sword, as in *Iris ensata*

Ensete
en-SET-eh
From Amharic (Ethiopia) *anset*, the
vernacular name (*Musaceae*)

ensifolius
en-see-FOH-lee-us
ensifolia, ensifolium
With leaves shaped like a sword, as in
Kniphofia ensifolia

ensiformis
en-see-FOR-mis
ensiformis, ensiforme
In the shape of a sword, as in *Pteris
ensiformis*

Eomecon

ee-oh-MEE-kon

From Greek *eos*, meaning "dawn" or "the east," and *mecon*, meaning "poppy," because this genus originates in China (*Papaveraceae*)

Epacris

ep-AH-kris

From Greek *epi*, meaning "upon," and *acris*, meaning "summit," an allusion to the high altitude habitat of some species (*Ericaceae*)

Ephedra

uh-FED-ruh

From Greek *epi*, meaning "upon," and *hedraios*, meaning "sitting," because the stem segments appear to sit one atop another; the classical name for the similar-looking, but unrelated *Equisetum* (*Ephedraceae*)

Epidendrum

eh-pi-DEN-dron

From Greek *epi*, meaning "upon," and *dendron*, meaning "tree," because these orchids are epiphytic (*Orchidaceae*)

Epigaea

eh-pi-JEE-uh

From Greek *epi*, meaning "upon," and *gaia*, meaning "earth"; the plant has a creeping habit (*Ericaceae*)

Epilobium

eh-pi-LOH-bee-um

From Greek *epi*, meaning "upon," and *lobos*, meaning "pod," because the corolla is at the tip of the ovary (*Onagraceae*)

Epimedium [1]

eh-pi-MEE-dee-um

From Greek *epimedion*, a vernacular name for an unknown plant (*Berberidaceae*)

Epipactis

eh-pi-PAK-tis

From Greek *epipaktis*, a vernacular name for a plant that is said to have curdled milk, perhaps *E. helleborine* (*Orchidaceae*)

Epiphyllum

eh-pi-FIL-um

From Greek *epi*, meaning "upon," and *phyllon*, "a leaf"; the flowers appear to develop on the leaves, but this plant is leafless and the stems are foliose (*Cactaceae*)

epiphyllus

ep-ih-FIL-us

epiphylla, epiphyllum

On the leaf (for example, flowers), as in *Saxifraga epiphylla*

epiphyticus

ep-ih-FIT-ih-kus

epiphytica, epiphyticum

Growing on another plant, as in *Cyrtanthus epiphyticus*

Epipremnum

eh-pi-PREM-num

From Greek *epi*, meaning "upon," and *premnon*, meaning "trunk," because this climber clings to tree trunks (*Araceae*)

Episcia

eh-PIS-ee-uh

From Greek *epi*, meaning "upon," and *skia*, meaning "shadow," because plants prefer shade (*Gesneriaceae*)

Epithelantha

eh-pi-thu-LAN-thu

From Greek *epi* ("upon"), *thele* ("nipple"), and *anthos* ("flower"), because the blooms develop atop the stem tubercles (*Cactaceae*)

equestris

e-KWES-tris

equestris, equestre

—

equinus

e-KWEE-nus

equina, equinum

Relating to horses, equestrian, as in *Phalaenopsis equestris*

1

Epimedium hybridum, E. versicolor, and *E. sulphureum*

2

Equisetum sylvaticum

3

Erica cerinthoides

equisetifolius
ek-wih-set-ih-FOH-lee-us
equisetifolia, equisetifolium
—
equisetiformis
eck-kwiss-ee-tih-FOR-mis
equisetiformis, equisetiforme
Resembling horsetail (*Equisetum*), as in
Russelia equisetiformis

Equisetum [2]
ek-wi-SEE-tum
From Latin *equis*, meaning "horse," and *seta*,
meaning "bristle," because the stems and/or
roots of some species resemble horses' tails
(*Equisetaceae*)

Eragrostis
eh-ruh-GROS-tis
From Greek *eros*, meaning "love," and
agrostis, meaning "grass," because the
spikelets have a heart shape (*Poaceae*)

Eranthemum
eh-ran-thu-mum
From Greek *er*, meaning "spring," and
anthemon, meaning "flower," suggesting an
early-blooming plant (*Acanthaceae*)

Eranthis
eh-RAN-this
From Greek *er*, meaning "spring," and
anthos, meaning "flower"; winter aconite
(*E. hyemalis*) is one of the earliest spring-
flowering perennials (*Ranunculaceae*)

Ercilla
er-SIH-luh
Named after Alonso de Ercilla y Zúñiga
(1533–94), Spanish soldier and poet
(*Phytolaccaceae*)

erectus
ee-RECK-tus
erecta, erectum
Erect; upright, as in *Trillium erectum*

Eremophila
eh-ruh-MOF-il-uh
From Greek *erema*, meaning "desert," and
philos, meaning "loving," because many
species live in arid parts of Australia
(*Scrophulariaceae*)

Eremurus
eh-ruh-MEW-rus
From Greek *erema*, meaning "desert," and
ouros, meaning "tail," referring to both the
long, narrow inflorescence and the preferred
habitat (*Asphodelaceae*)

eri-
Used in compound words to denote woolly

Eria
EAR-ee-uh
From Greek *erion*, for "wool," because some
species have woolly flowers (*Orchidaceae*)

eriantherus
er-ee-AN-ther-uz
erianthera, eriantherum
With woolly anthers, as in *Penstemon
eriantherus*

erianthus
er-ee-AN-thus
eriantha, erianthum
With woolly flowers, as in *Kohleria eriantha*

Erica [3]
Eh-ri-kuh
From Greek *ereiko*, meaning "to break,"
because the stems are brittle; the classical
Latin name for *E. arborea* (*Ericaceae*)

ericifolius
er-ik-ih-FOH-lee-us
ericifolia, ericifolium
With leaves like heather (*Erica*), as in
Banksia ericifolia

ericoides
er-ik-OY-deez
Resembling heather (*Erica*), as in *Aster
ericoides*

Erigeron
uh-RIG-uh-ron
The classical Latin name for groundsel or
old-man-in-the-Spring (*Senecio vulgaris*);
derived from Greek *er* ("spring"), *erion*
("wool"), and *geron* ("old man"), because
some species are early flowering and, once
the blooms fade, the pappus is like the hair
of an old man (*Asteraceae*)

Erinacea
er-in-AY-see-uh
From Latin *erinaceus*, meaning "hedgehog,"
because these plants are spiny (*Fabaceae*)

erinaceus
er-in-uh-SEE-us
erinacea, erinaceum
Like a hedgehog, as in *Dianthus erinaceus*

Erinus
EH-ri-nus
From Greek *erinos*, a vernacular name for
another plant (*Plantaginaceae*)

Eriobotrya
eh-ree-oh-BOT-ree-uh
From Greek *erion*, meaning "wool," and
botrys, meaning "a bunch of grapes," because
the edible fruit (loquats) is clustered and
somewhat woolly (*Rosaceae*)

eriocarpus
er-ee-oh-KAR-pus
eriocarpa, eriocarpum
With woolly fruit, as in *Pittosporum
eriocarpum*

Eryngium

Most members of the carrot family (*Apiaceae*) are known for their ethereal flower clusters (umbels), where numerous small flowers sit atop wiry stalks all radiating out from a point in the center. Often long-lasting, these umbrella-like blooms provide border interest all year round. In sea hollies (*Eryngium*) however, it is not just the flowers that cause a stir; they are often subtended by elaborate spiny bracts enameled in silver or blue. Cut the flowering stems before they fade and hang them to dry; they will last a long time indoors and make attractive decorations.

Eryngium campestre

eriocephalus
er-ri-oh-SEF-uh-lus
eriocephala, eriocephalum
With a woolly head, as in *Lamium eriocephalum*

Eriogonum
eh-ree-oh-GOH-num
From Greek *erion*, meaning "wool," and *gonia*, meaning "knee," in reference to *E. tomentosum*, which is a woolly plant with jointed stems (*Polygonaceae*)

Eriophorum
eh-ree-OF-oh-room
From Greek *erion*, meaning "wool," and *phorus*, meaning "bearing," because the seed is in cottony clusters, thus the common name cotton grass (*Cyperaceae*)

Eriophyllum
eh-ree-oh-FIL-um
From Greek *erion*, meaning "wool," and *phyllon*, meaning "leaf" (*Asteraceae*)

eriostemon
er-ree-oh-STEE-mon
With woolly stamens, as in *Geranium eriostemon*

Erodium
eh-ROH-dee-um
From Greek *erodios*, meaning "heron," because the fruit is beaklike (*Geraniaceae*)

erosus
e-ROH-sus
erosa, erosum
Jagged, as in *Cissus erosa*

erubescens
er-oo-BESS-enz
Becoming red; blushing, as in *Philodendron erubescens*

Eruca
eh-ROO-kuh
This is the Latin name used by Pliny for this or a similar plant; the derivation is obscure, but perhaps based on Greek *uro*, meaning "burn," alluding to the peppery flavor of *E. vesicaria* subsp. *sativa* (arugula), or perhaps *eruco*, meaning "caterpillar" (*Brassicaceae*)

Eryngium
eh-RING-gee-um
From Greek *eryggion*, a bristly plant, possibly referring to *E. campestre* (*Apiaceae*)

Erysimum
eh-RIS-i-mum
From Greek *erysimon*, a kind of mustard (*Brassicaceae*)

Erythrina
eh-rith-REE-nuh
From Greek *erythros*, meaning "red," the color of the flowers and seed (*Fabaceae*)

erythro-
Used in compound words to denote red

erythrocarpus
er-ee-throw-KAR-pus
erythrocarpa, erythrocarpum
With red fruit, as in *Actinidia erythrocarpa*

Erythronium
eh-rith-ROH-nee-um
From Greek *erythros*, meaning "red," referring to the flowers of *E. dens-canis*; the Greek *erythronion* was a vernacular name for another plant (*Liliaceae*)

erythropodus
er-ee-THROW-pod-us
erythropoda, erythropodum
With a red stem, as in *Alchemilla erythropoda*

erythrosorus
er-rith-roh-SOR-us
erythrosora, erythrosorum
Having red spore cases, as in *Dryopteris erythrosora*

Escallonia
es-kuh-LOH-nee-uh
Named after Antonio José Escallón y Flóres (1739–1819), Spanish explorer in Colombia (*Escalloniaceae*)

Eucalyptus

Perhaps best known for the fragrant oil extracted from the leaves, *Eucalyptus* is a large genus of trees and shrubs centered in Australia. The name is derived from Greek and refers to the cap that covers the flowers in bud. In most flowers, the sepals surround the flower buds to protect them, while the petals are showy and attractive. In eucalypts, the sepals and/or petals are fused together to form a cap; when ready to open, the cap pops off and a dense tuft of stamens is revealed. These pollen-producing structures have taken on the role of petals and so they are profuse and sometimes colorful.

Eucalyptus globulus

Eschscholzia

esh-OLT-zee-uh
Named after Johann Friedrich von Eschscholtz (1793–1831), Baltic German physician and entomologist (*Papaveraceae*)

Escobaria

es-koh-BAR-ee-uh
Named after brothers Rómulo (1882–1946) and Numa Escobar Zerman (1874–1949), Mexican agronomists (*Cactaceae*)

esculentus

es-kew-LEN-tus
esculenta, esculentum
Edible, as in *Colocasia esculenta*

Espostoa

es-pos-TOH-uh
Named after Nicolas Esposto (1877–?), Italian-born Peruvian pharmacist and botanist (*Cactaceae*)

Etlingera

et-LIN-ger-uh
Named after Andreas Ernst Etlinger (ca. 1756–85), German botanist and artist (*Zingiberaceae*)

etruscus

ee-TRUSS-kus
estrusca, estruscum
Connected with Tuscany, Italy, as in *Crocus etruscus*

eucalyptifolius

yoo-kuh-lip-tih-FOH-lee-us
eucalyptifolia, eucalyptifolium
With leaves like *Eucalyptus*, as in *Leucadendron eucalyptifolium*

Eucalyptus

yoo-kuh-LIP-tus
From Greek *eu*, meaning "good/well," and *kalyptos*, meaning "covered," because in the flowers, the sepals (and sometimes petals) are fused together creating a lid (*Myrtaceae*)

Eucharis

YOO-kuh-ris
From Greek *eukharis*, meaning "gracious," as befits this fragrant and attractive perennial (*Amaryllidaceae*)

euchlorus

YOO-klor-us
euchlora, euchlorum
A healthy green, as in *Tilia × euchlora*

Eucomis

yoo-KOH-mis
From Greek *eu*, meaning "good," and *kome*, meaning "hair," a beautiful head of hair, referring to the tuft of leaves atop the inflorescence (*Asparagaceae*)

Eucommia

yoo-KOH-mee-uh
From Greek *eu*, meaning "good," and *kommi*, meaning "gum," because latex can be extracted from the tree (*Eucommiaceae*)

Eucryphia

yoo-KRI-fee-uh
From Greek *eu*, meaning "good," and *kryphios*, meaning "hidden"; the sepals are joined at the tips of the flower buds, initially concealing the petals (*Cunoniaceae*)

Eugenia

yoo-JEE-nee-uh
Named after Prince Eugene of Savoy (1663–1736), soldier and politician in the Holy Roman Empire (*Myrtaceae*)

eugenioides

yoo-jee-nee-OY-deez
Resembling the genus *Eugenia*, as in *Pittosporum eugenioides*

Eulophia
yoo-LOH-fee-uh
From Greek *eu*, meaning "good," and *lophus*, meaning "crest," because the flower lip is crested (*Orchidaceae*)

Euodia
yoo-OH-dee-uh
From Greek *euodia*, meaning "sweet fragrance," because the foliage is scented (*Rutaceae*)

Euonymus [1]
yoo-ON-ee-mus
From Greek *eu*, meaning "good," and *onoma*, meaning "name," but the reason for this positive appellation is uncertain (*Celastraceae*)

eupatorioides
yoo-puh-TOR-ee-oy-deez
Resembling *Eupatorium*, as in *Agrimonia eupatoria*

Eupatorium
ew-puh-TOR-ee-um
Named after Mithridates Eupator, King of Pontus, comprising much of modern-day Turkey and the Black Sea coast (132–63 BC) (*Asteraceae*)

Euphorbia [2]
yoo-FOR-bee-uh
Named after Euphorbus (dates unknown), physician to Juba II, King of Numidia and Mauretania (*Euphorbiaceae*)

euphorbioides
yoo-for-bee-OY-deez
Resembling spurge (*Euphorbia*), as in *Neobuxbaumia euphorbioides*

Euptelea
yoop-TEE-lee-uh
From Greek *eu*, meaning "good," and *ptelea*, meaning "elm," which they resemble (*Eupteleaceae*)

europaeus
yoo-ROH-pay-us
europaea, europaeum
Connected with Europe, as in *Euonymus europaeus*

Eurya
YUR-ee-uh
From Greek *eurys*, meaning "large/wide," but allusion unclear (*Pentaphylacaceae*)

Euryale
yur-ee-AH-lee
Named after the gorgon Euryale from Greek mythology, perhaps a reference to the writhing petioles and spiny-coated leaf surfaces (*Nymphaeaceae*)

Eurybia
yoo-RY-bee-uh
From Greek *eurys*, meaning "wide," and *baios*, meaning "few," because the capitulae have few, but broad ray florets (*Asteraceae*)

1

Euonymus alatus

Euryops
YOO-ry-ops
From Greek *eurys*, meaning "wide," and *opsis*, meaning "eye," alluding to the large flower heads (*Asteraceae*)

Eustoma
yoo-STOH-muh
From Greek *eu*, meaning "good," and *stoma*, meaning "mouth," a reference to the attractive tubular flowers (*Gentianaceae*)

Euthamia
yoo-THAM-ee-uh
From Greek *eu*, meaning "good," and *thama*, meaning "crowded," an allusion to the crowded branches (*Asteracaeae*)

Eutrochium
yoo-TROH-kee-um
From Greek *eu*, meaning "good," and *trocho*, meaning "wheel," because the leaves are whorled (*Asteraceae*)

evansianus
eh-vanz-ee-AH-nus
evansiana, evansianum
—

evansii
eh-VANS-ee-eye
Named after various people called Evans, including Thomas Evans (1751–1814), as in *Begonia grandis* subsp. *evansiana*

Euphorbia characias

Evolvulus arbuscula

Evolvulus [3]

ee-VOL-voo-lus

From Latin *evolvo*, meaning "to untwine," because unlike climbing *Convolvulus*, these plants do not twine (*Convolvulaceae*)

Exacum

eks-AH-kum

From Celtic Gaulish *exacon*, the vernacular name for *Centaurium*, a plant in the same family (*Gentianaceae*)

exaltatus

eks-all-TAH-tus

exaltata, exaltatum

Very tall, as in *Nephrolepis exaltata*

exaratus

ex-a-RAH-tus

exarata, exaratum

Engraved; furrowed, as in *Agrostis exarata*

excavatus

ek-ska-VAH-tus

excavata, excavatum

Hollowed out, as in *Calochortus excavatus*

excellens

ek-SEL-lenz

Excellent, as in *Sarracenia* × *excellens*

excelsior

eks-SEL-see-or

Taller, as in *Fraxinus excelsior*

excelsus

ek-SEL-sus

excelsa, excelsum

Tall, as in *Araucaria excelsa*

excisus

eks-SIZE-us

excisa, excisum

Cut away; cutout, as in *Adiantum excisum*

excorticatus

eks-kor-tih-KAH-tus

excorticata, excorticatum

Lacking or stripped of bark, as in *Fuchsia excorticata*

exiguus

eks-IG-yoo-us

exigua, exiguum

Very little; poor, as in *Salix exigua*

eximius

eks-IM-mee-us

eximia, eximium

Distinguished, as in *Eucalyptus eximia*

Exochorda

eks-oh-KOR-duh

From Greek *exo*, meaning "outside," and *chorde*, meaning "string," referring to fibers on the ovary outer wall (*Rosaceae*)

exoniensis

eks-oh-nee-EN-sis

exoniensis, exoniense

From Exeter, England, as in *Passiflora* × *exoniensis*

expansus

ek-SPAN-sus

expansa, expansum

Expanded, as in *Catasetum expansum*

exsertus

ek-SER-tus

exserta, exsertum

Protruding, as in *Acianthus exsertus*

extensus

eks-TEN-sus

extensa, extensum

Extended, as in *Acacia extensa*

eyriesii

eye-REE-see-eye

Named after Alexander Eyries, French nineteenth-century cactus collector, as in *Echinopsis eyriesii*

fabaceus

fab-AY-see-us

fabacea, fabaceum

Like a fava bean, as in *Marah fabacea*

Fabiana

fah-bee-AH-nuh

Named after Francisco Fabián y Fuero
(1719–1801), Spanish cleric (*Solanaceae*)

facetus

fa-CEE-tus

faceta, facetum

Elegant, as *Rhododendron facetum*

fagifolius

fag-ih-FOH-lee-us

fagifolia, fagifolium

With leaves like beech (*Fagus*), as in *Clethra
fagifolia*

Fagopyrum

fay-goh-PY-rum

From Latin *fagus*, meaning "beech," and
Greek *pyros*, meaning "wheat," because the
seed resembles beech mast (*Polygonaceae*)

Fagus [1]

FAY-gus

The classical Latin name for beech, possibly
derived from Greek *fagein*, meaning "to eat,"
because the seed is edible (*Fagaceae*)

falcatus

fal-KAH-tus

falcata, falcatum

Shaped like a sickle, as in *Cyrtanthus falcatus*

falcifolius

fal-sih-FOH-lee-us

falcifolia, falcifolium

With leaves in the shape of a sickle, as in
Allium falcifolium

falciformis

fal-sif-FOR-mis

falciformis, falciforme

Shaped like a sickle, as in *Falcatifolium
falciforme*

falcinellus

fal-sin-NELL-us

falcinella, falcinellum

Like a small sickle, as in *Polystichum
falcinellum*

fallax

FAL-laks

Deceptive; false, as in *Crassula fallax*

1

Fagus sylvatica

Fallopia

fuh-LOH-pee-uh

Named after Gabriele Falloppio (1523–62),
Italian anatomist who described the
Fallopian tube (*Polygonaceae*)

Fallugia

fuh-LOO-jee-uh

Named after Virgilio Fallugi (1627–1707),
Italian cleric and botanist (*Rosaceae*)

Farfugium

far-FOO-jee-um

From Latin *farfaria*, vernacular name for
related *Tussilago farfara* (*Asteraceae*)

Fargesia

far-JEE-see-uh

Named after Paul Guillaume Farges
(1844–1912), French missionary (*Poaceae*)

farinaceus

far-ih-NAH-kee-us

farinacea, farinaceum

Producing starch; mealy, like flour, as in
Salvia farinacea

farinosus

far-ih-NOH-sus

farinosa, farinosum

Mealy; powdery, as in *Rhododendron
farinosum*

farnesianus

far-nee-zee-AH-nus

farnesiana, farnesianum

Connected with the Farnese Gardens,
Rome, Italy, as in *Acacia farnesiana*

farreri

far-REY-ree

Named after Reginald Farrer (1880–1920),
British plant hunter and botanist, as in
Viburnum farreri

fasciatus

fash-ee-AH-tus

fasciata, fasciatum

Bound together, as in *Aechmea fasciata*

Fascicularia

fah-sik-ew-LAIR-ee-uh

From Latin *fasciculus*, meaning "bundle,"
and *arius*, meaning "pertaining to," because
the habit is clustered (*Bromeliaceae*)

fascicularis

fas-sik-yoo-LAH-ris

fascicularis, fasciculare

—

fasciculatus

fas-sik-yoo-LAH-tus

fasciculata, fasciculatum

Clustered or grouped together in bundles, as in *Ribes fasciculatum*

fastigiatus

fas-tij-ee-AH-tus

fastigiata, fastigiatum

With erect, upright branches, often creating the effect of a column, as in *Cotoneaster fastigiatus*

fastuosus

fast-yoo-OH-sus

fastuosa, fastuosum

Proud, as in *Cassia fastuosa*

Fatsia [1]

FAT-see-uh

From Japanese *yatsude*, meaning "eight fingers," because the leaves typically have eight fingerlike lobes (*Araliaceae*)

fatuus

FAT-yoo-us

fatua, fatuum

Insipid; poor quality, as in *Avena fatua*

Faucaria

faw-KAH-ree-uh

From Latin *fauces*, meaning "animal mouth," because the toothed leaves look like jaws (*Aizoaceae*)

febrifugus [2]

feb-ri-FEW-gus

febrifuga, febrifugum

Can reduce fever, as in *Dichroa febrifuga*

fecundus

feh-KUN-dus

fecunda, fecundum

Fertile; fruitful, as in *Aeschynanthus fecundus*

fejeensis

fee-jee-EN-sis

fejeensis, fejeense

From the Fiji Islands, South Pacific, as in *Davallia fejeensis*

Felicia

fuh-LIS-ee-uh

Named after the mysterious Herr Felix, a German official who died in 1846 (*Asteraceae*)

Fendlera

FEND-lur-uh

Named after August Fendler (1813–83), Prussian-born American naturalist (*Hydrangeaceae*)

1

Fatsia japonica

2

Dichroa febrifuga

fenestralis

fen-ESS-tra-lis

fenestralis, fenestrale

With openings like a window, as in *Vriesea fenestralis*

Fenestraria

fen-us-TRAIR-ee-uh

From Latin *fenestra*, meaning "window," because each leaf has a transparent tip (*Aizoaceae*)

fennicus

FEN-nih-kus

fennica, fennicum

Connected with Finland, as in *Picea fennica*

ferax

FER-aks

Fruitful, as in *Fargesia ferax*

Ferocactus

fer-oh-KAK-tus

From Latin *ferus*, meaning "fierce," plus cactus (*Cactaceae*)

ferox

FER-oks

Ferocious; thorny, as in *Datura ferox*

Ferraria

fuh-RAIR-ee-uh

Named after Giovanni Baptista Ferrari (1584–1655), Italian cleric and botanist (*Iridaceae*)

ferreus

FER-ee-us

ferrea, ferreum

Connected with iron; hard as iron, as in *Caesalpinia ferrea*

ferrugineus [3]

fer-oo-GIN-ee-us

ferruginea, ferrugineum

The color of rust, as in *Digitalis ferruginea*

fertilis

fer-TIL-is

fertilis, fertile

With plenty of fruit; with many seed, as in *Robinia fertilis*

Ferula [4]

FE-roo-luh

From Latiun *ferula*, meaning "rod," because the stem was used to make walking sticks and splints (*Apiaceae*)

festalis

FES-tuh-lis

festalis, festale

—

festivus

fes-TEE-vus

festiva, festivum

Festive; bright, as in *Hymenocallis × festalis*

Festuca

fes-TOO-kuh

From Latin *festuca*, meaning "straw" or "stalk" (*Poaceae*)

fibrillosus

fy-BRIL-oh-sus

fibrillosa, fibrillosum

—

fibrosus

fy-BROH-sus

fibrosa, fibrosum

Fibrous, as in *Dicksonia fibrosa*

Ficaria

fy-KAIR-ee-uh

Resembling *Ficus*, as the tubers resemble figs (*Ranunculaceae*)

3

Digitalis ferruginea

4

Ferula assa-foetida

ficifolius
fik-ee-FOH-lee-us
ficifolia, ficifolium
With figlike leaves, as in *Cucumis ficifolius*

ficoides
fy-KOY-deez
—

ficoideus
fy-KOY-dee-us
ficoidea, ficoideum
Resembling a fig (*Ficus*), as in *Senecio ficoides*

Ficus
FY-kus
From Latin *ficus*, the edible fig, *F. carica* (*Moraceae*)

filamentosus
fil-uh-men-TOH-sus
filamentosa, filamentosum
—

filarius
fil-AH-ree-us
filaria, filarium
With filaments or threads, as in *Yucca filamentosa*

fili-
Used in compound words to denote threadlike

filicaulis
fil-ee-KAW-lis
filicaulis, filicaule
With a threadlike stem, as in *Alchemilla filicaulis*

filicifolius
fil-ih-see-FOH-lee-us
filicifolia, filicifolium
With leaves like a fern, as in *Polyscias filicifolia*

filicinus
fil-ih-SEE-nus
filicina, filicinum
—

filiculoides
fil-ih-kyu-LOY-deez
Resembling a fern, as in *Asparagus filicinus*

Filipendula
fi-li-PEN-doo-luh
From Latin *filus*, meaning "thread," and *pendulus*, meaning "hanging," because the tubers of *F. vulgaris* are connected by threadlike roots (*Rosaceae*)

filipendulus
fil-ih-PEN-dyoo-lus
filipendula, filipendulum
Like meadowsweet (*Filipendula*), as in *Oenanthe filipendula*

filipes
fil-EE-pays
With threadlike stalks, as in *Rosa filipes*

fimbriatus
fim-bry-AH-tus
fimbriata, fimbriatum
Fringed, as in *Silene fimbriata*

firmatus
fir-MAH-tus
firmata, firmatum
—

firmus
fir-MUS
firma, firmum
Strong, as in *Abies firma*

Firmiana
fur-me-AH-nuh
Named after Karl Joseph von Firmian (1716–82), Austrian politician (*Malvaceae*)

fissilis
FISS-ill-is
fissilis, fissile
—

fissus
FISS-us
fissa, fissum
—

fissuratus
fis-zhur-RAH-tus
fissurata, fissuratum
With a split, as in *Alchemilla fissa*

fistulosus
fist-yoo-LOH-sus
fistulosa, fistulosum
Hollow, as in *Asphodelus fistulosus*

Fittonia
fih-TOH-nee-uh
Named after sisters Elizabeth (dates unknown) and Sarah Mary Fitton (ca. 1796–1874), Irish botanists (*Acanthaceae*)

Fitzroya
fitz-ROY-uh
Named after Robert FitzRoy (1805–65), British sailor, politician, and scientist; captained HMS *Beagle* while Charles Darwin was onboard naturalist (*Cupressaceae*)

flabellatus
fla-bel-AH-tus
flabellata, flabellatum
Like an open fan, as in *Aquilegia flabellata*

flabellifer
fla-BEL-lif-er
flabellifera, flabelliferum
Bearing a fanlike structure, as in *Borassus flabellifer*

flabelliformis
fla-bel-ih-FOR-mis
flabelliformis, flabelliforme
Shaped like a fan, as in *Erythrina flabelliformis*

flaccidus
FLA-sih-dus
flaccida, flaccidum
Weak: soft; feeble, as in *Yucca flaccida*

flagellaris
fla-gel-AH-ris
flagellaris, flagellare
—

flagelliformis
fla-gel-ih-FOR-mis
flagelliformis, flagelliforme
Like a whip; with long, thin shoots, as in *Celastrus flagellaris*

flammeus
FLAM-ee-us
flammea, flammeum
A flame color; flamelike, as in *Tigridia flammea*

Ficus carica

flavens

flav-ENZ

—

flaveolus

fla-VEE-oh-lus

flaveola, flaveolum

—

flavescens

flav-ES-enz

—

flavidus

FLA-vid-us

flavida, flavidum

Various kinds of yellow, as in *Anigozanthos flavidus*

flavicomus

flay-vih-KOH-mus

flavicoma, flavicomum

With yellow hair, as in *Euphorbia flavicoma*

flavissimus

flav-ISS-ih-mus

flavissima, flavissimum

Deep yellow, as in *Zephyranthes flavissima*

flavovirens

fla-voh-VY-renz

Greenish yellow, as in *Callistemon flavovirens*

flavus

FLA-vus

flava, flavum

Pure yellow, as in *Crocus flavus*

flexicaulis

fleks-ih-KAW-lis

flexicaulis, flexicaule

With a supple stem, as in *Strobilanthes flexicaulis*

flexilis

FLEKS-il-is

flexilis, flexile

Pliant, as in *Pinus flexilis*

flexuosus

fleks-yoo-OH-sus

flexuosa, flexuosum

Indirect; zigzagging, as in *Corydalis flexuosa*

floccigerus

flok-KEE-jer-us

floccigera, floccigerum

floccosus

flok-KOH-sus

floccosa, floccosum

With a woolly texture, as in *Rhipsalis floccosa*

florentinus

flor-en-TEE-nus

florentina, florentinum

Connected with Florence, Italy, as in *Malus florentina*

flore-pleno

FLOR-ee PLEE-no

With double flowers, as in *Aquilegia vulgaris* var. *flore-pleno*

floribundus

flor-ih-BUN-dus

floribunda, floribundium

—

floridus

flor-IH-dus

florida, floridum

Very free-flowering, as in *Wisteria floribunda*

GENUS SPOTLIGHT

Foeniculum

Fennel has an identity crisis. This English name is applied to several different plants, some related, others not. *Foeniculum* is the main fennel genus and it, like the English name, derives from the Latin word for hay. Both Florence fennel, with its edible bulb (actually a collection of leaf bases), and the herb fennel (cultivated for edible leaves and seed) are varieties of *Foeniculum vulgare*. Giant fennel is *Ferula vulgare*, sea fennel (or rock samphire) is *Crithmum maritimum*, while hog fennel (or masterwort) is *Peucedanum ostruthium*, all in the same family as *Foeniculum*. Dog's fennel—aka stinking chamomile (*Anthemis cotula*) or white heat aster (*Symphyotrichum ericoides*)—and fennel flower (or black cumin; *Nigella sativa*), definitely are not.

Foeniculum vulgare

floridanus

flor-ih-DAH-nus

floridana, floridanum

Connected with Florida, as in *Illicium floridanum*

floriferus

flor-IH-fer-us

florifera, floriferum

Especially free-flowering, as in *Townsendia florifera*

flos

flos

Used in combination to denote flower, as in *Lychnis flos-cuculi* (meaning cuckooflower)

fluitans

FLOO-ih-tanz

Floating, as in *Glyceria fluitans*

fluminensis

floo-min-EN-sis

fluminensis, fluminense

From Rio de Janeiro, Brazil, as in *Tradescantia fluminensis*

fluvialis

floo-vee-AHL-is

fluvialis, fluviale

—

fluviatilis

floo-vee-uh-TIL-is

fluviatilis, fluviatile

Growing in a river or running water, as in *Isotoma fluviatilis*

Fockea

FOK-ee-uh

Named after Hendrik Charles Focke (1802–56), Dutch botanist (*Apocynaceae*)

foeniculaceus

fen-ee-kul-ah-KEE-us

foeniculacea, foeniculaceum

Like fennel (*Foeniculum*), as in *Argyranthemum foeniculaceum*

Foeniculum

fen-IK-ew-lum

From Greek *fenum*, meaning "hay," an allusion to the scent of the leaves of fennel (*F. vulgare*), which supposedly resemble hay (*Apiaceae*)

foetidissimus

fet-uh-DISS-ih-mus

foetidissima, foetidissimum

With a really bad smell, as in *Iris foetidissima*

Forsythia viridissima

foetidus

FET-uh-dus

foetida, foetidum

With a bad smell, as in *Vestia foetida*

Fokienia

foh-kee-EN-ee-uh

Named after Fukien (now Fujian) Province in China (*Cupressaceae*)

foliaceus

foh-lee-uh-SEE-us

foliacea, foliaceum

Like a leaf, as in *Aster foliaceus*

foliatus

fol-ee-AH-tus

foliata, foliatum

With leaves, as in *Aletris foliata*

foliolotus

foh-lee-oh-LOH-tus

foliolota, foliolotum

—

foliolosus

foh-lee-oh-LOH-sus

foliolosa, foliolosum

With leaflets, as in *Thalictrum foliolosum*

foliosus

foh-lee-OH-sus

foliosa, foliosum

With many leaves; leafy, as in *Dactylorhiza foliosa*

follicularis

fol-lik-yoo-LAY-ris

follicularis, folliculare

With follicles, as in *Cephalotus follicularis*

Trachycarpus fortunei

fontanus

FON-tah-nus

fontana, fontanum

Growing in fast-running water, as in *Cerastium fontanum*

formosanus

for-MOH-sa-nus

formosana, formosanum

Connected with Taiwan (formerly Formosa), as in *Pleione formosana*

formosus

for-MOH-sus

formosa, formosum

Handsome; beautiful, as in *Pieris formosa*

forrestianus

for-rest-ee-AH-nus

forrestiana, forrestianum

—

forrestii

for-rest-EE-eye

Named after George Forrest (1873–1932), Scottish plant collector, as in *Hypericum forrestii*

Forsythia [1]

for-SYTH-ee-uh

Named after William Forsyth (1737–1804), Scottish botanist and a founding member of the Royal Horticultural Society (*Oleaceae*)

fortunei [2]

for-TOO-nee-eye

Named after Robert Fortune (1812–80), Scottish plant hunter and horticulturist, as in *Trachycarpus fortunei*

1

2

3

Fothergilla gardenii

Fragaria virginiana

Fuchsia magellanica

Fothergilla  [1]
foth-ur-GIL-uh
Named after John Fothergill (1712–80), British physician and naturalist (*Hamamelidaceae*)

Fouquieria
foo-kee-AIR-ee-uh
Named after Pierre Éloi Fouquier (1776–1850), French physician (*Fouquieriaceae*)

foveolatus
foh-vee-oh-LAH-tus
foveolata, foveolatum
With slight pitting, as in *Chionanthus foveolatus*

fragantissimus
fray-gran-TISS-ih-mus
fragrantissima, fragrantissimum
Very fragrant, as in *Lonicera fragrantissima*

Fragaria [2]
fruh-GAIR-ee-uh
From Latin *fraga*, meaning "fragrant," as is the fruit of strawberry (*Rosaceae*)

fragarioides
fray-gare-ee-OY-deez
Resembling strawberry (*Fragaria*), as in *Waldsteinia fragarioides*

fragilis
FRAJ-ih-lis
fragilis, fragile
Brittle; quick to wilt, as in *Salix fragilis*

fragrans
FRAY-granz
Fragrant, as in *Osmanthus fragrans*

Frailea
FRAY-lee-uh
Named after Manuel Fraile (1850–?), Spanish horticulturist (*Cactaceae*)

Francoa
fran-KOH-uh
Named after Francisco Franco (sixteenth century), Spanish physician (*Francoaceae*)

Frangula
FRANG-ew-luh
From Latin *frango*, meaning "to break," because they have brittle branches (*Rhamnaceae*)

Franklinia
frank-LIN-ee-uh
Named after Benjamin Franklin (1706–90), American statesman (*Theaceae*)

fraseri
FRAY-zer-ee
Named after John Fraser (1750–1811), Scottish plant collector and nurseryman, as in *Magnolia fraseri*

fraxineus
FRAK-si-nus
fraxinea, fraxineum
Like ash (*Fraxinus*), as in *Blechnum fraxineum*

fraxinifolius
fraks-in-ee-FOH-lee-us
fraxinifolia, fraxinifolium
With leaves like ash (*Fraxinus*), as in *Pterocarya fraxinifolia*

Fraxinus
FRAKS-in-us
From the classical Latin name for ash tree (*Oleaceae*)

Freesia
FREE-shuh
Named after Friedrich Heinrich Theodor Freese (1797–1876), German physician (*Iridaceae*)

Fremontodendron
fre-mont-oh-DEN-dron
Named after John Charles Frémont (1813–90), American soldier and politician, plus Greek *dendron*, meaning "tree" (*Malvaceae*)

frigidus
FRIH-jih-dus
frigida, frigidum
Growing in cold regions, as in *Artemisia frigida*

Fritillaria
fri-tuh-LAIR-ee-uh
From Latin *fritillus*, meaning "checkered," as in the tepals of *F. meleagris* (*Liliaceae*)

frondosus
frond-OH-sus
frondosa, frondosum
Especially leafy, as in *Primula frondosa*

frutescens
froo-TESS-enz
—

fruticans
FROO-tih-kanz
—

fruticosus
froo-tih-KOH-sus
fruticosa, fruticosum
Shrubby; bushy, as in *Argyranthemum frutescens*

fruticola
froo-TIH-koh-luh
Growing in bushy places, as in *Chirita fruticola*

fruticulosus
froo-tih-koh-LOH-sus
fruticulosa, fruticulosum
Dwarf and shrubby, as in *Matthiola fruticulosa*

fucatus
few-KAH-tus
fucata, fucatum
Painted; dyed, as in *Crocosmia fucata*

Fuchsia [3]
FEW-shuh or FOOK-see-uh
Named after Leonhart Fuchs (1501–66), German physician (*Onagraceae*)

fuchsioides
few-shee-OY-deez
Resembling *Fuchsia*, as in *Iochroma fuchsioides*

fugax
FOO-gaks
Withering quickly; fleeting, as in *Urginea fugax*

fulgens
FUL-jenz
—

fulgidus
FUL-jih-dus
fulgida, fulgidum
Shining; glistening, as in *Rudbeckia fulgida*

fuliginosus
few-lih-gin-OH-sus
fuliginosa, fuliginosum
A dirty brown or sooty color, as in *Carex fuliginosa*

fulvescens
ful-VES-enz
Becoming tawny in color, as in *Masdevallia fulvescens*

fulvidus
FUL-vee-dus
fulvida, fulvidum
Slightly tawny in color, as in *Cortaderia fulvida*

fulvus
FUL-vus
fulva, fulvum
Tawny orange in color, as in *Hemerocallis fulva*

Fumaria
foo-MAIR-ee-uh
From Latin *fumus*, meaning "smoke," which is what the roots smell of (*Papaveraceae*)

fumariifolius
foo-mar-ee-FOH-lee-us
fumariifolia, fumariifolium
With leaves like fumitory (*Fumaria*), as in *Scabiosa fumariifolia*

funebris
fun-EE-bris
funebris, funebre
Connected to graveyards, as in *Cupressus funebris*

fungosus
fun-GOH-sus
fungosa, fungosum
Like fungus; spongy, as in *Borinda fungosa*

furcans
fur-kanz
—

furcatus
fur-KA-tus
furcata, furcatum
Forked, as in *Pandanus furcatus*

Furcraea
fur-KREE-uh
Named after Antoine François de Fourcroy (1755–1809), French chemist (*Asparagaceae*)

fuscatus
fus-KA-tus
fuscata, fuscatum
Brownish in color, as in *Sisyrinchium fuscatum*

fuscus
FUS-kus
fusca, fuscum
A dusky or swarthy brown, as in *Nothofagus fusca*

futilis
FOO-tih-lis
futilis, futile
Without use, as in *Salsola futilis*

gaditanus

gad-ee-TAH-nus

gaditana, gaditanum

Connected with Cadiz, Spain, as in *Narcissus gaditanus*

Gagea

GAGE-ee-uh

Named after Thomas Gage (1781–1820), British botanist (*Liliaceae*)

Gaillardia

gay-LAR-dee-uh

Named after M. Gaillard de Charentonneau (eighteenth century), French magistrate and patron of botany (*Asteraceae*)

galacifolius

guh-lay-sih-FOH-lee-us

galacifolia, galacifolium

With leaves like beetleweed or wandflower (*Galax*), as in *Shortia galacifolia*

Galanthus

gah-LAN-thus

From Greek *gala*, meaning "milk," and *anthos*, meaning "flower," referring to the white flowers of snowdrops (*Amaryllidaceae*)

Galax

GAY-lax

From Greek *gala*, meaning "milk," because the flowers are white (*Diapensiaceae*)

galeatus

ga-le-AH-tus

galeata, galeatum

—

galericulatus

gal-er-ee-koo-LAH-tus

galericulata, galericulatum

Shaped like a helmet, as in *Sparaxis galeata*

Galega

GA-leh-guh

From Greek *gala*, meaning "milk," because this fodder was said to improve milk yield from goats (*Fabaceae*)

galegifolius

guh-lee-gih-FOH-lee-us

galegifolia, galegifolium

With leaves like goat's rue or professor weed (*Galega*), as in *Swainsona galegifolia*

Campanula garganica

Galium

GA-lee-um

From Greek *galion*, the vernacular name for a bedstraw, originating in *gala*, meaning "milk," because the flowers of *G. verum* were said to curdle milk, while the stems of *G. aparine* were used to strain milk (*Rubiaceae*)

gallicus

GAL-ih-kus

gallica, gallicum

Connected with France, as in *Rosa gallica*

Galtonia

gawl-TOH-nee-uh

Named after Francis Galton (1822–1911), British statistician and psychologist (*Amaryllidaceae*)

gangeticus

gan-GET-ih-kus

gangetica, gangeticum

Of the Ganges regions of India and Bangladesh, as in *Asystasia gangetica*

Gardenia

gar-DEE-nee-uh

Named after Alexander Garden (1730–91), Scottish physician and biologist (*Rubiaceae*)

garganicus [1]

gar-GAN-ih-kus

garganica, garganicum

Connected with Monte Gargano, Italy, as in *Campanula garganica*

Garrya

GA-ree-uh

Named after Nicholas Garry (ca. 1782–1856), British-born deputy governor of the Hudson's Bay Company (*Garryaceae*)

Gasteria

gas-TEER-ee-uh

From Greek *gaster*, meaning "stomach," which the flowers resemble (*Asphodelaceae*)

Gaultheria

gawl-THEER-ee-uh

Named after Jean-François Gaultier (1708–56), Canadian botanist and physician (*Ericaceae*)

Geranium

When is a geranium not a *Geranium*? When it is a *Pelargonium*! Cranesbills are members of the genus *Geranium*, but although many gardeners prefer common names over Latin, "geranium" is now so widely known that it has become a common name. It is also the common name for *Pelargonium*, a close relation of *Geranium*. But this causes confusion, because most true *Geranium* are hardy herbaceous perennials, while most *Pelargonium* are tender shrubs used as bedding. To avoid confusion, perhaps we should use the name "cranesbill" for the former and "storksbill" for the latter?

Geranium pratense

Gaura
GOW-ruh
From Greek *gauros*, meaning "superb" (*Onagraceae*)

Gaylussacia
gay-loo-SAY-see-uh
Named after Louis Joseph Gay-Lussac (1778–1850), French chemist (*Ericaceae*)

Gazania
guh-ZAY-nee-uh
From Greek *gaza*, meaning "riches," in reference to the great show of flowers; or named after Theodorus Gaza (1398–1478), Greek scholar (*Asteraceae*)

Geissorhiza
guy-so-RY-zuh
From Greek *geissos*, meaning "tile," and *rhiza*, meaning "root," due to the overlapping corm tunics (*Iridaceae*)

gelidus
JEL-id-us
gelida, gelidum
Connected with ice-cold regions, as in *Rhodiola gelida*

Gelsemium
jel-SEE-mee-um
From Italian *gelsomino*, meaning "jasmine" (*Gelsemiaceae*)

gemmatus
jem-AH-tus
gemmata, gemmatum
Bejeweled, as in *Wikstroemia gemmata*

gemmiferus
jem-MIH-fer-us
gemmifera, gemmiferum
With buds, as in *Primula gemmifera*

generalis
jen-er-RAH-lis
generalis, generale
Normal, as in *Canna* × *generalis*

genevensis
gen-EE-ven-sis
genevensis, genevense
From Geneva, Switzerland, as in *Ajuga genevensis*

geniculatus
gen-ik-yoo-LAH-tus
geniculata, geniculatum
With a sharp bend like a knee, as in *Thalia geniculata*

Genista
jen-IS-tuh
The classical Latin name for broom, from which the English royal House of Plantagenet gets its name, *planta-genista* (*Fabaceae*)

genistifolius
jih-nis-tih-FOH-lee-us
genistifolia, genistifolium
With leaves like broom (*Genista*), as in *Linaria genistifolia*

Gentiana
jen-shee-AH-nuh
Named after King Gentius of Illyria (second century), who supposedly discovered the medicinal properties of *G. lutea* (*Gentianaceae*)

Gentianopsis
jen-shee-ah-NOP-sis
Resembling the related genus *Gentiana* (*Gentianaceae*)

geoides
jee-OY-deez
Like avens (*Geum*), as in *Waldsteinia geoides*

geometrizans
jee-oh-MET-rih-zanz
With markings in a formal pattern, as in *Myrtillocactus geometrizans*

georgianus
jorj-ee-AH-nus
georgiana, georgianum
Connected with the U.S. state of Georgia, as in *Quercus georgiana*

georgicus

JORJ-ih-kus

georgica, georgicum

Connected with Georgia (Eurasia), as in *Pulsatilla georgica*

geranioides

jer-an-ee-OY-deez

Resembling *Geranium*, as in *Saxifraga geranioides*

Geranium

juh-RAY-nee-um

From Greek *geranos*, meaning "crane," because the fruit resembles this bird's beak, thus the common name cranesbill (*Geraniaceae*)

Gerbera

JER-bur-uh

Named after Traugott Gerber (1710–43), German physician and botanist (*Asteraceae*)

germanicus

jer-MAN-ih-kus

germanica, germanicum

Connected with Germany, as in *Iris germanica*

Gesneria

jez-NEER-ee-uh

Named after Konrad Gessner (1516–65), Swiss biologist (*Gesneriaceae*)

Geum [1]

JEE-oom

From Greek *geuo*, meaning "to give relish," because the roots of *G. urbanum* are said to taste good (*Rosaceae*)

Gevuina

geh-voo-AY-nuh

From Mapuche (Chile) *guevin*, the vernacular name (*Proteaceae*)

Gibbaeum

gib-AY-um

From Latin *gibbus*, meaning "hunched," because the succulent leaves are swollen and rounded (*Aizoaceae*)

gibberosus

gib-er-OH-sus

gibberosa, gibberosum

With a hump on one side, as in *Scaphosepalum gibberosum*

gibbiflorus

gib-bih-FLOR-us

gibbiflora, gibbiflorum

With flowers that have a hump on one side, as in *Echeveria gibbiflora*

gibbosus

gib-OH-sus

gibbosa, gibbosum

—

gibbus

gib-us

gibba, gibbum

With a swelling on one side, as in *Fritillaria gibbosa*

gibraltaricus

jib-ral-TAH-rih-kus

gibraltarica, gibraltaricum

Connected with Gibraltar, Europe, as in *Iberis gibraltarica*

giganteus

jy-GAN-tee-us

gigantea, giganteum

Unusually tall or large, as in *Stipa gigantea*

giganthus

jy-GAN-thus

gigantha, giganthum

With large flowers, as in *Hemsleya gigantha*

Gilia

GIL-ee-uh

Named after Filippo Luigi Gilii (1756–1821), Italian naturalist (*Polemoniaceae*)

Gillenia

gil-EN-ee-uh

Named after Arnold Gille (1586–1633), German alchemist (*Rosaceae*)

gilvus

GIL-vus

gilva, gilvum

Dull yellow, as in *Echeveria × gilva*

1

Geum urbanum

Ginkgo
GINK-goh
From Chinese *yinxing*, meaning "silver apricot," in reference to the edible seed; when translated into Japanese, it is *ginkyo*, but Linnaeus misread the name, switching the "y" for a "g," giving the current spelling (*Ginkgoaceae*)

glabellus
gla-BELL-us
glabella, glabellum
Smooth, as in *Epilobium glabellum*

glaber
gla-ber
glabra, glabrum
Smooth, hairless, as in *Bougainvillea glabra*

glabratus
GLAB-rah-tus
glabrata, glabratum
—

glabrescens
gla-BRES-senz
—

glabriusculus
gla-bree-US-kyoo-lus
glabriuscula, glabriusculum
Hairless, as in *Corylopsis glabrescens*

glacialis
glass-ee-AH-lis
glacialis, glaciale
Connected with ice-cold, glacial regions, as in *Dianthus glacialis*

gladiatus
glad-ee-AH-tus
gladiata, gladiatum
Like a sword, as in *Coreopsis gladiata*

Gladiolus
gla-dee-OH-lus
From Latin *gladiolus*, meaning "little sword," the shape of the leaves (*Iridaceae*)

Glandora
glan-DOR-uh
From Latin *glandulosus*, meaning "glandular," and *Lithodora*, a related genus; these plants differ from *Lithodora* in that they have glands inside the corolla (*Boraginaceae*)

Glandularia
glan-dew-LAIR-ee-uh
From Latin *glandula*, meaning "small acorn," a reference to the shape of the fruit, or *glandulosus*, meaning "glandular," as in the stigmas (*Verbenaceae*)

glanduliferus
glan-doo-LIH-fer-us
glandulifera, glanduliferum
With glands, as in *Impatiens glandulifera*

glanduliflorus
gland-yoo-LIH-flor-us
glanduliflora, glanduliflorum
With glandular flowers, as in *Stapelia glanduliflora*

glandulosus
glan-doo-LOH-sus
glandulosa, glandulosum
Glandular, as in *Erodium glandulosum*

glaucescens
glaw-KES-enz
With a bloom; blue-green in color, as in *Ferocactus glaucescens*

Glaucidium
glaw-KID-ee-um
The Greek *glaucidium* means "owl," but in this case, the name refers to the poppy genus *Glaucium*, which has similar foliage (*Ranunculaceae*)

glaucifolius
glau-see-FOH-lee-us
glaucifolia, glaucifolium
With gray-green leaves; with leaves with a bloom, as in *Diospyros glaucifolia*

Glaucium
GLAW-kee-um
From Greek *glaukon*, meaning "blue-gray," the color of the leaves (*Papaveraceae*)

glaucophyllus
glaw-koh-FIL-us
glaucophylla, glaucophyllum
With gray-green leaves, or with a bloom, as in *Rhododendron glaucophyllum*

GENUS SPOTLIGHT

Ginkgo

The maidenhair is a lonely tree. Known scientifically as *Ginkgo biloba*, it is a genus of a single species, the only member of its family, order, class, and division. This has not always been the case—the fossil record suggests that *Ginkgo* and its relatives were once common and widely distributed. The remaining species, which has earned the title "living fossil," occurs naturally only in a handful of sites in China. Given its long history, it is not surprising that this tree is a survivor and grows well as a city street tree, despite high levels of environmental pollution.

Ginkgo biloba

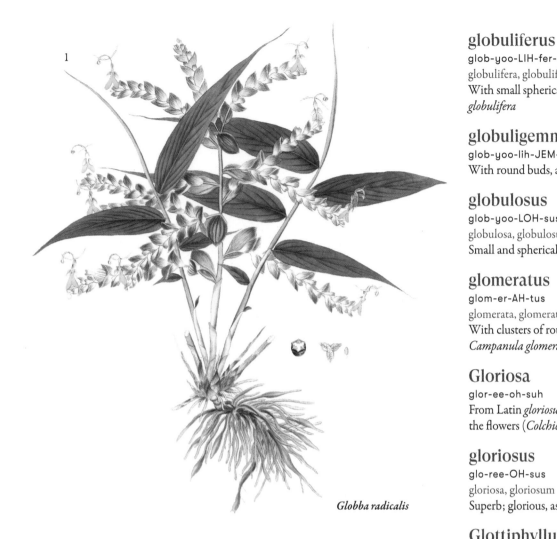

1

Globba radicalis

glaucus
GLAW-kus
glauca, glaucum
With a bloom on the leaves, as in *Festuca glauca*

Glebionis
gle-bee-OH-nis
From Greek *gleba*, meaning "soil," and *ionis*, meaning "characteristic of," perhaps in reference to how these plants can appear in plowed fields (*Asteraceae*)

Glechoma
gle-KOH-muh
From Greek *glechon*, meaning "mint" or "thyme" (*Lamiaceae*)

Gleditsia
gleh-DIT-zee-uh
Named after Johann Gottlieb Gleditsch (1714–86), German physician and botanist (*Fabaceae*)

Globba [1]
GLOH-buh
From Ambonese (Indonesia) *galoba*, the vernacular name for a spice (*Zingiberaceae*)

globiferus
glo-BIH-fer-us
globifera, globiferum
With spherical clusters of small globes, as in *Pilularia globifera*

globosus
glo-BOH-sus
globosa, globosum
Round, as in *Buddleja globosa*

Globularia
gloh-bew-LAIR-ee-uh
From Latin *globosus*, meaning "like a small ball," from the shape of the flower heads (*Plantaginaceae*)

globularis
glob-YOO-lah-ris
globularis, globulare
Relating to a small sphere, as in *Carex globularis*

globuliferus
glob-yoo-LIH-fer-us
globulifera, globuliferum
With small spherical clusters, as in *Saxifraga globulifera*

globuligemma
glob-yoo-lih-JEM-uh
With round buds, as in *Aloe globuligemma*

globulosus
glob-yoo-LOH-sus
globulosa, globulosum
Small and spherical, as in *Hoya globulosa*

glomeratus
glom-er-AH-tus
glomerata, glomeratum
With clusters of rounded heads, as in *Campanula glomerata*

Gloriosa
glor-ee-oh-suh
From Latin *gloriosus*, meaning "glorious," like the flowers (*Colchicaceae*)

gloriosus
glo-ree-OH-sus
gloriosa, gloriosum
Superb; glorious, as in *Yucca gloriosa*

Glottiphyllum
glot-i-FIL-um
From Greek *glotta*, meaning "tongue," and *phyllon*, meaning "leaf" (*Aizoaceae*)

Gloxinia
gloks-IN-ee-uh
Named after Benjamin Peter Gloxin (1765–94), German physician and botanist (*Gesneriaceae*)

gloxinioides
gloks-in-ee-OY-deez
Resembling *Gloxinia*, as in *Penstemon gloxinioides*

glumaceus
gloo-MA-see-us
glumacea, glumaceum
With glumes (the bracts that enclose the flowers of grasses), as in *Dendrochilum glumaceum*

Glumicalyx
gloo-me-KAY-lix
From Latin *gluma*, meaning "husk," because the calyx resemble a grass husk (*Scrophulariaceae*)

Gossypium arboreum

glutinosus

loo-tin-OH-sus

glutinosa, glutinosum

sticky; glutinous, as in *Eucryphia glutinosa*

Glyceria

gly-SEER-ee-uh

From Greek *glykeros*, meaning "sweet," in reference to the tasty seed (*Poaceae*)

Glycine

gly-SEE-nee

From Greek *glykeros*, meaning "sweet," an allusion to the sweet-tasting tubers of *Apios americana*, which used to belong to this genus, as *G. apios* (*Fabaceae*)

glycinoides

gly-sin-OY-deez

Like soybean (*Glycine*), as in *Clematis glycinoides*

Glycyrrhiza (also glycyrrhiza)

gli-sur-RY-zuh

From Greek *glykeros*, meaning "sweet," and *rhiza*, meaning "root," because the roots of *G. glabra* are used to make licorice (*Fabaceae*)

gnaphaloides

naf-fal-OY-deez

Like cudweed (*Gnaphalium*), as in *Senecio gnaphaloides*

Gomphocarpus

gom-foh-KAR-pus

From Greek *gomphos*, meaning "bolt" or "club," and *karpos*, meaning "fruit," an allusion to the inflated fruit (*Apocynaceae*)

Gomphrena

gom-FREE-nuh

From Latin *gromphaena*, the name for this or a related plant (*Amaranthaceae*)

gongylodes

GON-jih-loh-deez

Swollen; roundish, as in *Cissus gongylodes*

goniocalyx

gon-ee-oh-KAL-iks

Calyx with corners or angles, as in *Eucalyptus goniocalyx*

Goniolimon

goh-nee-oh-LIM-on

From Latin *gonio*, meaning "angle," and *Limonium*, a related genus, because the inflorescence branches are angled (*Plumbaginaceae*)

Goodyera

good-ee-AIR-uh

Named after John Goodyer (1592–1664), British botanist (*Orchidaceae*)

Gordonia

gor-DOH-nee-uh

Named after James Gordon (1708–80), British nurseryman (*Theaceae*)

gossypinus

goss-ee-PEE-nus

gossypina, gossypinum

Like cotton (*Gossypium*), as in *Strobilanthes gossypina*

Gossypium

gos-IP-ee-um

From Greek *gossypion*, meaning "cotton," possibly derived from Arabic *goz*, meaning a "silky substance"; commercial cotton is harvested from *G. hirsutum* and other species (*Malvaceae*)

gracilentus

grass-il-EN-tus

gracilenta, gracilentum

Graceful; slender, as in *Rhododendron gracilentum*

graciliflorus

grass-il-ih-FLOR-us

graciliflora, graciliflorum

With slender or graceful flowers, as in *Pseuderanthemum graciliflorum*

gracilipes

gra-SIL-i-peez

With a slender stalk, as in *Mahonia gracilipes*

gracilis

GRASS-il-is

gracilis, gracile

Graceful; slender, as in *Geranium gracile*

graecus

GRAY-kus

graeca, graecum

Greek, as in *Fritillaria graeca*

gramineus

gram-IN-ee-us

graminea, gramineum

Like grass, as in *Iris graminea*

graminifolius

gram-in-ee-FOH-lee-us

graminifolia, graminifolium

With grasslike leaves, as in *Stylidium graminifolium*

granadensis

gran-uh-DEN-sis

granadensis, granadense

From Granada, Spain, or Colombia, South America, as in *Drimys granadensis*

grandiceps

GRAN-dee-keps

With large head, as in *Leucogenes grandiceps*

grandicuspis

gran-dih-KUS-pis

grandicuspis, grandicuspe

With big points, as in *Sansevieria grandicuspis*

grandidentatus

gran-dee-den-TAH-tus

grandidentata, grandidentatum

With big teeth, as in *Thalictrum grandidentatum*

grandiflorus

gran-dih-FLOR-us

grandiflora, grandiflorum

With large flowers, as in *Platycodon grandiflorus*

grandifolius

gran-dih-FOH-lee-us

grandifolia, grandifolium

With large leaves, as in *Haemanthus grandifolius*

grandis

gran-DIS

grandis, grande

Big; showy, as in *Licuala grandis*

graniticus

gran-NY-tih-kus

granitica, graniticum

Growing on granite and rocks, as in *Dianthus graniticus*

granulatus

gran-yoo-LAH-tus

granulata, granulatum

Bearing grainlike structures, as in *Saxifraga granulata*

granulosus

gran-yool-OH-sus

granulosa, granulosum

Made of small grains, as in *Centropogon granulosus*

Graptopetalum

grap-toh-PET-uh-lum

From Greek *graptos*, meaning "inscribed," and *petalon*, meaning "petal," because the flowers of some species are curiously marked (*Crassulaceae*)

Graptophyllum

grap-toh-FIL-um

From Greek *graptos*, meaning "inscribed," and *phyllon*, meaning "leaf," for the often variegated foliage (*Acanthaceae*)

gratianopolitanus

grat-ee-an-oh-pol-it-AH-nus

gratianopolitana, gratianopolitanum

Connected with Grenoble, France, as in *Dianthus gratianopolitanus*

gratissimus

gra-TIS-ih-mus

gratissima, gratissimum

Especially pleasing, as in *Luculia gratissima*

gratus

GRAH-tus

grata, gratum

Giving pleasure, as in *Conophytum gratum*

graveolens [1]

grav-ee-OH-lenz

With a heavy scent, as in *Ruta graveolens*

Grevillea

gre-VILL-ee-uh

Named after Charles Francis Greville (1749–1809), British politician, art collector, and scientist, and a founding member of the Royal Horticultural Society (*Proteaceae*)

Grewia

GROO-ee-uh

Named after Nehemiah Grew (1641–1712), British plant anatomist (*Malvaceae*)

Greyia

GREY-ee-uh

Named after George Grey (1812–98), British soldier, explorer, and colonial administrator (*Francoaceae*)

Grindelia

grin-DEEL-ee-uh

Named after David Hieronymus Grindel (1776–1836), Latvian botanist (*Asteraceae*)

Griselinia

griz-uh-LIN-ee-uh

Named after Francesco Griselini (1717–87), Italian botanist (*Griseliniaceae*)

1

Ruta graveolens

2

Guzmania monostachya

griseus

GREE-see-us

grisea, griseum

Gray, as in *Acer griseum*

grosseserratus

groseser-AH-tus

grosseserrata, grosseserratum

With large saw teeth, as in *Clematis occidentalis* subsp. *grosseserrata*

grossus

GROSS-us

grossa, grossum

Particularly large, as in *Betula grossa*

Grusonia

groo-SOH-nee-uh

Named after Hermann Gruson (1821–95), German engineer (*Cactaceae*)

guianensis

gee-uh-NEN-sis

guianensis, guianense

From Guiana, South America, as in *Couroupita guianensis*

guineensis

gin-ee-EN-sis

guineensis, guineense

From the Guinea coast, West Africa, as in *Elaeis guineensis*

gummifer

GUM-mif-er

gummifera, gummiferum

Producing gum, as in *Seseli gummiferum*

gummosus

gum-MOH-sus

gummosa, gummosum

Gummy, as in *Ferula gummosa*

Gunnera

GUH-ner-uh

Named after Johan Ernst Gunner (1718–73), Norwegian cleric and botanist (*Gunneraceae*)

guttatus

goo-TAH-tus

guttata, guttatum

With spots, as in *Mimulus guttatus*

Guzmania [2]

gooz-MAN-ee-uh

Named after Anastasio Guzmán (unknown–1807), Spanish pharmacist and naturalist (*Bromeliaceae*)

3

Gypsophila elegans

Gymnocalycium

jim-noh-kay-LIS-ee-um

From Greek *gymnos*, meaning "naked," and calyx, the ring of sepals that protect the flower bud; these cacti have no spines on the calyx (*Cactaceae*)

Gymnocarpium

jim-noh-KAR-pee-um

From Greek *gymnos*, meaning "naked," and *karpos*, for "fruit," because the sporangia are not covered with a membrane (*Woodsiaceae*)

gymnocarpus

jim-noh-KAR-pus

gymnocarpa, gymnocarpum

With naked, uncovered fruit, as in *Rosa gymnocarpa*

Gymnocladus

jim-noh-KLAY-dus

From Greek *gymnos*, meaning "naked," and *klados*, meaning "branch"; the type species (*G. dioica*) is deciduous and slow to produce new leaves in spring and shed old leaves in fall (*Fabaceae*)

Gynura

jy-NOOR-uh

From Greek *gyne*, meaning "female," and *oura*, meaning "tail," perhaps an allusion to the female floral part, which has a branched style (*Asteraceae*)

Gypsophila [3]

jip-SOF-i-luh

From Greek *gypsos*, meaning "gypsum/chalk," and *philos*, meaning "loving," because some species prefer to grow on lime-rich soil (*Caryophyllaceae*)

Haageocereus

hah-jee-oh-SER-ee-us

Named after the Haage family of Germany, famous cactus growers, plus *Cereus*, a related genus (*Cactaceae*)

Haastia

HAHS-tee-uh

Named after Sir Johann Franz Julius von Haast (1822–87), German-born New Zealand geologist and explorer (*Asteraceae*)

haastii

HAAS-tee-eye

Named after Sir Johann Franz Julius von Haast (1822–87), German explorer and geologist, as in *Olearia × haastii*

Habenaria

ha-ben-AIR-ee-uh

From Latin *habena*, meaning "reins," an allusion to the flower spur (*Orchidaceae*)

Haberlea

hah-bur-LEE-uh

Named after Karl Konstantin Christian Haberle (1764–1832), Austrian scientist (*Gesneriaceae*)

Habranthus

ha-BRAN-thus

From Greek *habros*, meaning "delicate," and *anthos*, meaning "flower" (*Amaryllidaceae*)

Hacquetia

ha-KET-ee-uh

Named after Balthazar Hacquet (1739–1815), French-born Austrian scientist (*Apiaceae*)

hadriaticus

had-ree-AT-ih-kus

hadriatica, hadriaticum

Connected with the shores of the Adriatic Sea, Europe, as in *Crocus hadriaticus*

Haemanthus

HE-man-thus

From Greek *haimatos*, meaning "blood," and *anthos*, meaning "flower," because many species have scarlet blooms (*Amaryllidaceae*)

haematocalyx

hem-at-oh-KAL-icks

With a blood-red calyx, as in *Dianthus haematocalyx*

haematochilus

hem-mat-oh-KY-lus

haematochila, haematochilum

With a blood-red lip, as in *Oncidium haematochilum*

haematodes

hem-uh-TOH-deez

Blood-red, as in *Rhododendron haematodes*

Hakea

HAH-kee-uh

Named after Christian Ludwig von Hake (1745–1818), German administrator (*Proteaceae*)

hakeoides

hak-ee-OY-deez

Resembles *Hakea*, as in *Berberis hakeoides*

Hakonechloa

ha-kon-uh-KLOH-uh

Named for the town of Hakone in Japan, plus Greek *chloa*, "a blade of grass" (*Poaceae*)

Halesia

ha-LEE-see-uh, HAIL-see-uh

Named after Stephen Hales (1677–1761), British cleric and botanist (*Styracaceae*)

Halimium

ha-LIM-ee-um

From Greek *halimos*, the name for *Atriplex halimus*, which has similar leaves (*Cistaceae*)

Halimodendron

ha-lim-oh-DEN-dron

From Greek *halos*, meaning "the sea," and *dendron*, meaning "tree," because this genus thrives on saline soil (*Fabaceae*)

halophilus

hal-oh-FIL-ee-us

halophila, halophilum

Salt-loving, as in *Iris spuria* subsp. *halophila*

Hamamelis

ha-muh-MEL-is

From Greek *hama*, meaning "with," and *melon*, meaning "apple," because plants have both flowers and fruit at the same time; the medlar (*Mespilus germanica*) was known as *hamamelis* for this reason (*Hamamelidaceae*)

hamatus

ham-AH-tus

hamata, hamatum

—

hamosus

ham-UH-sus

hamosa, hamosum

Hooked, as in *Euphorbia hamata*

Haplopappus

hap-loh-PAP-us

From Greek *haplo*, meaning "simple," and *pappos*, meaning "fluff," because each fruit has a single ring of bristles (*Asteraceae*)

Hardenbergia

har-den-BURG-ee-uh

Named after Franziska von Hardenberg (unknown–1853), Austrian patron of botany (*Fabaceae*)

harpophyllus

harp-oh-FIL-us

harpophylla, harpophyllum

With leaves shaped like sickles, as in *Laelia harpophylla*

Harrisia

ha-RIS-ee-uh

Named after William Harris (1860–1920), Irish botanist (*Cactaceae*)

hastatus

hass-TAH-tus

hastata, hastatum

Shaped like a spear, as in *Verbena hastata*

hastilabius

hass-tih-LAH-bee-us

hastilabia, hastilabium

With a spear-shaped lip, as in *Oncidium hastilabium*

hastulatus

hass-TOO-lat-tus

hastulata, hastulatum

Shaped somewhat like a spear, as in *Acacia hastulata*

Hatiora

hat-ee-OR-uh

An anagram of the genus *Hariota*, named after Thomas Harriot (1560–1621), British mathematician and astronomer (*Cactaceae*)

Haworthia

huh-WORTH-ee-uh
Named after Adrian Hardy Haworth (1768–1833), British botanist and entomologist (*Asphodelaceae*)

Hebe

HE-bee
Named after Hebe, the Greek goddess of youth, although the allusion is unclear (*Plantaginaceae*)

hebecarpus

hee-be-KAR-pus
hebecarpa, hebecarpum
With down-covered fruit, as in *Senna hebecarpa*

hebephyllus

hee-bee-FIL-us
hebephylla, hebephyllum
With down-covered leaves, as in *Cotoneaster hebephyllus*

Hechtia

HECH-tee-uh
Named after Julius Gottfried Conrad Hecht (1771–1837), Prussian botanist and politician (*Bromeliaceae*)

Hedera

HED-uh-ruh
From Latin *hedera*, the vernacular name, possibly originating in Celtic *hedra*, meaning "cord" (*Araliaceae*)

hederaceus

hed-er-AYE-see-us
hederacea, hederaceum
Like ivy (*Hedera*), as in *Glechoma hederacea*

hederifolius

hed-er-ih-FOH-lee-us
hederifolia, hederifolium
With leaves like ivy (*Hedera*), as in *Veronica hederifolia*

Hedychium

he-dik-ee-um
From Greek *hedys*, meaning "sweet," and *chios*, meaning "snow," referring to the white, fragrant flowers of *H. coronarium* (*Zingiberaceae*)

Helenium (also helenium)

huh-LEE-nee-um
Named after Helen of Troy in Greek mythology, because the flowers were said to spring up from her teardrops (*Asteraceae*)

Heliamphora

he-lee-am-FOR-uh
From Greek *helos*, meaning "marsh," and *amphora*, meaning "vessel/container," because these pitcher plants live in bogs (*Sarraceniaceae*)

Helianthemum

he-lee-ANTH-uh-mum
From Greek *helios*, meaning "sun," and *anthemis*, for "flower," because the blooms open only in sunny weather (*Cistaceae*)

helianthoides

hel-ih-anth-OH-deez
Resembling sunflower (*Helianthus*), as in *Heliopsis helianthoides*

Helianthus

he-lee-ANTH-us
From Greek *helios*, meaning "sun," and *anthos*, meaning "flower," because the flower heads resemble the sun and the flower buds follow it (*Asteraceae*)

Helichrysum

he-li-KRY-sum
From Greek *helios*, meaning "sun," and *khrysos*, meaning "gold," because some species have yellow flowers (*Asteraceae*)

GENUS SPOTLIGHT

Helianthus

The sunflower (*Helianthus annuus*) is a favorite among children due to its ease of cultivation and vigorous growth—the tallest recorded was over 30 feet tall. Both its English and Latin names refer to the flower head, which looks like the sun, an inflorescence composed of small brown disk florets surrounded by yellow ray florets. Contrary to popular belief, mature sunflowers do not follow the sun (a process known as heliotropism) and generally face eastward. However, the developing flower buds can track the sun, returning to an easterly position each night.

Helianthus annuus

Hedera helix

1

Helleborus orientalis

2

Helonias bullata

3

Hemerocallis dumortieri

Heliconia
he-li-KOH-nee-uh
Named after Mount Helicon in Greece, home of the Muses (*Heliconiaceae*)

Helictotrichon
he-lik-to-TRY-kon
From Greek *helictos*, meaning "twisted," and *trichon*, meaning "hair," because the seed has twisted awns (*Poaceae*)

Heliopsis
he-lee-OP-sis
From Greek *helios*, meaning "sun," and *opsis*, meaning "like," because the flower heads resemble the sun (*Asteraceae*)

Heliotropium
he-lee-oh-TROH-pee-um
From Greek *helios*, meaning "sun," and *tropos*, meaning "to turn," from the mistaken idea that these flowers follow the sun (*Boraginaceae*)

helix
HEE-licks
In a spiral shape; applied to twining plants, as in *Hedera helix*

Helleborus [1]
he-lee-BORE-us
From Greek *helein*, meaning "to injure," and *bora*, meaning "food," an allusion to the hellebore's poisonous properties (*Ranunculaceae*)

hellenicus
hel-LEN-ih-kus
hellenica, hellenicum
Connected with Greece, as in *Linaria hellenica*

helodes
hel-OH-deez
From boggy land, as in *Drosera helodes*

Helonias [2]
huh-LOW-nee-as
From Greek *helos*, meaning "marsh," a common habitat (*Melanthiaceae*)

Heloniopsis
huh-low-nee-OP-sis
Resembling the related genus *Helonias* (*Melanthiaceae*)

helveticus
hel-VET-ih-kus
helvetica, helveticum
Connected with Switzerland, as in *Erysimum helveticum*

helvolus
HEL-vol-us
helvola, helvolum
Reddish yellow, as in *Vanda helvola*

Helwingia
hel-WING-ee-uh
Named after Georg Andreas Helwing (1666–1748), Prussian cleric and botanist (*Helwingiaceae*)

Hemerocallis [3]
hem-er-oh-KAH-lis, hem-uh-ROK-uh-lis
From Greek *hemeros*, meaning "day," and *kallos*, meaning "beauty," because daylily flowers last only one day (*Asphodelaceae*)

Hemigraphis
he-mee-GRAF-is
From Greek *hemi*, meaning "half," and *graphis*, meaning "brush," because the outer stamens are coated in hairs but the inner ones are not (*Acanthaceae*)

Hemionitis
hem-ee-oh-NY-tis
From Greek *hemionos*, meaning "mule," because a similar plant was said to ward off pregnancy, and mules are sterile (*Pteridaceae*)

Hemiptelea

hem-ip-TEE-lee-uh

From Greek *hemi*, meaning "half," and *ptelea*, meaning "elm," because this genus is related to true elms, *Ulmus* (*Ulmaceae*)

hemisphaericus

hem-is-FEER-ih-kus

hemisphaerica, hemisphaericum

In the shape of half a sphere, as in *Quercus hemisphaerica*

henryi

HEN-ree-eye

Named after Augustine Henry (1857–1930), Irish plant collector, as in *Lilium henryi*

Hepatica

he-PAT-i-kuh

From Greek *hepatos*, meaning "liver," which the leaves resemble (*Ranunculaceae*)

hepaticifolius

hep-at-ih-sih-FOH-lee-us

hepaticifolia, hepaticifolium

With leaves like liverwort (*Hepatica*), as in *Cymbalaria hepaticifolia*

hepaticus

hep-AT-ih-kus

hepatica, hepaticum

Dull brown; the color of liver, as in *Anemone hepatica*

hepta-

Used in compound words to denote seven

Heptacodium

hep-tuh-KOH-dee-um

From Greek *hepta*, meaning "seven," and *codeia*, meaning "poppy," because the flowers were said to occur in clusters of seven, although the bloom in the center is actually the bud of a developing flower cluster (*Caprifoliaceae*)

heptaphyllus

hep-tah-FIL-us

heptaphylla, heptaphyllum

With seven leaves, as in *Parthenocissus heptaphylla*

heracleifolius

hair-uh-klee-ih-FOH-lee-us

heracleifolia, heracleifolium

With leaves like hogweed (*Heracleum*), as in *Begonia heracleifolia*

Heracleum

huh-RAK-lee-um

Named after Hercules/Heracles, hero of Greek mythology, a reference to the great size of some hogweeds (*Apiaceae*)

herbaceus

her-buh-KEE-us

herbacea, herbaceum

Herbaceous, that is, not woody, as in *Salix herbacea*

Herbertia

her-BUR-tee-uh

Named after William Herbert (1778–1847), British botanist, cleric, and politician (*Iridaceae*)

Herpolirion

her-poh-LIR-ee-on

From Greek *herp*, meaning "to creep," and *lirion*, meaning "lily," in reference to the creeping habit (*Asphodelaceae*)

Hesperaloe

hes-per-AL-oh

From Greek *hesperos*, meaning "western" plus the genus *Aloe*; distant relatives of *Aloe* from western North America (*Asparagaceae*)

Hesperantha

hes-puh-RANTH-uh

From Greek *hesperos*, meaning "evening," and *anthos*, meaning "flower," because some species bloom in the evening (*Iridaceae*)

Hesperis

HES-puh-ris

From Greek *hesperos*, meaning "evening," when the flowers are most strongly scented (*Brassicaceae*)

Hesperocallis

hes-per-oh-KAL-is

From Greek *hesperos*, meaning "western," and *kallos*, meaning "beauty" (*Asparagaceae*)

heter-, hetero-

Used in compound words to denote various or diverse

OFFICER AND GENTLEMAN

The discovery of new plant species has long stimulated the imagination. While professional botanists are responsible for many such discoveries, people from other walks of life have also contributed to our understanding of the world's flora. Soldiers posted to far-flung stations were well positioned. John C. Frémont collected *Fremontodendron* while exploring the western United States, while *Colquhounia* was found by Sir Robert Colquhoun, stationed in Nepal. *Briggsia* was named in commemoration of a soldier, Munro Briggs Scott. A promising young botanist at Kew, London, he joined the army during World War I and was killed in France at the age of 28.

Heuchera

Plant breeders continue to develop new *Heuchera* cultivars with leaves in a range of vivid shades. They expanded the color palette by crossing plants of different colors and selecting the best-looking offspring. *Heuchera* will also crossbreed with the related genus *Tiarella*, and the offspring are known as intergeneric hybrids. These hybrids are given a unique genus name that combines both parental names: *Heuchera* plus *Tiarella* equals x *Heucherella*. Several hybrid genera are widely grown, including x *Fatshedera* (*Fatsia* x *Hedera)* and x *Cuprocyparis* (*Cupressus* x *Xanthocyparis*). Intergeneric hybrids may be vigorous, as in Leyland cypress (x *Cuprocyparis leylandii*), and can combine the best characteristics of each parent.

Heuchera sanguinea

heteracanthus
het-er-a-KAN-thus
heteracantha, heteracanthum
With various or diverse spines, as in *Agave heteracantha*

heteranthus
het-er-AN-thus
heterantha, heteranthum
With various or diverse flowers, as in *Indigofera heterantha*

heterocarpus
het-er-oh-KAR-pus
heterocarpa, heterocarpum
With various or diverse fruits, as in *Ceratocapnos heterocarpa*

heterodoxus
het-er-oh-DOKS-us
heterodoxa, heterodoxum
Differing from the type of the genus, as in *Heliamphora heterodoxa*

Heteromeles
het-uh-roh-ME-leez
From Greek *heteros*, meaning "different," and *melon*, meaning "apple"; either a comparison with apple (*Malus*) fruit or an allusion to the flowers, which only have ten stamens each, while those of an apple have fifteen-plus each (*Rosaceae*)

heteropetalus
het-er-oh-PET-uh-lus
heteropetala, heteropetalum
With various or diverse petals, as in *Erepsia heteropetala*

heterophyllus
het-er-oh-FIL-us
heterophylla, heterophyllum
With various or diverse leaves, as in *Osmanthus heterophyllus*

heteropodus
het-er-oh-PO-dus
heteropoda, heteropodum
With various or diverse stalks, as in *Berberis heteropoda*

Heterotheca
het-uh-roh-THEE-kuh
From Greek *heteros*, meaning "different," and *theke*, meaning "container," an allusion to the different types of fruit produced by ray and disk florets (*Asteraceae*)

Heuchera
HOOK-uh-ruh
Named after Johann Heinrich von Heucher (1677–1747), Austrian physician and botanist (*Saxifragaceae*)

hexa-
Used in compound words to denote six

hexagonopterus
heks-uh-gon-OP-ter-us
hexagonoptera, hexagonopterum
With six-angled wings, as in *Phegopteris hexagonoptera*

hexagonus
hek-sa-GON-us
hexagona, hexagonum
With six angles, as in *Cereus hexagonus*

hexandrus
heks-AN-drus
hexandra, hexandrum
With six stamens, as in *Sinopodophyllum hexandrum*

hexapetalus
heks-uh-PET-uh-lus
hexapetala, hexapetalum
With six petals, as in *Ludwigia grandiflora* subsp. *hexapetala*

hexaphyllus
heks-uh-FIL-us
hexaphylla, hexaphyllum
With six leaves or leaflets, as in *Stauntonia hexaphylla*

hians
HY-anz
Gaping, as in *Aeschynanthus hians*

Hibbertia
hi-BER-tee-uh
Named after George Hibbert (1757–1837), British merchant, politician, and botanist (*Dilleniaceae*)

hibernicus
hy-BER-nih-kus
hibernica, hibernicum
Connected with Ireland, as in *Hedera hibernica*

Hibiscus [1]
hi-BIS-kus
From Greek *hibiskos*, the vernacular name for related marshmallow (*Althaea officinalis*); perhaps deriving from *ibiskum*, suggesting these plants live with ibis, marshland birds (*Malvaceae*)

hiemalis
hy-EH-mah-lis
hiemalis, hiemale
Of the winter; winter flowering, as in *Leucojum hiemale*

Hieracium
hy-er-AY-see-um
From Greek *hierax*, meaning "hawk," but the etymology is obscure (*Asteraceae*)

Hierochloe
hy-er-oh-KLO-ee
From Greek *hieros*, meaning "sacred," and *chloa*, "a blade of grass"; used in religious ceremonies (*Poaceae*)

hierochunticus
hi-er-oh-CHUN-tih-kus
hierochuntica, hierochunticum
Connected with Jericho, as in *Anastatica hierochuntica*

himalaicus
him-al-LAY-ih-kus
himalaica, himalaicum
Connected with the Himalayas, as in *Stachyurus himalaicus*

Himalayacalamus
him-uh-lay-oh-KAL-uh-mus
From the Himalayas, plus Greek *kalamos*, meaning "reed," a bamboo (*Poaceae*)

himalayensis
him-uh-lay-EN-is
himalayensis, himalayense
From the Himalayas, as in *Geranium himalayense*

Hippeastrum
hip-ee-AS-trum
From Greek *hippeus*, meaning "horseman," and *astros*, meaning "star," perhaps because the flowers resemble a weapon or ornament worn by a knight, although the etymology obscure (*Amaryllidaceae*)

Hippocrepis
hip-oh-KREP-is
From Greek *hippos*, meaning "horse," and *krepis*, meaning "slipper," because the seed pods resemble horseshoes (*Fabaceae*)

Hippophae
HIP-oh-fee
From Greek *hippos*, meaning "horse," and *phaos*, meaning "shining," because the foliage was fed to horses to improve their health and appearance (*Elaeagnaceae*)

Hippuris
hip-ew-ris
From Greek *hippos*, meaning "horse," and *oura*, meaning "tail," which the plant resembles (*Plantaginaceae*)

hircinus
her-SEE-nus
hircina, hircinum
Goatlike, or with a goatlike odor, as in *Hypericum hircinum*

hirsutissimus
her-soot-TEE-sih-mus
hirsutissima, hirsutissimum
Especially hairy, as in *Clematis hirsutissima*

hirsutus
her-SOO-tus
hirsuta, hirsutum
Hairy, as in *Lotus hirsutus*

hirsutulus
her-SOOT-oo-lus
hirsutula, hirsutulum
Somewhat hairy, as in *Viola hirsutula*

hirtellus
her-TELL-us
hirtella, hirtellum
Hairy, as in *Plectranthus hirtellus*

hirtiflorus
her-tih-FLOR-us
hirtiflora, hirtiflorum
With hairy flowers, as in *Passiflora hirtiflora*

hirtipes
her-TYE-pees
With hairy stems, as in *Viola hirtipes*

hirtus
HER-tus
hirta, hirtum
Hairy, as in *Columnea hirta*

1

Hibiscus syriacus

hispanicus

his-PAN-ih-kus

hispanica, hispanicum

Connected with Spain, as in *Narcissus hispanicus*

hispidus

HISS-pih-dus

hispida, hispidum

With bristles, as in *Leontodon hispidus*

Hohenbergia

hoh-un-BUR-ge-uh

Named after the Prince of Württemberg, called Hohenberg, a German or Austrian noble, although his exact identity is uncertain (*Bromeliaceae*)

Hoheria

ho-HEER-ee-uh

From Maori (New Zealand) *houhere*, the vernacular name (*Malvaceae*)

Holboellia [1]

hol-BOWL-ee-uh

Named after Frederik Ludvig Holbøll (1765–1829), Danish botanist (*Lardizabalaceae*)

Holcus

HOL-kus

From Greek *holkos*, meaning "a kind of grain," or possibly, meaning "strap," a reference to the straplike leaves of this grass (*Poaceae*)

hollandicus

hol-LAN-dih-kus

hollandica, hollandicum

Connected with Holland, as in *Allium hollandicum*

Holmskioldia

holm-ski-OLD-ee-uh

Named after Johan Theodor Holmskjold (1731–93), Danish botanist and courtier (*Lamiaceae*)

holo-

Used in compound words to denote completely

holocarpus

ho-loh-KAR-pus

holocarpa, holocarpum

With complete fruit, as in *Staphylea holocarpa*

1

Holboellia latifolia

holochrysus

ho-loh-KRIS-us

holochrysa, holochrysum

Completely golden, as in *Aeonium holochrysum*

Holodiscus

ho-loh-DIS-kus

From Greek *holos*, meaning "whole," and *diskus*, meaning "disk," because each flower has a circular, unlobed nectar-producing structure (*Rosaceae*)

holosericeus

ho-loh-ser-ee-KEE-us

holosericea, holosericeum

With silky hairs all over, as in *Convolvulus holosericeus*

Homalocladium

hom-uh-loh-KLAY-dee-um

From Greek *homalos*, meaning "flat" or "smooth," and *klados*, meaning "branch," in reference to the flat stems that give the plant the name "tapeworm plant" (*Polygonaceae*)

Hoodia

HOOD-ee-uh

Named after Van Hood (nineteenth century), British horticulturist (*Apocynaceae*)

Hordeum

HOR-dee-um

From the classical Latin name for barley, *H. vulgare* (*Poaceae*)

horizontalis

hor-ih-ZON-tah-lis

horizontalis, horizontale

Close to the ground; horizontal, as in *Cotoneaster horizontalis*

Horminum (also horminum)

hor-MINE-um

From the classical Latin name for sage, possibly *Salvia horminum* (*Lamiaceae*)

horridus

HOR-id-us

horrida, horridum

With many prickles, as in *Euphorbia horrida*

hortensis

hor-TEN-sis

hortensis, hortense

—

hortorum

hort-OR-rum

—

hortulanus

hor-tew-LAH-nus

hortulana, hortulanum

Relating to gardens, *Lysichiton × hortensis*

Hydrangea macrophylla

Hosta
HOS-tuh
Named after Nicolaus Thomas Host (1761–1834), Croatian botanist and physician (*Asparagaceae*)

Hottonia
huh-TOH-nee-uh
Named after Petrus Houttuyn (1648–1709), Dutch physician and botanist (*Primulaceae*)

Houstonia
hoo-STOW-nee-uh
Named after William Houstoun (ca. 1695–1733), Scottish botanist (*Rubiaceae*)

Houttuynia
how-TEE-nee-uh
Named after Maarten Houttuyn (1720–98), Dutch physician and naturalist (*Saururaceae*)

Hovenia
hoh-VEE-nee-uh
Named after David ten Hove (1724–87), Dutch politician (*Rhamnaceae*)

Howea
HOW-ee-uh
Named after Lord Howe Island, Australia, where both species reside; the island is named after Richard Howe (1726–99), British naval officer (*Arecaceae*)

Hoya
HOY-uh
Named after Thomas Hoy (ca. 1750–1822), British horticulturist and botanist (*Apocynaceae*)

Huernia
HER-nee-uh
Named after Justin Heurnius (1587–1652), Dutch missionary (*Apocynaceae*)

hugonis
hew-GO-nis
Named after Father Hugh Scallon, missionary in China in the late nineteenth and early twentieth centuries, as in *Rosa hugonis*

humifusus
hew-mih-FEW-sus
humifusa, humifusum
Of a sprawling habit, as in *Opuntia humifusa*

humilis
HEW-mil-is
humilis, humile
Low growing, dwarfish, as in *Chamaerops humilis*

Humulus
HUM-ew-lus
From German *humela*, an old name for hops, *H. lupulus* (*Cannabaceae*)

hungaricus
hun-GAR-ih-kus
hungarica, hungaricum
Connected with Hungary, as in *Colchicum hungaricum*

Hunnemannia
hoon-MAN-ee-uh
Named after John Hunnemann (1760–1839), British botanist (*Papaveraceae*)

hunnewellianus
hun-ee-we-el-AH-nus
hunnewelliana, hunnewellianum
Named after the Hunnewell family of the Hunnewell Arboretum, Wellesley, Massachusetts, as in *Rhododendron hunnewellianum*

hupehensis
hew-pay-EN-sis
hupehensis, hupehense
From Hupeh (Hubei), China, as in *Sorbus hupehensis*

Hyacinthella
hy-uh-sinth-ELL-uh
From *Hyacinthus*, a related genus, plus diminutive *ella* (*Asparagaceae*)

hyacinthinus
hy-uh-sin-THEE-nus
hyacinthina, hyacinthinum
—

Hyacinthoides
hy-uh-sinth-OY-deez
Resembling the related genus *Hyacinthus* (*Asparagaceae*)

Hyacinthus (also hyacinthus)
hy-uh-SIN-thus
hyacintha, hyacinthum
Dark purple-blue, or like a hyacinth, as in *Triteleia hyacinthina*

hyalinus
hy-yuh-LEE-nus
hyalina, hyalinum
Transparent; almost transparent, as in *Allium hyalinum*

hybridus
hy-BRID-us
hybrida, hybridum
Mixed; hybrid, as in *Helleborus × hybridus*

Hydrangea [2]
hy-DRANE-juh
From Greek *hydor*, meaning "water," and *angeion*, meaning "small vessel," a reference to the cuplike fruit (*Hydrangeaceae*)

hydrangeoides
hy-drain-jee-OY-deez
Resembling *Hydrangea*, as in *Schizophragma hydrangeoides*

Hydrocharis
hy-DRO-kar-is
From Greek *hydor*, meaning "water," and *chari*, meaning "grace," because these are delicate floating aquatic plants (*Hydrocharitaceae*)

Hydrocleys
hy-DROK-lee-is
From Greek *hydor*, meaning "water," and *clavis*, meaning "club," a reference to the club-shaped female floral parts (*Alismataceae*)

Hydrocotyle
hr-droh-KOT-il-ee
From Greek *hydor*, meaning "water," and *kotyle*, meaning "cup," because this aquatic herb may have cup-shaped leaves (*Araliaceae*)

Hydrophyllum
hy-droh-FIL-um
From Greek *hydor*, meaning "water," and *phyllon*, meaning "leaf," because the leaf spots look like drops of water (*Boraginaceae*)

hyemalis [1]
hy-EH-mah-lis
hyemalis, hyemale
Relating to winter; winter flowering, as in *Eranthis hyemalis*

Hygrophila
hy-GRO-fil-uh
From Greek *hygro*, meaning "damp," and *phila*, meaning "loving" (*Acanthaceae*)

hylaeus
hy-la-ee-us
hylaea, hylaeum
From the woods, as in *Rhododendron hylaeum*

Hylocereus
hy-loh-SER-ee-us
From Greek *hyle*, meaning "wood," and *Cereus*, a related genus, because these cacti climb trees (*Cactaceae*)

Hylomecon
hy-loh-ME-kon
From Greek *hyle*, meaning "wood," and *mekon*, meaning "poppy," a plant of woodlands (*Papaveraceae*)

Hylotelephium
hy-loh-tel-EFF-ee-um
From Greek *hyle*, meaning "wood," and *Telephium*, another genus; wild plants can be found on woodland margins (*Crassulaceae*)

hymen-
Used in compound words to denote membranous

hymenanthus
hy-men-AN-thus
hymenantha, hymenanthum
With flowers with a membrane, as in *Trichopilia hymenantha*

Hymenocallis [2]
hy-men-OK-uh-lis
From Greek *hymen*, meaning "membrane," and *kallis*, meaning "beauty," for the membranous corona in the center of the flower (*Amaryllidaceae*)

hymenorrhizus
hy-men-oh-RY-zus
hymenorrhiza, hymenorrhizum
With membranous roots, as in *Allium hymenorrhizum*

hymenosepalus
hy-men-no-SEP-uh-lus
hymenosepala, hymenosepalum
With membranous sepals, as in *Rumex hymenosepalus*

Hymenosporum
hy-men-oh-SPORE-um
From Greek *hymen*, meaning "membrane," and *spora*, meaning "seed," because the seed have membranous wings (*Pittosporaceae*)

Hymenoxys
hy-men-OX-is
From Greek *hymen*, meaning "membrane," and *oxys*, meaning "sharp," because the scales surrounding each flower are sharp-tipped (*Asteraceae*)

Hyophorbe
hy-oh-FOR-bee
From Greek *hyos*, meaning "swine," and *phorba*, meaning "food," because the fruit were fed to pigs (*Arecaceae*)

Eranthis hyemalis

Hymenocallis littoralis

Hypericum perforatum

Hyoscyamus

hy-oh-SY-ah-mus

From Greek *hyos*, meaning "swine," and *kyamos*, meaning "bean"; this is henbane (*H. niger*), which is poisonous to humans but supposedly not to pigs (*Solanaceae*)

hyperboreus

hy-puh-BOR-ee-us

hyperborea, hyperboreum

Connected with the far north, as in *Sparganium hyperboreum*

hypericifolius

hy-PER-ee-see-FOH-lee-us

hypericifolia, hypericifolium

With leaves like St. John's wort (*Hypericum*), as in *Melalaeuca hypericifolia*

hypericoides

hy-per-ih-KOY-deez

Resembling *Hypericum*, as in *Ascyrum hypericoides*

Hypericum [3]

hy-PER-ik-um

From Greek *hyper*, meaning "above," and *eikon*, meaning "icon," because the plant was often hung above religious figures (*Hypericaceae*)

hypnoides

hip-NO-deez

Resembling moss, as in *Saxifraga hypnoides*

hypo-

Used in compound words to denote under

Hypocalymma

hy-poh-ka-LIM-uh

From Greek *hypo*, meaning "under," and *kalymma*, meaning "veil," because the calyx forms a cap over the flower bud (*Myrtaceae*)

Hypochaeris

hy-poh-KEER-is

From Greek *hypo*, meaning "under," and *choiras*, meaning "pig," suggesting that swine consume the roots, although the name may also derive from *kharieis*, meaning "beauty" (*Asteraceae*)

hypochondriacus

hy-po-kon-dree-AH-kus

hypochondriaca, hypochondriacum

With a melancholy appearance; with flowers of a dull color, as in *Amaranthus hypochondriacus*

Hypoestes

hy-poh-ES-teez

From Greek *hypo*, meaning "below," and *estia*, meaning "house," an allusion to the flowers, which emerge from within several bracts (*Acanthaceae*)

hypogaeus

hy-poh-JEE-us

hypogaea, hypogaeum

Underground; developing in the soil, as in *Copiapoa hypogaea*

hypoglaucus

hy-poh-GLAW-kus

hypoglauca, hypoglaucum

Glaucous underneath, as in *Cissus hypoglauca*

hypoglottis

hh-poh-GLOT-tis

Underside of the tongue, from the shape of the pods, as in *Astragalus hypoglottis*

hypoleucus

hy-poh-LOO-kus

hypoleuca, hypoleucum

White underneath, as in *Centaurea hypoleuca*

hypophyllus

hy-poh-FIL-us

hypophylla, hypophyllum

Underneath the leaf, as in *Ruscus hypophyllum*

hypopitys

hi-po-PY-tees

Growing under pines, as in *Monotropa hypopitys*

Hypoxis

hy-POKS-iss

From Greek *hypo*, meaning "below," and *oxys*, meaning "sharp," a reference to the fruiting capsules, which have a pointed base (*Hypoxidaceae*)

hyrcanus

hyr-KAH-nus

hyrcana, hyrcanum

Connected with the region of the Caspian Sea (Hyrcania in antiquity), as in *Hedysarum hyrcanum*

hyssopifolius

hiss-sop-ih-FOH-lee-us

hyssopifolia, hyssopifolium

With leaves like hyssop (*Hyssopus*), as in *Cuphea hyssopifolia*

Hyssopus

his-OP-us

Classical Greek name for an aromatic herb (*Lamiaceae*)

hystrix

HIS-triks

Bristly; like a porcupine, as in *Colletia hystrix*

ibericus

eye-BEER-ih-kus

iberica, ibericum

Connected with Iberia (Spain and Portugal), as in *Geranium ibericum*

iberidifolius

eye-beer-id-ih-FOH-lee-us

iberidifolia, iberidifolium

With leaves that resemble *Iberis*, as in *Brachyscome iberidifolia*

Iberis

AY-ber-is

From Greek *Iberes*, meaning "of Iberia," because several species come from Spain (*Brassicaceae*)

Ibervillea

ay-ber-VILL-ee-uh

Possibly named after Iberville Parish, Louisiana, because the type species was collected in that state (*Cucurbitaceae*)

Ichtyoselmis

ik-tee-oh-SEL-mis

From Greek *ichthys*, meaning "fish," and *selmis*, meaning "fishing line," because the flowers look like fish hooked on a line (*Papaveraceae*)

icos-

Used in compound words to denote twenty

icosandrus

eye-koh-SAN-drus

icosandra, icosandrum

With twenty stamens, as in *Phytolacca icosandra*

idaeus

eye-DAY-ee-us

idaea, idaeum

Connected with Mount Ida, Crete, as in *Rubus idaeus*

Idesia

ay-DEE-see-uh

Named after Eberhard Ysbrants Ides (1657–1708), Dutch diplomat and explorer (*Salicaceae*)

ignescens

ig-NES-enz

—

igneus

ig-NE-us

ignea, igneum

Fiery red, as in *Cuphea ignea*

ikariae

eye-KAY-ree-ay

Of the island of Ikaria, in the Aegean Sea, as in *Galanthus ikariae*

Ilex (also ilex)

AY-leks

From the classical Latin name *ilex* for hollly oak (*Quercus ilex*), because both are evergreen trees with spiny leaves (*Aquifoliaceae*)

ilicifolius

il-liss-ee-FOH-lee-us

ilicifolia, ilicifolium

With leaves like holly (*Ilex*), as in *Itea ilicifolia*

illecebrosus

il-lee-see-BROH-sus

illecebrosa, illecebrosum

Enticing; charming, as in *Tigridia illecebrosa*

Illicium

i-LISS-ee-um

From Latin *illicere*, meaning "to allure," an allusion to the scented flowers and foliage (*Schisandraceae*)

GENUS SPOTLIGHT

Ilex

The holly genus *Ilex* is often incorporated into other Latin names to indicate that a plant has spiny, hollylike leaves, such as in *Itea ilicifolia* and *Olearia ilicifolia*. This is somewhat ironic, because hollies were named *Ilex* due to their similarity to the spiny leaves of holm oak (*Quercus ilex*). Furthermore, most hollies do not have the traditional spiny leaves you find in English (*I. aquifolium*) or American (*I. opaca*) holly. And even in these species, not all of the leaves are spiny. As hollies mature, leaves near the top of the tree are often spine-free—why waste energy forming spines when no herbivore can reach them?

Ilex aquifolium

1

2

3

Gladiolus imbricatus

Impatiens balsamina

Imperata cylindrica

illinitus
il-lin-EYE-tus
illinita, illinitum
Smeared; smirched, as in *Escallonia illinita*

illinoinensis
il-ih-no-in-EN-sis
illinoinensis, illinoinense
From Illinois, as in *Carya illinoinensis*

illustris
il-LUS-tris
illustris, illustre
Brilliant; lustrous, as in *Amsonia illustris*

illyricus
il-LEER-ih-kus
illyrica, illyricum
Connected with Illyria, the name for an area of the western Balkan
Peninsula in antiquity, as in *Gladiolus illyricus*

ilvensis
il-VEN-sis
ilvensis, ilvense
Of Elba, Italy, or the Elbe River, as in *Woodsia ilvensis*

imberbis
IM-ber-bis
imberbis, imberbe
Without spines or beard, as in *Rhododendron imberbe*

imbricans
IM-brih-KANS
—
imbricatus [1]
IM-brih-KA-tus
imbricata, imbricatum
With elements that overlap in a regular pattern, as in *Gladiolus
imbricatus*

immaculatus
im-mak-yoo-LAH-tus
immaculata, immaculatum
Spotless, as in *Aloe immaculata*

immersus
im-MER-sus
immersa, immersum
Growing under water, as in *Pleurothallis immersa*

Impatiens [2]
im-PAT-ee-enz, im-PAY-shunz
From Latin *impatiens*, meaning "impatient," because the ripe fruit
suddenly bursts open (*Balsaminaceae*)

Imperata [3]
im-puh-RAH-tuh
Named after Ferrante Imperato (1550–1625), Italian apothecary
(*Poaceae*)

imperialis
im-peer-ee-AH-lis
imperialis, imperiale
Very fine; showy, as in *Fritillaria imperialis*

Dactylorhiza incarnata

Narcissus incomparabilis

implexus
im-PLECK-sus
implexa, implexum
Tangled, as in *Kleinia implexa*

impressus
im-PRESS-us
impressa, impressum
With impressed or sunken surfaces, as in *Ceanothus impressus*

inaequalis
in-ee-KWA-lis
inaequalis, inaequale
Unequal, as in *Geissorhiza inaequalis*

incanus
in-KAN-nus
incana, incanum
Gray, as in *Geranium incanum*

incarnatus [4]
in-kar-NAH-tus
incarnata, incarnatum
The color of flesh, as in *Dactylorhiza incarnata*

Incarvillea
in-kar-VILL-ee-uh
Named after Pierre Nicolas Le Chéron d'Incarville (1706–57),
French missionary and botanist (*Bignoniaceae*)

incertus
in-KER-tus
incerta, incertum
Doubtful; uncertain, as in *Draba incerta*

incisus
in-KYE-sus
incisa, incisum
With deeply cut and irregular incisions, as in *Prunus incisa*

inclaudens
in-KLAW-denz
Not closing, as in *Erepsia inclaudens*

inclinatus
in-klin-AH-tus
inclinata, inclinatum
Bent downward, as in *Moraea inclinata*

incomparabilis [5]
in-kom-par-RAH-bih-lis
incomparabilis, incomparabile
Incomparable, as in *Narcissus* × *incomparabilis*

incomptus
in-KOMP-tus
incompta, incomptum
Without adornment, as in *Verbena incompta*

inconspicuus
in-kon-SPIK-yoo-us
inconspicua, inconspicuum
Inconspicuous, as in *Hoya inconspicua*

A
B
C
D
E
F
G
H
I
J
K
L
M
N
O
P
Q
R
S
T
U
V
W
X
Y
Z

i

Masdevallia infracta

incrassatus

in-kras-SAH-tus

incrassata, incrassatum

Thickened, as in *Leucocoryne incrassata*

incurvatus

in-ker-VAH-tus

incurvata, incurvatum

—

incurvus

in-ker-VUS

incurva, incurvum

Bent inward, as in *Carex incurva*

indicus

IN-dih-kus

indica, indicum

Connected with India; may also apply to plants originating from the East Indies or China, as in *Lagerstroemia indica*

Indigofera

in-di-GOF-er-uh

From Latin *indigo*, a blue dye, and *fera*, meaning "to bear," because this important commercial dye was extracted from *I. tinctoria* and a few other species (*Fabaceae*)

indivisus

in-dee-VEE-sus

indivisa, indivisum

Without divisions, as in *Cordyline indivisa*

Indocalamus

in-doh-KAL-uh-mus

From Latin *Indos*, meaning "the Indies," and Greek *kalamos*, a reed; this bamboo genus occurs primarily in China, reaching into Indochina (*Poaceae*)

induratus

in-doo-RAH-tus

indurata, induratum

Hard, as in *Cotoneaster induratus*

inebrians

in-ee-BRI-enz

Intoxicating, as in *Ribes inebrians*

inermis

IN-er-mis

inermis, inerme

Without arms, for example, without prickles, as in *Acaena inermis*

infaustus

in-FUS-tus

infausta, infaustum

Unfortunate (sometimes therefore used of poisonous plants), unlucky, as in *Colletia infausta*

infectorius

in-fek-TOR-ee-us

infectoria, infectorium

Dyed; colored, as in *Quercus infectoria*

infestus

in-FES-tus

infesta, infestum

Dangerous; troublesome, as in *Melilotus infestus*

inflatus

in-FLAH-tus

inflata, inflatum

Swollen up, as in *Codonopsis inflata*

infortunatus

in-for-tu-NAH-tus

infortunata, infortunatum

Unfortunate (of poisonous plants), as in *Clerodendrum infortunatum*

infractus

in-FRAC-tus

infracta, infractum

Curving inward, as in *Masdevallia infracta*

infundibuliformis

in-fun-dih-bew-LEE-for-mis

infundibuliformis, infundibuliforme

In the shape of a funnel or trumpet, as in *Crossandra infundibuliformis*

infundibulus

in-fun-DIB-yoo-lus

infundibula, infundibulum

A funnel, as in *Dendrobium infundibulum*

ingens

IN-genz

Enormous, as in *Tulipa ingens*

inodorus

in-oh-DOR-us

inodora, inodorum

Without scent, as in *Hypericum* × *inodorum*

inornatus

in-or-NAH-tus

inornata, inornatum

Without ornament, as in *Boronia inornata*

inquinans

in-KWIN-anz

Polluted; stained; defiled, as in *Pelargonium inquinans*

insignis

in-SIG-nis

insignis, insigne

Distinguished; remarkable, as in *Rhododendron insigne*

insititius

in-si-tih-TEE-us

insititia, insititium

Grafted, as in *Prunus insititia*

insulanus

in-su-LAH-nus

insulana, insulanum

—

insularis

in-soo-LAH-ris

insularis, insulare

Relating to an island, as in *Tilia insularis*

integer

IN-teg-er

integra, integrum

Entire, as in *Cyananthus integer*

integrifolius

in-teg-ree-FOH-lee-us

integrifolia, integrifolium

With leaves that are complete, uncut, as in *Meconopsis integrifolia*

intermedius

in-ter-MEE-dee-us

intermedia, intermedium

Intermediate in color, form, or habit, as in *Forsythia* × *intermedia*

interruptus

in-ter-UP-tus

interrupta, interruptum

Interrupted; not continuous, as in *Bromus interruptus*

intertextus

in-ter-TEKS-tus

intertexta, intertextum

Intertwined, as in *Matucana intertexta*

intortus

in-TOR-tus

intorta, intortum

Twisted, as in *Melocactus intortus*

intricatus

in-tree-KAH-tus

intricata, intricatum

Tangled, as in *Asparagus intricatus*

intumescens

in-tu-MES-enz

Swollen, as in *Carex intumescens*

intybaceus

in-tee-BAK-ee-us

intybacea, intybaceum

Like chicory (*Cichorium intybus*), as in
Hieracium intybaceum

Inula

IN-ew-luh

From Greek *inaein*, meaning "to clean," an
allusion to medicinal properties (*Asteraceae*)

inversus

in-VERS-us

inversa, inversum

Turned over, as in *Quaqua inversa*

involucratus

in-vol-yoo-KRAH-tus

involucrata, involucratum

With a ring of bracts surrounding several
flowers, as in *Cyperus involucratus*

involutus

in-vol-YOO-tus

involuta, involutum

Rolled inward, as in *Gladiolus involutus*

Iochroma

eye-oh-KRO-muh

From Greek *ios*, meaning "violet," and
chroma, meaning "color," because some
species have flowers of a violet color
(*Solanaceae*)

ioensis

eye-oh-EN-sis

ioensis, ioense

From Iowa, as in *Malus ioensis*

ionanthus

eye-oh-NAN-thus

ionantha, ionanthum

With flowers of a violet color, as in
Saintpaulia ionantha

ionopterus

eye-on-OP-ter-us

ionoptera, ionopterum

With violet wings, as in *Koellensteinia
ionoptera*

Ipomoea

eye-poh-MEE-uh

From Greek *ipos*, meaning "worm," and
homoios, meaning "resembling," alluding
to the twining stems (*Convolvulaceae*)

Ipomopsis

eye-poh-MOP-sis

Resembling the genus *Ipomoea*
(*Polemoniaceae*)

Iresine

eye-ri-SEE-nee

From Greek *eiresione*, meaning "a staff
wrapped in wool," because in some species
there are woolly hairs on the sepals
surrounding the flowers (*Amaranthaceae*)

iridescens

ir-id-ES-enz

Iridescent, as in *Phyllostachys iridescens*

iridiflorus

ir-id-uh-FLOR-us

iridiflora, iridiflorum

With flowers like *Iris*, as in *Canna iridiflora*

iridifolius

ir-id-ih-FOH-lee-us

iridifolia, iridifolium

With leaves like *Iris*, as in *Billbergia iridifolia*

iridioides

ir-id-ee-OY-deez

Resembling *Iris*, as in *Dietes iridioides*

Iris [1]

EYE-ris

Named after the Greek goddess of the
rainbow, because the flowers are colorful
(*Iridaceae*)

irregularis

ir-reg-yoo-LAH-ris

irregularis, irregulare

With parts of different sizes, as in *Primula
irregularis*

irriguus

ir-EE-gyoo-us

irrigua, irriguum

Watered, as in *Pratia irrigua*

Isatis

eye-SAH-tis

From Greek *isatis*, a dye plant, because
I. tinctoria (woad) provides a blue dye
(*Brassicaceae*)

1

Iris japonica

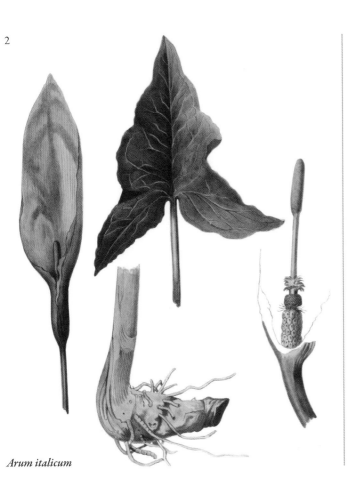

Arum italicum

Ixora coccinea

Isolepis
eye-so-LEP-is
From Greek *isos*, meaning "equal," and *lepis*, meaning "scale," because the glumes in the inflorescence are all of similar size (*Cyperaceae*)

isophyllus
eye-so-FIL-us
isophylla, isophyllum
With leaves of the same size, as in *Penstemon isophyllus*

Isopogon
eye-soh-POH-gon
From Greek *isos*, meaning "equal," and *pogon*, for "beard," referring to fringed flowers or some species that have hairy fruit (*Proteaceae*)

Isopyrum
eye-soh-PY-rum
From Greek *isos*, meaning "equal," and *pyros*, meaning "wheat," an allusion to the grainlike fruit (*Ranunculaceae*)

Isotoma
eye-soh-TOH-muh
From Greek *isos*, meaning "equal," and *tome*, meaning "division," because the petal tubes are divided into sections of similar size (*Campanulaceae*)

italicus [2]
ee-TAL-ih-kus
italica, italicum
Connected with Italy, as in *Arum italicum*

Itea
eye-TEE-uh
From Greek *itea*, meaning "willow," because some species have willowy leaves (*Iteaceae*)

Ixia
IKS-ee-uh
From Greek *ixos*, meaning "mistletoe," a plant with sticky berries and a reference to the viscid sap of some *Ixia*; an alternative etymology is Greek *ixias*, meaning "chameleon," because this genus produces so many different flower colors (*Iridaceae*)

ixioides
iks-ee-OY-deez
Resembling African iris or corn lily (*Ixia*), as in *Libertia ixioides*

Ixiolirion
iks-ee-oh-LIR-ee-on
From *Ixia*, another genus, and Greek *lirion*, meaning "lily" (*Ixioliriaceae*)

ixocarpus
iks-so-KAR-pus
ixocarpa, ixocarpum
With sticky fruit, as in *Physalis ixocarpa*

Ixora [3]
iks-OR-uh
From Sanskrit *isvara*, meaning "ruler" or "lord," perhaps from the use of the flowers in Hindu ceremonies (*Rubiaceae*)

Jaborosa

jab-uh-ROH-suh

From Arabic *jaborose*, vernacular name for mandrake (*Mandragora*), a plant in the same family (*Solanaceae*)

Jacaranda [1]

jah-kur-AN-duh

From a Guarani (Brazil), meaning "fragrant," converted to Latin via Portuguese (*Bignoniaceae*)

jackii

JAK-ee-eye

Named after John George Jack (1861–1949), a Canadian dendrologist at the Arnold Arboretum, Boston, Massachusetts, as in *Populus* × *jackii*

Jacobaea

jak-oh-BEE-uh

Named after Saint James, because *J. vulgaris* flowers on or around his feast day (*Asteraceae*)

jacobaeus

jak-oh-BAY-ee-us

jacobaea, jacobaeum

Named for Saint James, or for Santiago (Cape Verde), as in *Senecio jacobaea*

1

Jacaranda puberula

Jacquemontia

jak-uh-MON-tee-uh

Named after Victor Jacquemont (1801–32), French botanist and geologist (*Convolvulaceae*)

jalapa [2]

juh-LAP-a

Connected with Xalapa, Mexico, as in *Mirabilis jalapa*

jamaicensis

ja-may-KEN-sis

jamaicensis, jamaicense

From Jamaica, as in *Brunfelsia jamaicensis*

Jamesbrittenia

james-bri-TEEN-ee-uh

Named after James Britten (1846–1924), British botanist (*Scrophulariaceae*)

Jamesia

JAME-zee-uh

Named after Edwin P. James (1797–1861), American physician, geologist, and naturalist (*Hydrangeaceae*)

japonicus

juh-PON-ih-kus

japonica, japonicum

Connected with Japan, as in *Cryptomeria japonica*

Jasione

jaz-ee-OH-nee

From Greek *jasione*, the, meaning of which is unclear, although it may refer to a *Campanula* or *Convolvulus* (*Campanulaceae*)

jasmineus

jaz-MIN-ee-us

jasminea, jasmineum

Like jasmine (*Jasminum*), as in *Daphne jasminea*

2

Mirabilis jalapa

GENUS SPOTLIGHT

Jasminum

The word "jasmine" evokes two images, a vigorous flowering vine and a sweetly scented flower, which is why the name is so common in both Latin and English plant names. True jasmines belong to the genus *Jasminum*, and while many are vines with fragrant blooms, both shrubby (for example, *J. parkeri*) and unscented (for example, *J. nudiflorum*) species occur. Outside *Jasminum*, other climbing jasmines can be found in *Trachelospermum* (Star jasmine), *Stephanotis* (Madagascar jasmine), and *Gelsemium* (Carolina jessamine), while *Gardenia jasminoides* and *Rhododendron jasminiflorum* are both nonclimbing shrubs with scented flowers.

1

Jasminum didymum

jasminiflorus
jaz-min-IH-flor-us
jasminiflora, jasminiflorum
With flowers like jasmine (*Jasminum*), as in *Rhododendron jasminiflorum*

jasminoides
jaz-min-OY-deez
Resembling jasmine (*Jasminum*), as in *Trachelospermum jasminoides*

Jasminum [1]
JAZ-min-um
From Persian-Farsi *yasamin*, meaning "gift from God," and now the name of the plant (*Oleaceae*)

Jatropha
juh-TRO-fuh
From Greek *iatros*, meaning "doctor," and *trophe*, meaning "food," because the seed of *J. curcas* (Barbados nut or physic nut) was a popular purgative (*Euphorbiaceae*)

javanicus
juh-VAHN-ih-kus
javanica, javanicum
Connected with Java, as in *Rhododendron javanicum*

Jeffersonia
jef-ur-SOH-nee-uh
Named after Thomas Jefferson (1743–1826), third president of the United States (*Berberidaceae*)

jejunus
jeh-JOO-nus
jejuna, jejunum
Small, as in *Eria jejuna*

Jovellana
joh-vuh-LAH-nuh
Named after Gaspar Melchor de Jovellanos (1744–1811), Spanish politician and author (*Calceolariaceae*)

Jovibarba
joh-vee-BAR-buh
From Latin *Jovis*, meaning "Jupiter," and *barba*, meaning "beard," an allusion to the fringed petals (*Crassulaceae*)

Juanulloa
wan-oo-LOH-uh
Named after Jorge Juan y Santacilia (1713–73) and Antonio Ulloa y de la Torre-Giral (1716–95), Spanish scientists and explorers (*Solanaceae*)

Jubaea
joo-BEE-uh
Named after Juba II, King of Numidia and Mauretania (ca. 52 BC–AD 23), patron of the arts and sciences (*Arecaceae*)

170

Juniperus communis

Juglans cinerea

jubatus

joo-BAH-tus

jubata, jubatum

With awns, as in *Cortaderia jubata*

jucundus

joo-KUN-dus

jucunda, jucundum

Agreeable; pleasing, as in *Osteospermum jucundum*

jugalis

joo-GAH-lis

jugalis, jugale

—

jugosus

joo-GOH-sus

jugosa, jugosum

Yoked, as in *Pabstia jugosa*

Juglans [3]

JUG-lunz

From classical Latin name for English walnut, *J. regia*, meaning "Jupiter's nut" (*Juglandaceae*)

julaceus

joo-LA-see-us

julacea, julaceum

With catkins, as in *Leucodon julaceus*

junceus

JUN-kee-us

juncea, junceum

Like a rush, as in *Spartium junceum*

Juncus

JUN-kus

From classical Latin name for rush, deriving from a word meaning "to bind," because the leaves were used to make rope, etc. (*Juncaceae*)

juniperifolius

joo-nip-er-ih-FOH-lee-us

juniperifolia, juniperifolium

With leaves like a juniper (*Juniperus*), as in *Armeria juniperifolia*

juniperinus

joo-nip-er-EE-nus

juniperina, juniperinum

Like a juniper (*Juniperus*); blue-black, as in *Grevillea juniperina*

Juniperus [2]

joo-NIP-er-us

From classical Latin name for juniper, originating in Latin *junis*, meaning "youth," and *pario*, meaning "to birth"; etymologies include the evergreen foliage, the production of new berries while old ones remain on the tree, or Jupiter the father (*Jovis pater*), as the plant was used in religious ceremonies (*Cupressaceae*)

Justicia

just-ISH-ee-uh

Named after James Justice (1698–1763), Scottish horticulturist (*Acanthaceae*)

juvenilis

joo-VEE-nil-is

juvenilis, juvenile

Young, as in *Draba juvenilis*

Kalanchoe flammea

Kalmia latifolia

Kadsura
kad-SOOR-uh
From Japanese vernacular name for *K. japonica* (*Schisandraceae*)

Kalanchoe [1]
kal-an-KOH-ee
From Chinese *kalanchauhuy*, the vernacular name reported by missionary Georg Joseph Kamel (*Crassulaceae*)

Kalimeris
kal-i-MER-is
From Greek *kallos*, meaning "beautiful," and *meris*, meaning "parts" (*Asteraceae*)

Kalmia [2]
KAL-me-uh
Named after Pehr Kalm (1716–79), Swedish botanist (*Ericaceae*)

Kalmiopsis
kal-me-OP-sis
Resembling the related genus *Kalmia* (*Ericaceae*)

Kalopanax
kal-oh-PAN-aks
From Greek *kallos*, meaning "beautiful," and *Panax*, a related genus (*Araliaceae*)

kamtschaticus
kam-SHAY-tih-kus
kamtschatica, kamtschaticum
Connected with Kamchatka, Russia, as in *Sedum kamtschaticum*

kansuensis
kan-soo-EN-sis
kansuensis, kansuense
From Kansu (Gansu), China, as in *Malus kansuensis*

karataviensis
kar-uh-taw-vee-EN-sis
karataviensis, karataviense
From the Karatau mountains, Kazakhstan, as in *Allium karataviense*

kashmirianus
kash-meer-ee-AH-nus
kashmiriana, kashmirianum
Connected with Kashmir, as in *Actaea kashmiriana*

Keckiella
kek-ee-EL-uh
Named after David Daniels Keck (1903–95), American botanist, plus the diminutive *ella* (*Plantaginaceae*)

Kelseya

KEL-see-uh
Named after Francis Duncan Kelsey (1849–1905), American botanist (*Rosaceae*)

Kennedia [1]

KEN-uh-dee-uh
Named after John Kennedy (1759–1842), Scottish horticulturist (*Fabaceae*)

kermesinus

ker-mes-SEE-nus
kermesina, kermesinum
Crimson, as in *Passiflora kermesina*

Kerria

KER-ee-uh
Named after William Kerr (?–1814), Scottish horticulturist (*Rosaceae*)

Keteleeria

ket-uh-LEER-ee-uh
Named after Jean Baptiste Keteleer (1813–1903), French horticulturist (*Pinaceae*)

kewensis

kew-EN-sis
kewensis, kewense
From Kew Gardens, London, England, as in *Primula kewensis*

Kigelia

ki-JEE-lee-uh
From Mozambican indigenous name *kigeli-keia* (*Bignoniaceae*)

Kirengeshoma [2]

ki-reng-i-SHOH-muh
From Japanese vernacular name, based on *ki* ("yellow"), *renge* ("lotus blossom"), and *shoma* ("hat"), (*Hydrangeaceae*)

kirkii

KIR-kee-eye
Named after Thomas Kirk (1828–98), renowned botanist of New Zealand flora; or after his son Harry Bower Kirk (1903–44), professor of biology at the University of New Zealand; or after Sir John Kirk (1832–1922), botanist and British Consul to Zanzibar, as in *Coprosma × kirkii* (after Thomas Kirk)

Kitaibela

kit-AY-bel-uh
Named after Pál Kitaibel (1757–1817), Hungarian botanist and chemist (*Malvaceae*)

kiusianus

key-oo-see-AH-nus
kiusiana, kiusianum
Connected with Kyushu, Japan, as in *Rhododendron kiusianum*

Kleinia

KLY-nee-uh
Named after Jacob Theodor Klein (1685–1759), German diplomat, historian, and scientist (*Asteraceae*)

1

Kennedia carinata

2

Kirengeshoma palmata

Knautia

NAWT-ee-uh

Named after brothers Christian (1659–1716) and Christoph Knaut (1638–94), German physicians and botanists (*Caprifoliaceae*)

Knightia

NITE-ee-uh

Named after Thomas Andrew Knight (1759–1838), British botanist and horticulturist, second president of the Royal Horticultural Society (*Proteaceae*)

Kniphofia [3]

ny-FOH-fee-uh

Named after Johann Hieronymus Kniphof (1704–63), German physician and botanist (*Asphodelaceae*)

Koeleria

koh-LEER-ee-uh

Named after Georg Ludwig Koeler (1764–1807), German botanist (*Poaceae*)

Koelreuteria

kol-roy-TEER-ee-uh

Named after Joseph Gottlieb Kölreuter (1733–1806), German botanist (*Sapindaceae*)

Kohleria

koh-LEER-ee-uh

Named after Michael Kohler (dates unknown), Swiss educator (*Gesneriaceae*)

Kolkwitzia

kol-KWITZ-ee-uh

Named after Richard Kolkwitz (1873–1956), German botanist (*Caprifoliaceae*)

koreanus

kor-ee-AH-nus

koreana, koreanum

Connected with Korea, as in *Abies koreana*

Kunzea

KUN-zee-uh

Named after Gustave Kunze (1793–1851), German zoologist and botanist (*Myrtaceae*)

kurdicus

KUR-dih-kus

kurdica, kurdicum

Referring to the Kurdish homeland in western Asia, as in *Astragalus kurdicus*

3

Kniphofia triangularis

labiatus

la-bee-AH-tus

labiata, labiatum

Lipped, as in *Cattleya labiata*

labilis

LA-bih-lis

labilis, labile

Slippery; unstable, as in *Celtis labilis*

labiosus

la-bee-OH-sus

labiosa, labiosum

Lipped, as in *Besleria labiosa*

Lablab

LAB-lab

From Arabic *lablab*, vernacular name for morning glory (*Fabaceae*)

Laburnum [1]

lah-BUR-num

The classical Latin name for this plant (*Fabaceae*)

laburnifolius

lah-bur-nih-FOH-lee-us

laburnifolia, laburnifolium

With leaves like *Laburnum*, as in *Crotalaria laburnifolia*

lacerus

LASS-er-us

lacera, lacerum

Cut into fringelike sections, as in *Costus lacerus*

Lachenalia

lak-uh-NAY-lee-uh

Named after Werner de Lachenal (1736–1800), Swiss botanist (*Asparagaceae*)

laciniatus

la-sin-ee-AH-tus

laciniata, laciniatum

Divided into narrow sections, as in *Rudbeckia laciniata*

lacrimans

LAK-ri-manz

Weeping, as in *Eucalyptus lacrimans*

lacteus

lak-TEE-us

Milk white, as in *Cotoneaster lacteus*

lacticolor

lak-tee-KOL-or

Milk white, as in *Protea lacticolor*

lactiferus

lak-TEE-fer-us

lactifera, lactiferum

Producing a milky sap, as in *Gymnema lactiferum*

lactiflorus

lak-tee-FLOR-us

lactiflora, lactiflorum

With milk-white flowers, as in *Campanula lactiflora*

Lactuca

lak-TOO-kuh

From Latin *lac*, meaning "milk," due to the milky sap (*Asteraceae*)

lacunosus

lah-koo-NOH-sus

lacunosa, lacunosum

With deep holes or pits, as in *Allium lacunosum*

lacustris

lah-KUS-tris

lacustris, lacustre

Relating to lakes, as in *Iris lacustris*

ladaniferus

lad-an-IH-fer-us

ladanifera, ladaniferum

—

ladanifer

lad-an-EE-fer

Producing ladanum, a fragrant gum resin, as in *Cistus ladanifer*

Laelia

LAY-lee-uh

Named after Laelia, one of the Vestal Virgins in ancient Rome, priestesses of the goddess of the hearth (*Orchidaceae*)

laetevirens

lay-tee-VY-renz

Vivid green, as in *Parthenocissus laetevirens*

laetiflorus

lay-tee-FLOR-us

laetiflora, laetiflorum

With bright flowers, as in *Helianthus × laetiflorus*

laetus

LEE-tus

laeta, laetum

Bright; vivid, as in *Pseudopanax laetum*

1

Laburnum anagyroides

laevigatus

lee-vih-GAH-tus

laevigata, laevigatum

—

laevis

LEE-vis

laevis, laeve

Smooth, as in *Crocus laevigatus*

Lagarosiphon

lah-gur-oh-SY-fon

From Greek *lagaros*, meaning "lax," and *siphon*, meaning "tube," probably referring to the long, tubular hypanthium that carries female flowers to the water surface (*Hydrocharitaceae*)

Lagarostrobos

lah-gur-oh-STROH-bus

From Greek *lagaros*, meaning "lax," and *strobilus*, meaning "cone" (*Podocarpaceae*)

Lagerstroemia

lah-guh-STROH-mee-uh

Named after Magnus von Lagerström (ca. 1696–1759), Swedish merchant and naturalist (*Lythraceae*)

lagodechianus

lah-go-chee-AH-nus

lagodechiana, lagodechianum

Connected with Lagodekhi, Georgia, as in *Galanthus lagodechianus*

Lagunaria

lah-goo-NAIR-ee-uh

Named after Andrés de Laguna (1499–1559), Spanish botanist and physician (*Malvaceae*)

Lagurus

lag-GOOR-us

From Greek *lagus*, meaning "hare," and *ourus*, meaning "tail," a reference to the shape of the inflorescence (*Poaceae*)

lamellatus

la-mel-LAH-tus

lamellata, lamellatum

Layered, as in *Vanda lamellata*

Lamium

LAY-mee-um

From the classical Latin name for the plant, probably derived from Greek *laimos*, meaning "throat," a reference to the flower shape (*Lamiaceae*)

Lampranthus

lam-PRAN-thus

From Greek *lampros*, meaning "lamp," and *anthos*, meaning "flower," because the blooms are said to shine (*Aizoaceae*)

Lamprocapnos [1]

lam-pro-KAP-nos

From Greek *lampros*, meaning "bright," and *kapnos*, meaning "smoke"; the first part refers to the vivid flowers, while the second is common in members of the family *Fumariaceae* (now subsumed in *Papaveraceae*), which are said to have a smoky smell (*Papaveraceae*)

lanatus

la-NA-tus

lanata, lanatum

Woolly, as in *Lavandula lanata*

lanceolatus

lan-see-oh-LAH-tus

lanceolata, lanceolatum

lanceus

lan-SEE-us

lancea, lanceum

In the shape of a spear, as in *Tasmannia lanceolata*

lanigerus

lan-EE-ger-rus

lanigera, lanigerum

—

lanosus

LAN-oh-sus

lanosa, lanosum

lanuginosus

lan-oo-gih-NOH-sus

lanuginosa, lanuginosum

Woolly, as in *Leptospermum lanigerum*

Lantana (also lantana)

lan-TAH-nuh

From Latin name for a viburnum (possibly *V. lantana*), because both genera have similar inflorescences (*Verbenaceae*)

1

Lamprocapnos spectabilis

Lapageria
la-puh-JEER-ee-uh
Named after Marie Josèphe Rose Tascher de La Pagerie (1763–1814), first wife of Napoléon Bonaparte (*Philesiaceae*)

Lapeirousia
la-pay-ROO-see-uh
Named after Philippe-Isidore Picot de Lapeyrouse (1744–1818), French naturalist (*Iridaceae*)

lappa
LAP-ah
A burr (prickly seed case or flower head), as in *Arctium lappa*

lapponicus
Lap-PON-ih-kus
lapponica, lapponicum
—

lapponum
Lap-PON-num
Connected with Lapland, as in *Salix lapponum*

Laportea
la-POR-tee-uh
Named after François Louis Nompar de Caumont La Force, comte de Castelnau [also known as François Laporte] (1810–1880), French naturalist (*Urticaceae*)

Lapsana
lap-SAH-nuh
From Greek *lapsanae*, meaning "vegetable," possibly radish, which has similar leaves (*Asteraceae*)

Lardizabala
lar-di-ZAB-uh-luh
Named after Miguel de Lardizábal y Uribe (1744–1824), Spanish-born Mexican politician (*Lardizabalaceae*)

laricifolius
lah-ris-ih-FOH-lee-us
laricifolia, laricifolium
With leaves like a larch, as in *Penstemon laricifolius*

laricinus
lar-ih-SEE-nus
laricina, laricinum
Like a larch (*Larix*), as in *Banksia laricina*

Lathyrus odoratus

Larix
LA-riks
The classical Latin name for larch (*Pinaceae*)

lasi-
Used in compound words to denote woolly

lasiandrus
las-ee-AN-drus
lasiandra, lasiandrum
With woolly stamens, as in *Clematis lasiandra*

lasioglossus
las-ee-oh-GLOSS-us
lasioglossa, lasioglossum
With a rough tongue, as in *Lycaste lasioglossa*

Latania
luh-TAY-nee-uh
From French vernacular name *latanier*, used in Mauritius (*Arecaceae*)

lateralis
lat-uh-RAH-lis
lateralis, laterale
On the side, as in *Epidendrum laterale*

lateritius
la-ter-ee-TEE-us
lateritia, lateritium
Brick-red, as in *Kalanchoe lateritia*

Lathraea
LATH-ree-uh
From Greek *lathraios*, meaning "hidden," because this parasitic plant is largely concealed underground, only emerging to flower (*Orobanchaceae*)

Lathyrus [2]
LATH-uh-rus
From Greek *lathyros*, meaning "pea"; true peas belong to the genus *Pisum* (*Fabaceae*)

lati-
Used in compound words to denote broad

latiflorus
lat-ee-FLOR-us
latiflora, latiflorum
With broad flowers, as in *Dendrocalamus latiflorus*

latifolius
lat-ee-FOH-lee-us
latifolia, latifolium
With broad leaves, as in *Lathyrus latifolius*

latifrons
lat-ee-FRONS
With broad fronds, as in *Encephalartos latifrons*

latilobus

la-tih-LOH-bus

laptiloba, latilobum

With wide lobes, as in *Campanula latiloba*

latispinus

la-tih-SPEE-nus

latispina, latispinum

With wide thorns, as in *Ferocactus latispinus*

laudatus

law-DAH-tus

laudata, laudatum

Worthy of praise, as in *Rubus laudatus*

Laurelia

law-REE-lee-uh

Named for it resemblance to the unrelated genus *Laurus* (*Atherospermataceae*)

Laureliopsis

law-ree-lee-OP-sis

Resembling the related genus *Laurelia* (*Atherospermataceae*)

laurifolius

law-ree-FOH-lee-us

laurifolia, laurifolium

With leaves like bay (*Laurus*), as in *Cistus laurifolius*

laurinus

law-REE-nus

laurina, laurinum

Like a laurel or bay tree (*Laurus*), as in *Hakea laurina*

laurocerasus

law-roh-KER-uh-sus

From the Latin for cherry and laurel, as in *Prunus laurocerasus*

Laurus

LAW-rus

The classical Latin name for laurel, *L. nobilis* (*Lauraceae*)

Lavandula

la-VAN-dew-luh

From Latin *lavare*, meaning "to wash," because it is often used in soaps (*Lamiaceae*)

lavandulaceus

la-van-dew-LAY-see-us

lavandulacea, lavandulaceum

Like lavender (*Lavandula*), as in *Chirita lavandulacea*

lavandulifolius

la-van-dew-lih-FOH-lee-us

lavandulifolia, lavandulifolium

With leaves like lavender (*Lavandula*), as in *Salvia lavandulifolia*

Lavatera

la-vuh-TEAR-uh

Named after brothers Johann Heinrich (ca. 1611–91) and Johann Jacob Lavater (1594–1636), Swiss physicians and naturalists (*Malvaceae*)

Lawsonia

law-SOH-nee-uh

Named after Isaac Lawson (?–1747), Scottish physician (*Lythraceae*)

laxiflorus

laks-ih-FLO-rus

laxiflora, laxiflorum

With loose, open flowers, as in *Lobelia laxiflora*

laxifolius

laks-ih-FOH-lee-us

laxifolia, laxifolium

With loose, open leaves, as in *Athrotaxis laxifolia*

laxus

LAKS-us

laxa, laxum

Loose; open, as in *Freesia laxa*

Ledebouria

le-duh-BORE-ee-uh

Named after Carl Friedrich von Ledebour (1785–1851), German-Estonian botanist (*Asparagaceae*)

ledifolius

lee-di-FOH-lee-us

ledifolia, ledifolium

With leaves like *Ledum*, as in *Ozothamnus ledifolius*

Leea

LEE-uh

Named after James Lee (1715–95), Scottish horticulturist (*Vitaceae*)

Legousia

luh-GOO-see-uh

Named after Bénigne Le Gouz de Gerland (1695–1774), French academic (*Campanulaceae*)

leianthus

lee-AN-thus

leiantha, leianthum

With smooth flowers, as in *Bouvardia leiantha*

leichtlinii

leekt-LIN-ee-eye

Named after Max Leichtlin (1831–1910), German plant collector from Baden-Baden, Germany, as in *Camassia leichtlinii*

leiocarpus

lee-oh-KAR-pus

leiocarpa, leiocarpum

With smooth fruit, as in *Cytisus leiocarpus*

leiophyllus

lay-oh-FIL-us

leiophylla, leiophyllum

With smooth leaves, as in *Pinus leiophylla*

Lemna

LEM-nuh

From Greek name for a water plant, possibly *Callitriche verna*, which can resemble duckweed (*Araceae*)

Lens

LENZ

From Latin *lenticula*, meaning "lens," to which lentils (*L. culinaris*) resemble (*Fabaceae*)

lentiginosus

len-tig-ih-NOH-sus

lentiginosa, lentiginosum

With freckles, as in *Coelogyne lentiginosa*

lentus

LEN-tus

lenta, lentum

Tough but flexible, as in *Betula lenta*

leonis

le-ON-is

With the color of, or teeth like, a lion, as in *Angraecum leonis*

Leonotis

lee-oh-NOH-tis

From Greek *leon*, meaning "lion," and *otis*, meaning "ear," which the petals are said to resemble (*Lamiaceae*)

Lavatera phoenicea

Leontodon saxatilis

Leontopodium alpinum

Leontice
lee-on-TY-kee
From Greek *leontike*, a vernacular name for an unknown plant, often translated as "lion's foot" (*Berberidaceae*)

Leontodon [1]
lee-ON-toh-don
From Greek *leon*, meaning "lion," and *odons*, meaning "tooth"; the leaf margins are deeply toothed (*Asteraceae*)

leontoglossus
lee-on-toh-GLOSS-us
leontoglossa, leontoglossum
With a throat or tongue of a lion, as in *Masdevallia leontoglossa*

Leontopodium [2]
lee-on-toh-POH-dee-um
From Greek *leon*, meaning "lion," and *podion*, meaning "foot," because the inflorescence with its woolly bracts supposedly resembles a paw (*Asteraceae*)

leonurus
lee-ON-or-us
leonura, leonurum
Like a lion's tail, as in *Leonotis leonurus*

leopardinus
leh-par-DEE-nus
leopardina, leopardinum
With spots like a leopard, as in *Calathea leopardina*

Lepidium
le-PID-ee-um
From Greek *lepidion*, meaning "little scale," because the fruit is small and circular (*Brassicaceae*)

Lepidozamia
le-pid-oh-ZAY-mee-uh
From Greek *lepis*, meaning "scale," and *Zamia*, a related genus, because the trunk is covered in scalelike leaf bases (*Zamiaceae*)

lepidus
le-PID-us
lepida, lepidum
Graceful; elegant, as in *Lupinus lepidus*

lept-
Used in compound words to denote thin or slender

leptanthus
lep-TAN-thus
leptantha, leptanthum
With slender flowers, as in *Colchicum leptanthum*

Leptinella
lep-tin-EL-uh
From Greek *leptos*, meaning "slender," referring to the narrow ovary (*Asteraceae*)

leptocaulis
lep-toh-KAW-lis
leptocaulis, leptocaule
With thin stems, as in *Cylindropuntia leptocaulis*

leptocladus
lep-toh-KLAD-us
leptoclada, leptocladum
With thin branches, as in *Acacia leptoclada*

Leptospermum scoparium

4

Leucanthemum vulgare

leptophyllus
lep-toh-FIL-us
leptophylla, leptophyllum
With thin leaves, as in *Cassinia leptophylla*

leptopus
LEP-toh-pus
With thin stalks, as in *Antigonon leptopus*

leptosepalus
lep-toh-SEP-a-lus
leptosepala, leptosepalum
With thin sepals, as in *Caltha leptosepala*

Leptospermum [3]
lep-toh-SPER-mum
From Greek *leptos*, meaning "slender," and *sperma*, meaning "seed," alluding to the filamentous seed (*Myrtaceae*)

leptostachys
lep-toh-STAH-kus
With slender spikes, as in *Aponogeton leptostachys*

Leschenaultia
lesh-uh-NAWL-tee-uh
Named after Jean-Baptiste Louis Claude Théodore Leschenault de La Tour (1773–1826), French botanist and ornithologist (*Goodeniaceae*)

Lespedeza
les-puh-DEE-zuh
Named after Vicente Manuel de Céspedes y Velasco (ca. 1721–94), Spanish governor of East Florida (*Fabaceae*)

leuc-
Used in compound words to denote white

Leucadendron
lew-kuh-DEN-dron
From Greek *leukos*, meaning "white," and *dendron*, meaning "tree," due to the foliage that is covered in silvery silky hairs (*Proteaceae*)

Leucanthemum [4]
lew-KANTH-uh-mum
From Greek *leukos*, meaning "white," and *anthemon*, meaning "flower," because the inflorescences have white rays (*Asteraceae*)

leucanthus
loo-KAN-thus
leucantha, leucanthum
With white flowers, as in *Tulbaghia leucantha*

Leuchtenbergia
looch-ten-BURG-ee-uh
Named after Maximilian Joseph Eugene Auguste Napoleon de Beauharnais, 3rd Duke of Leuchtenberg (1817–52), Bavarian nobleman (*Cactaceae*)

leucocephalus
loo-koh-SEF-uh-lus
leucocephala, leucocephalum
With a white head, as in *Leucaena leucocephala*

leucochilus
loo-KOH-ky-lus
leucochila, leucochilum
With white lips, as in *Oncidium leucochila*

Leucocoryne

loo-koh-kor-AY-nee

From Greek *leukos*, meaning "white," and *koryne*, meaning "club," an allusion to the three sterile anthers (*Amaryllidaceae*)

leucodermis

loo-koh-DER-mis

leucodermis, leucoderme

With white skin, as in *Pinus leucodermis*

Leucogenes

loo-KOJ-en-eez

From Greek *leukos*, meaning "white"; variously interpreted as "white genus" or "generating white" or "white noble," the latter an allusion to the similar-looking edelweiss (*Leontopodium*), a German name, meaning "noble white" (*Asteraceae*)

Leucojum [1]

loo-KOH-jum

From Greek *leucoion*, meaning "white violet," a name applied perhaps to *Matthiola incana* or *Galanthus*; the flowers are white and smell of violets (*Amaryllidaceae*)

leuconeurus

loo-koh-NOOR-us

leuconeura, leuconeurum

With white nerves, as in *Maranta leuconeura*

leucophaeus

loo-koh-FAY-us

leucophaea, leucophaeum

Dusky white, as in *Dianthus leucophaeus*

leucophyllus

loo-koh-FIL-us

leucophylla, leucophyllum

With white leaves, as in *Sarracenia leucophylla*

Leucopogon

loo-koh-POH-gon

From Greek *leukos*, meaning "white," and *pogon*, meaning "beard," named for the typically white, bearded petals (*Ericaceae*)

leucorhizus

loo-koh-RYE-zus

leucorhiza, leucorhizum

With white roots, as in *Curcuma leucorhiza*

Leucosceptrum

loo-koh-SEP-trum

From Greek *leukos*, meaning "white," and *skeptron*, meaning "rod" or "staff," referring to the white, scepterlike inflorescence (*Lamiaceae*)

Leucospermum

loo-koh-SPER-mum

From Greek *leukos*, meaning "white," and *sperma*, meaning "seed," alluding to the white covering of the seed, an elaiosome (*Proteaceae*)

Leucothoe

loo-KOTH-oh-ee

Named for Leucothoe of Greek mythology, who was buried alive by her father, the King of Babylon, then transformed into a shrub by the sun god Helios (*Ericaceae*)

leucoxanthus

loo-koh-ZAN-thus

leucoxantha, leucoxanthum

Whitish yellow, as in *Sobralia leucoxantha*

leucoxylon

loo-koh-ZY-lon

With white wood, as in *Eucalyptus leucoxylon*

Levisticum

le-VISS-tik-um

From Latin *levo*, meaning "to relieve," alluding to medicinal properties; alternatively from Latin *ligusticum apium*, or "celery from Liguria" in northwest Italy (*Apiaceae*)

Lewisia

lew-IZ-ee-uh

Named after Meriwether Lewis (1774–1809), American soldier, politician, and explorer (*Montiaceae*)

Leycesteria

lay-ses-TEER-ee-uh, les-ter-EE-uh

Named after William Leycester (1775–1831), British judge and horticulturist (*Caprifoliaceae*)

Leymus

LAY-mus

An anagram of the related genus *Elymus* (*Poaceae*)

Liatris

lee-AH-tris

Derivation unknown (*Asteraceae*)

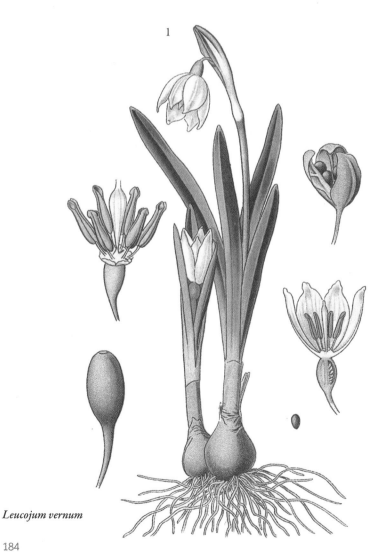

1

Leucojum vernum

Lilium maculatum

Lilium grayi

libani
LIB-an-ee
—
libanoticus
lib-an-OT-ih-kus
libanotica, libanoticum
Connected with Mount Lebanon, Lebanon, as in *Cedrus libani*

libericus
li-BEER-ih-kus
liberica, libericum
From Liberia, as in *Coffea liberica*

Libertia
li-BERT-ee-uh
Named after Marie-Anne Libert (1782–1865), Belgian botanist (*Iridaceae*)

Libocedrus
ly-boh-SEE-drus
From Greek *libos*, meaning "tears," and *Cedrus*, another conifer genus, in reference to resin droplets formed on the trunk (*Cupressaceae*)

liburnicus
li-BER-nih-kus
liburnica, liburnicum
Connected with Liburnia (now in Croatia), as in *Asphodeline liburnica*

Licuala
li-KWAH-luh
From Makassar (Indonesia) *leko wala*, the vernacular name for *L. spinosa* (*Arecaceae*)

lignosus
lig-NOH-sus
lignosa, lignosum
Woody, as in *Tuberaria lignosa*

Ligularia
lig-ew-LAIR-ee-uh
From Latin *ligula*, meaning "little tongue," a reference to the straplike ray florets (*Asteraceae*)

ligularis
lig-yoo-LAH-ris
ligularis, ligulare
—
ligulatus
lig-yoo-LAIR-tus
ligulata, ligulatum
Shaped like a strap, as in *Acacia ligulata*

ligusticifolius
lig-us-tih-kih-FOH-lee-us
ligusticifolia, ligusticifolium
With leaves like lovage, as in *Clematis ligusticifolia*

Ligusticum
li-GUST-ik-um
From Latin *ligusticum apium*, or "celery from Liguria" in northwest Italy (*Apiaceae*)

ligusticus
lig-US-tih-kus
ligustica, ligusticum
Connected with Liguria, Italy, as in *Crocus ligusticus*

ligustrifolius
lig-us-trih-FOH-lee-us
ligustrifolia, ligustrifolium
With leaves like privet (*Ligustrum*), as in *Hebe ligustrifolia*

ligustrinus
lig-us-TREE-nus
ligustrina, ligustrinum
Like privet (*Ligustrum*), as in *Ageratina ligustrina*

Ligustrum
lig-US-trum
From classical Latin name for privet, *L. vulgare* (*Oleaceae*)

lilacinus
ly-luc-SEE-nus
lilacina, lilacinum
Lilac, as in *Primula lilacina*

lili-
Used in compound words to denote lily

liliaceus
lil-lee-AY-see-us
liliacea, liliaceum
Like lily (*Lilium*), as in *Fritillaria liliacea*

liliiflorus
lil-lee-ih-FLOR-us
liliiflora, liliiflorum
With flowers like lily (*Lilium*), as in *Magnolia liliiflora*

liliifolius
lil-ee-eye-FOH-lee-us
liliifolia, liliifolium
With leaves like lily (*Lilium*), as in *Adenophora liliifolia*

Lilium [2, 3]
LIL-ee-um
From Greek *lirion*, meaning "lily" (*Liliaceae*)

1

Linaria vulgaris

2

Hakea linearis

limbatus
lim-BAH-tus
limbata, limbatum
Bordered, as in *Primula limbata*

limensis
lee-MEN-sis
limensis, limense
From Lima, Peru, as in *Haageocereus limensis*

limoniifolius
lim-on-ih-FOH-lee-us
limoniifolia, limoniifolium
With leaves like sea lavender (*Limonium*),
as in *Asyneuma limoniifolium*

Limonium
li-MOW-nee-um
From Greek *leimon*, meaning "meadow"
(*Plumbaginaceae*)

limosus
lim-OH-sus
limosa, limosum
Growing in marshy or muddy habitats,
as in *Carex limosa*

Linanthus
ly-NAN-thus
From *Linum*, the flax genus, and Greek
anthos, meaning "flower," suggesting
resemblance (*Polemoniaceae*)

Linaria [1]
li-NAIR-ee-uh
Resembling the genus *Linum*
(*Plantaginaceae*)

linariifolius
lin-ar-ee-FOH-lee-us
linariifolia, linariifolium
With leaves like toadflax (*Linaria*), as in
Melaleuca linariifolia

Lindera
LIN-der-uh
Named after Johann Linder (1676–1724),
Swedish physician and botanist (*Lauraceae*)

lindleyanus
lind-lee-AH-nus
lindleyana, lindleyanum
—
lindleyi
lind-lee-EYE
Named after John Lindley (1799–1865),
British botanist associated with the Royal
Horticultural Society, as in *Buddleja
lindleyana*

linearis [2]
lin-AH-ris
linearis, lineare
With narrow, almost parallel sides, as in
Ceropegia linearis

lineatus
lin-ee-AH-tus
lineata, lineatum
With lines or stripes, as in *Rubus lineatus*

lingua
LIN-gwa
A tongue or like a tongue, as in *Pyrrosia
lingua*

linguiformis
lin-gwih-FORM-is
linguiformis, linguiforme
—
lingulatus
lin-gyoo-LAH-tus
lingulata, lingulatum
Shaped like a tongue, as in *Guzmania
lingulata*

liniflorus
lin-ih-FLOR-us
liniflora, liniflorum
With flowers like flax (*Linum*), as in *Byblis
liniflora*

linifolius
lin-ih-FOH-lee-us
linifolia, linifolium
With leaves like flax (*Linum*), as in *Tulipa
linifolia*

Linnaea
lin-AY-uh
Named after Carl Linnaeus (1707–78),
Swedish botanist, zoologist, and father of
taxonomy; this was his favorite flower
(*Caprifoliaceae*)

linnaeanus
lin-ee-AH-nus
linnaeana, linnaeanum
—
linnaei
lin-ee-eye
Named after Carl Linnaeus (1707–78),
Swedish botanist, as in *Solanum linnaeanum*

linoides
li-NOY-deez
Resembling flax (*Linum*), as in *Monardella
linoides*

Linum
LY-num
From Latin *lin*, meaning "flax"; *Linum
usitatissimum* is the source of linseed oil
and linen (*Linaceae*)

Liquidambar

lik-wid-AM-bur

From Latin *liquidus*, meaning "liquid," and *ambar*, meaning "amber," because the trunk produces resin (*Altingiaceae*)

Liriodendron [3]

li-ree-oh-DEN-dron

From Greek *lirion*, meaning "lily," and *dendron*, meaning "tree," for the ornamental flowers (*Magnoliaceae*)

Liriope

li-ree-OH-pee

Named after Liriope, a nymph in Greek mythology and mother of Narcissus (*Asparagaceae*)

litangensis

lit-ang-EN-sis

litangensis, litangense

From Litang, China, as in *Lonicera litangensis*

Litchi

LY-chee

From Cantonese (China) *lai-zi* or Mandarin *lizhi*, the vernacular name for the fruit of *L. chinensis* (*Sapindaceae*)

Lithocarpus

lith-oh-KAR-pus

From Greek *lithos*, meaning "stone," and *karpos*, meaning "fruit," because the acorns have a hard shell (*Fagaceae*)

Lithodora

lith-oh-DOR-uh

From Greek *lithos*, meaning "stone," and *dorea*, meaning "gift," suggesting this pretty alpine is a welcome contribution from a rocky habitat (*Boraginaceae*)

lithophilus

lith-oh-FIL-us

lithophila, lithophilum

Growing in rocky habitats, as in *Anemone lithophila*

Lithophragma

lith-oh-FRAG-muh

From Greek *lithos*, meaning "stone," and *phragma* "fence" or "hedge," a reference to a rocky habitat; alternatively, an inaccurate Greek translation of *Saxifraga* (*Saxifragaceae*)

Lithops

LITH-ops

Resembling stones, from Greek *lithos*, meaning "stone" (*Aizoaceae*)

Litsea

LIT-see-uh

From Mandarin *litse*, meaning "small plum"; the fruit is small and fleshy (*Lauraceae*)

littoralis

lit-tor-AH-lis

littoralis, littorale

—

littoreus

lit-TOR-ee-us

littorea, littoreum

Growing by the sea, as in *Griselinia littoralis*

lividus

LI-vid-us

livida, lividum

Blue-gray; the color of lead, as in *Helleborus lividus*

Livistona

liv-i-STOW-nuh

Named after Patrick Murray, second Lord Elibank, who lived in Livingston Peel, west of Edinburgh (1632–71), Scottish horticulturist (*Arecaceae*)

Lloydia

LOY-dee-uh

Named after Edward Lhuyd (1660–1709), Welsh botanist and linguist (*Liliaceae*)

lobatus

low-BAH-tus

lobata, lobatum

With lobes, as in *Cyananthus lobatus*

Lobelia

loh-BEE-lee-uh

Named after Mathias de l'Obel (1538–1616), Flemish physician and botanist (*Campanulaceae*)

lobelioides

lo-bell-ee-OH-id-ees

Resembling *Lobelia*, as in *Wahlenbergia lobelioides*

lobophyllus

lo-bo-FIL-us

lobophylla, lobophyllum

With lobed leaves, as in *Viburnum lobophyllum*

Lobularia

lob-yoo-LAIR-ee-uh

From Latin *lobulus*, meaning "small lobe" or "pod," referring to the small, round fruit (*Brassicaceae*)

3

Liriodendron tulipifera

lobularis

lob-yoo-LAH-ris

lobularis, lobulare

With lobes, as in *Narcissus lobularis*

lobulatus

lob-yoo-LAH-tus

lobulata, lobulatum

With small lobes, as in *Crataegus lobulata*

Lodoicea

loh-doh-ISS-ee-uh

Named after Louis XV (1710–74), King of France (*Arecaceae*)

loliaceus

loh-lee-uh-SEE-us

loliacea, loliaceum

Like rye grass (*Lolium*), as in × *Festulolium loliaceum*

Lolium [1]

LOH-lee-um

From classical Latin name for a rye grass, probably *L. temulentum* (*Poaceae*)

Lomandra

loh-MAN-druh

From Greek *loma*, meaning "fringe," and *aner*, meaning "male," because in some species, the anthers have noticeable borders (*Asparagaceae*)

Lomatia

loh-MAH-tee-uh

From Greek *loma*, meaning "fringe," because the seed has a papery wing (*Proteaceae*)

Lomatium

loh-MAH-tee-um

From Greek *loma*, meaning "fringe," because the seed has a papery wing (*Apiaceae*)

longibracteatus

lon-jee-brak-tee-AH-tus

longibracteata, longibracteatum

With long bracts, as in *Pachystachys longibracteata*

longicaulis

lon-jee-KAW-lis

longicaulis, longicaule

With long stalks, as in *Aeschynanthus longicaulis*

longicuspis

lon-jee-KUS-pis

longicuspis, longicuspe

With long points, as in *Rosa longicuspis*

longiflorus

lon-jee-FLO-rus

longiflora, longiflorum

With long flowers, as in *Crocus longiflorus*

longifolius

lon-jee-FOH-lee-us

longifolia, longifolium

With long leaves, as in *Pulmonaria longifolia*

longilobus

lon-JEE-loh-bus

longiloba, longilobum

With long lobes, as in *Alocasia longiloba*

longipedunculatus

lon-jee-ped-un-kew-LAH-tus

longipedunculata, longipedunculatum

With a long peduncle, as in *Magnolia longipedunculata*

longipes

LON-juh-peez

With a long stalk, as in *Acer longipes*

longipetalus

lon-jee-PET-uh-lus

longipetala, longipetalum

With long petals, as in *Matthiola longipetala*

1

Lolium perenne

2

Lonicera hildebrandiana

longiracemosus

lon-jee-ray-see-MOH-sus

longiracemosa, longiracemosum

With long racemes, as in *Incarvillea longiracemosa*

longiscapus

lon-jee-SKAY-pus

longiscapa, longiscapum

With a long scape, as in *Vriesea longiscapa*

longisepalus

lon-jee-SEE-pal-us

longisepala, longisepalum

With long sepals, as in *Allium longisepalum*

longispathus

lon-jis-PAY-thus

longispatha, longispathum

With long spathes, as in *Narcissus longispathus*

longissimus

lon-JIS-ih-mus

longissima, longissimum

Especially long, as in *Aquilegia longissima*

longistylus

lon-jee-STY-lus

longistyla, longistylum

With a long style, as in *Arenaria longistyla*

3

Lotus corniculatus

longus

LONG-us

longa, longum

Long, as in *Cyperus longus*

Lonicera [2]

lo-NIS-er-uh

Named after Adam Lonitzer (1528–86), German physician and botanist; the family takes its name from genus *Caprifolium*, now a synonym of *Lonicera*, derived from Latin *capri*, meaning "goat," and *folium*, meaning "leaves," either because goats like to feed on the foliage, or because this vine's ability to climb trees matches that of goats (*Caprifoliaceae*)

lophanthus

low-FAN-thus

lophantha, lophanthum

With crested flowers, as in *Paraserianthes lophantha*

Lophomyrtus

lof-oh-MER-tus

From Greek *lophos*, meaning "crest," and *Myrtus*, a related genus, because within the flower's ovary are ribbonlike placenta lobes (*Myrtaceae*)

Lophophora

luh-FOF-er-uh

From Greek *lophos*, meaning "crest," and *phoreus*, meaning "bearer," because the areoles on the stem bear prominent tufts of hair (*Cactaceae*)

Lophospermum

lof-oh-SPER-mum

From Greek *lophos*, meaning "crest," and *sperma*, meaning "seed," alluding to the membranous wing around each seed (*Plantaginaceae*)

Loropetalum

lor-oh-PET-ah-lum

From Greek *loron*, meaning "strap," and *petalon*, meaning "petal," because this witch hazel relation has similar straplike petals (*Hamamelidaceae*)

lotifolius

lo-tif-FOH-lee-us

lotifolia, lotifolium

With leaves like *Lotus*, as in *Goodia lotifolia*

Lotus [3] (also lotus)

LOH-tus

From Greek *lotos*, a vernacular name for several different plants (*Fabaceae*)

louisianus

loo-ee-see-AH-nus

louisiana, louisianum

Connected with Louisiana, as in *Proboscidea louisiana*

lucens

LOO-senz

—

lucidus

LOO-sid-us

lucida, lucidum

Bright; shining; clear, as in *Ligustrum lucidum*

Luculia

luh-KOO-lee-uh

From Nepalese *luculi-swa*, the vernacular name for *L. gratissima* (*Rubiaceae*)

A
B
C
D
E
F
G
H
I
J
K
L
M
N
O
P
Q
R
S
T
U
V
W
X
Y
Z
i

1

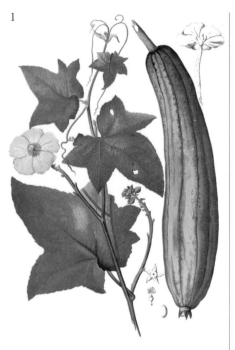

Luffa cylindrica

2

Lunaria annua

Ludisia

loo-DIZ-ee-uh

Said to derive from Ludis, whose widower composed a mournful elegy after her death, a nod perhaps to the dark foliage, although evidence for this etymology is scant (*Orchidaceae*)

ludovicianus

loo-doh-vik-ee-AH-nus

ludoviciana, ludovicianum

Connected with Louisiana, as in *Artemisia ludoviciana*

Ludwigia

lood-VIG-ee-uh

Named after Christian Gottlieb Ludwig (1709–73), German physician and botanist (*Onagraceae*)

Luffa [1]

LOO-fuh

From Arabic *lufah*, the vernacular name for *L. aegyptiaca* (*Cucurbitaceae*)

Luma (also luma)

LOO-muh

From Mapuche (Chile) vernacular name for this genus and/or *Amomyrtus luma* (*Myrtaceae*)

Lunaria [2]

loo-NAIR-ee-uh

From Latin *luna*, meaning "moon," because the fruit is circular (*Brassicaceae*)

lunatus

loo-NAH-tus

lunata, lunatum

—

lunulatus

loo-nu-LAH-tus

lunulata, lunulatum

Shaped like the crescent moon, as in *Cyathea lunulata*

Lupinus

loo-PIE-nus

From Latin *lupus*, meaning "wolf," because some weedy species are said to overrun fields and reduce fertility; the opposite is true, because this plant is a nitrogen fixer (*Fabaceae*)

lupulinus

lup-oo-LEE-nus

lupulina, lupulinum

Like hop (*Humulus lupulus*), as in *Medicago lupulina*

luridus

LEW-rid-us

lurida, luridum

Pale yellow, wan, as in *Moraea lurida*

lusitanicus

loo-si-TAN-ih-kus

lusitanica, lusitanicum

Connected with Lusitania (Portugal and some parts of Spain), as in *Prunus lusitanica*

luteolus

loo-tee-OH-lus

luteola, luteolum

Yellowish, as in *Primula luteola*

lutetianus

loo-tee-shee-AH-nus

lutetiana, lutetianum

Connected with Lutetia (Paris), France, as in *Circaea lutetiana*

luteus

LOO-tee-us

lutea, luteum

Yellow, as in *Calochortus luteus*

luxurians

luks-YOO-ee-anz

Luxuriant, as in *Begonia luxurians*

Luzula

luz-ew-luh

From Latin *luzulae*, meaning "little light," because the leaf hairs glisten with dew (*Juncaceae*)

Lycianthes

ly-see-AN-theez

From Greek *lykion*, meaning "thorny shrub," and *anthos*, meaning "flower" (*Solanaceae*)

Lycium

LY-see-um

From Greek *lykion*, meaning "thorny shrub from Lycia," now in Turkey; possibly referring to a species of *Rhamnus* (*Solanaceae*)

lycius

LY-cee-us

lycia, lycium

Connected with Lycia, now part of Turkey, as in *Phlomis lycia*

lycopersicum

ly-koh-PER-si-kum

'Wolf peach,' because in *Solanum lycopersicum*

lycopodioides

ly-kop-oh-dee-OY-deez

Resembling clubmoss (*Lycopodium*), as in *Cassiope lycopodioides*

Lycopodium

ly-koh-POH-dee-um

From Greek *lykos*, meaning "wolf," and *podion*, meaning "foot," which the stem tips are said to resemble (*Lycopodiaceae*)

Lycopus

ly-KOH-pus

From Greek *lykos*, meaning "wolf," and *pous*, meaning "foot," alluding to the shape of the leaves (*Lamiaceae*)

Lycoris

ly-KOR-is

After Lycoris (aka Cytheris), a Roman actress and lover of Mark Anthony (*Amaryllidaceae*)

lydius

LID-ee-us

lydia, lydium

Connected with Lydia, now part of Turkey, as in *Genista lydia*

Lygodium

ly-GOH-dee-um

From Greek *lygodes*, meaning "flexible," like the stems of this climbing fern (*Lygodiaceae*)

Lyonia

ly-OH-nee-uh

Named after John Lyon (1765–1814), Scottish-born American botanist (*Ericaceae*)

Lyonothamnus

ly-on-oh-THAM-nus

Named after William Scrugham Lyon (1851–1916), American botanist, plus Greek *thamnos*, meaning "shrub" (*Rosaceae*)

Lysichiton

ly-si-KY-ton

From Greek *lysis*, meaning "to dissolve," and *chiton*, meaning "tunic," referring to the prominent spathe that soon withers (*Araceae*)

Lysimachia

ly-si-MAK-ee-uh

From Greek *lysis*, meaning "to dissolve," and *mache*, meaning "strife," alluding to its medicinal properties, as in "loosestrife"; alternatively, after Lysimachus (ca. 360–281 BC), King of Thrace, Macedonia ,and Asia Minor or Lysimachus of Kos, a fourth- or fifth-century-BC physician (*Primulaceae*)

lysimachioides

ly-see-mak-ee-OY-deez

Resembling loosestrife (*Lysimachia*), as in *Hypericum lysimachioides*

Lythrum

LITH-rum

From Greek *lythron*, meaning "blood," a reference to flower color (*Lythraceae*)

Lycopersicon

The tomato (*Solanum lycopersicum*) was initially described in the same genus as the potato (*S. tuberosum*), but it was quickly moved into its own genus (as *Lycopersicon esculentum*). However, just as old styles come back into fashion, so DNA data has suggested that tomato should rejoin potato. The word *lycopersicon* is derived from Greek, meaning "wolf peach," and is a name endowed with suspicion. Although indigenous South Americans have long eaten the fruit, Europeans were initially skeptical of them. Tomatoes are members of a family that includes henbane (*Hyoscyamus*), deadly nightshade (*Atropa*), and mandrake (*Mandragora*), so they had reason to worry. Thankfully, tomatoes have shaken off their poisonous reputation.

Solanum lycopersicum

Maackia [1]

MAH-kee-uh

Named after Richard Otto Maack (1825–86), Russian naturalist and anthropologist (*Fabaceae*)

Macadamia

mak-uh-DAME-ee-uh

Named after John Macadam (1827–65), Scottish-born Australian chemist and politician (*Proteaceae*)

macedonicus

mas-eh-DON-ih-kus

macedonica, macedonicum

Connected with Macedonia, as in *Knautia macedonica*

Machaeranthera

mak-uh-RAN-thu-ruh

From Greek *macha*, meaning "sword," and *anthera*, meaning "anther" (*Asteraceae*)

macilentus

mas-il-LEN-tus

macilenta, macilentum

Thin; lean, as in *Justicia macilenta*

Mackaya

muh-KY-uh

Named after James Townsend Mackay (1775–1862), Scottish botanist (*Acanthaceae*)

Macleania

muh-KLAY-nee-uh

Named after John Maclean (eighteenth century), British merchant in Peru (*Ericaceae*)

Macleaya

muh-KLAY-uh

Named after Alexander Macleay (1767–1848), Scottish colonial official in New South Wales and entomologist (*Papaveraceae*)

Maclura

ma-KLUR-uh

Named after William Maclure (1763–1840), Scottish-born American geologist (*Moraceae*)

macro-

Used in compound words to denote either long or large

macracanthus

mak-ra-KAN-thus

macracantha, macracanthum

With large spines, as in *Acacia macracantha*

macrandrus

mak-RAN-drus

macrandra, macrandrum

With large anthers, as in *Eucalyptus macrandra*

macranthus

mak-RAN-thus

macrantha, macranthum

With large flowers, as in *Hebe macrantha*

macrobotrys

mak-roh-BOT-rees

With large grape-like clusters, as in *Strongylodon macrobotrys*

macrocarpus [2]

mak-roh-KAR-pus

macrocarpa, macrocarpum

With large fruit, as in *Cupressus macrocarpa*

1

Maackia amurensis

2

Begonia maculata

macrocephalus

mak-roh-SEF-uh-lus

macrocephala, macrocephalum

With large heads, as in *Centaurea macrocephala*

macrodontus

mak-roh-DON-tus

macrodonta, macrodontum

With large teeth, as in *Olearia macrodonta*

macromeris

mak-roh-MER-is

With many or large parts, as in *Coryphantha macromeris*

macrophyllus

mak-roh-FIL-us

macrophylla, macrophyllum

With large or long leaves, as in *Hydrangea macrophylla*

macropodus

mak-roh-POH-dus

macropoda, macropodum

With stout stalks, as in *Daphniphyllum macropodum*

macrorrhizus

mak-roh-RY-zus

macrorrhiza, macrorrhizum

With large roots, as in *Geranium macrorrhizum*

macrospermus

mak-roh-SPERM-us

macrosperma, macrospermum

With large seed, as in *Senecio macrospermus*

macrostachyus

mak-roh-STAH-kus

macrostachya, macrostachyum

With long or large spikes, as in *Setaria macrostachya*

Macrozamia

mak-roh-ZAY-mee-uh

From Greek *makros*, meaning "large," and *Zamia*, a related genus (*Zamiaceae*)

maculatus

mak-yuh-LAH-tus

maculata, maculatum

—

maculosus

mak-yuh-LAH-sus

maculosa, maculosum

With spots, as in *Begonia maculata*

madagascariensis

mad-uh-gas-KAR-ee-EN-sis

madagascariensis, madagascariense

From Madagascar, as in *Buddleja madagascariensis*

Magnolia

Many plant names were created to honor people, and *Magnolia* is no exception. Pierre Magnol was a French botanist who is credited with creating the plant family, a unit of classification still in use today. Once seen as exclusive, magnolias (note the lowercase "m") have become ubiquitous in towns and cities, so much so that their name is now appended to one of the most popular colors of paint for home interiors. As *Magnolia* became magnolia, it lost some of its luster, and few people familiar with the tree (or the color) have heard of poor Monsieur Magnol.

Magnolia campbellii

maderensis

ma-der-EN-sis

maderensis, maderense

From Madeira, as in *Geranium maderense*

Madia

MAY-dee-uh

From Mapuche (Chile) name *madi*, for *M. sativa* (*Asteraceae*)

Maesa

MAY-zuh

From Arabic *maas*, the vernacular name for *M. lanceolata* (*Primulaceae*)

magellanicus

ma-jell-AN-ih-kus

megallanica, megallanicum

Connected with the Straits of Magellan, South America, as in *Fuchsia magellanica*

magellensis

mag-ah-LEN-sis

magellensis, magellense

Of the Maiella massif, Italy, as in *Sedum magellense*

magnificus

mag-NIH-fih-kus

magnifica, magnificum

Splendid, magnificent, as in *Geranium* × *magnificum*

Magnolia

mag-NOH-lee-uh

Named after Pierre Magnol (1638–1715), French botanist (*Magnoliaceae*)

magnus

MAG-nus

magna, magnum

Great; big, as in *Alberta magna*

Mahonia

muh-HOH-nee-uh

Named after Bernard McMahon (1775–1816), Irish-born American horticulturist (*Berberidaceae*)

Maianthemum

my-ANTH-uh-mum

From Greek *Maios*, meaning "May," and *anthemon*, meaning "flower," alluding to the bloom period (*Asparagaceae*)

Maihuenia

my-WEN-ee-uh

From Mapuche (Chile) name *maihuén*, for *M. poeppigii* (*Cactaceae*)

majalis

maj-AH-lis

majalis, majale

Flowering in May, as in *Convallaria majalis*

GENUS SPOTLIGHT

Mahonia

Taxonomists are often unpopular with gardeners, because of their annoying habit of changing the name by which we know our favorite plant. Contrary to popular belief, such changes are enacted because new evidence arises revealing relationships that contradict current taxonomy. The popular genus *Mahonia* is a confusing case in point. Taxonomists have long disagreed as to whether it should be combined with its close relative *Berberis*. DNA evidence suggests the two should be merged, an unpopular move with many gardeners. One recently mooted solution is to hive off a handful of mahonias into new genera (*Moranothamnus*, *Alloberberis*) and let the rest retain their well-known name and independent status. It will be many years before we know whether the taxonomic community has embraced this option.

Mahonia napaulensis

major
MAY-jor
major, majus
Bigger; larger, as in *Astrantia major*

malabaricus
mal-uh-BAR-ih-kus
malabarica, malabaricum
Connected with the Malabar coast, India, as in *Bauhinia malabarica*

malacoides
mal-a-koy-deez
Soft, as in *Erodium malacoides*

malacospermus
mal-uh-ko-SPER-mus
malacosperma, malacospermum
With soft seeds, as in *Hibiscus malacospermus*

maliformis
ma-lee-for-mees
maliformis, maliforme
Shaped like an apple, as in *Passiflora maliformis*

Malope [1]
muh-LOH-pee
From Greek *malos*, meaning "tender," because the leaves are soft (*Malvaceae*)

Malpighia
mal-PIG-ee-uh
Named after Marcello Malpighi (1628–94), Italian biologist (*Malpighiaceae*)

Malus
MA-lus
From Latin *malus*, meaning "apple tree" (*Rosaceae*)

Malva
MAL-vuh
From Greek *malaco*, meaning "to soften," a reference to the use of some species as emollients (*Malvaceae*)

malvaceus
mal-VAY-see-us
malvacea, malvaceum
Like mallow (*Malva*), as in *Physocarpus malvaceus*

Malvastrum
mal-VAS-trum
From *Malva*, a related genus, plus Greek *astrum*, meaning "incomplete resemblance" (*Malvaceae*)

Malvaviscus
mal-vuh-VIS-kus
From *Malva*, a related genus plus Latin *viscidus*, meaning "sticky," like the sap (*Malvaceae*)

malviflorus
mal-VEE-flor-us
malviflora, malviflorum
With flowers like mallow (*Malva*), as in *Geranium malviflorum*

malvinus
mal-VY-nus
malvina, malvinum
Mauve, as in *Plectranthus malvinus*

Mammillaria [2]
mam-uh-LAIR-ee-uh
From Latin *mamilla*, meaning "nipple," referencing the shape of the spine-bearing tubercles, which in some species produce white latex (*Cactaceae*)

mammillatus
mam-mil-LAIR-tus
mammillata, mammillatum
—

mammillaris
mam-mil-LAH-ris
mammillaris, mammillare
—

mammosus
mam-OH-sus
mammosa, mammosum
Bearing nipple- or breastlike structures, as in *Solanum mammosum*

1

Malope trifida

2

Mammillaria elongata

Mandragora officinarum

Mangifera indica

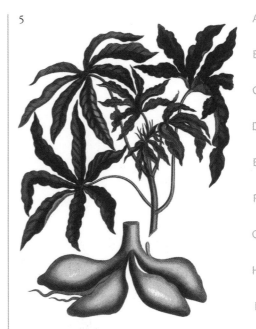

Manihot esculenta

Mandevilla
man-duh-VIL-uh
Named after Henry John Mandeville (1773–1861), British diplomat (*Apocynaceae*)

Mandragora [3]
man-druh-GOR-uh
From Greek *mandragoras*, meaning "mandrake," itself deriving from a proto-Greek or Persian word (*Solanaceae*)

mandshuricus
mand-SHEU-rih-kus
mandshurica, mandshuricum

—

manshuricus
man-SHEU-rih-kus
manshurica, manshuricum
Connected with Manchuria, northeast Asia, as in *Tilia mandshurica*

Manettia
muh-NET-ee-uh
Named after Saverio Manetti (1723–85), Italian physician, ornithologist, and botanist (*Rubiaceae*)

Mangifera [4]
man-GIF-er-uh
From Tamil (India) *mangai*, meaning "unripe mango fruit" (*manpalam* is Tamil for "ripe mango fruit"), adopted by Portuguese sailors, becoming the standard word for the fruit and plant; Latin *iferus* means "bearing" (*Anacardiaceae*)

manicatus
mah-nuh-KAH-tus
manicata, manicatum
With long sleeves, as in *Gunnera manicata*

Manihot [5]
MAN-ee-hot
From Brazilian indigenous *manioc*, the name for cassava, *M. esculenta* (*Euphorbiaceae*)

Maranta
ma-RAN-tuh
Named after Bartolomea Maranti (?–1751), Venetian botanist and physician (*Marantaceae*)

margaritaceus
mar-gar-ee-tuh-KEE-us
margaritacea, margaritaceum

—

margaritus
mar-gar-ee-tus
margarita, margaritum
Relating to pearls, as in *Anaphalis margaritacea*

margaritiferus
mar-guh-rih-TIH-fer-us
margaritifera, margaritiferum
With pearls, as in *Haworthia margaritifera*

marginalis
mar-gin-AH-lis
marginalis, marginale

—

marginatus
mar-gin-AH-tus
marginata, marginatum
Margined, as in *Saxifraga marginata*

Margyricarpus
mar-jir-ee-KAR-pus
From Greek *margaron*, meaning "pearl," and *karpos*, meaning "fruit," because the fruits are opalescent (*Rosaceae*)

marianus
mar-ee-AH-nus
mariana, marianum
Of the Virgin Mary (or sometimes Maryland), as in *Silybum marianum*

marilandicus
mar-i-LAND-ih-kus
marilandica, marilandicum
Connected with Maryland, as in *Quercus marilandica*

maritimus
muh-RIT-tim-mus
maritima, maritimum
Relating to the sea, as in *Armeria maritima*

1

Marrubium vulgare

2

Matthiola sinuata

marmoratus
mar-mor-RAH-tus
marmorata, marmoratum
—
marmoreus
mar-MOH-ree-us
marmorea, marmoreum
Marbled; mottled, as in *Kalanchoe marmorata*

maroccanus
mar-oh-KAH-nus
maroccana, maroccanum
Connected with Morocco, as in *Linaria maroccana*

Marrubium [1]
muh-ROO-bee-um
From Latin *marrubium*, the vernacular name for horehound, *M. vulgare* (*Lamiaceae*)

Marsilea
mar-si-LAY-uh
Named after Luigi Fernando Marsigli (1656–1730), Italian scholar, soldier, and emissary (*Marsileaceae*)

martagon
MART-uh-gon
A word of uncertain origin, thought in *Lilium martagon* to refer to the turban-ike flower

martinicensis
mar-teen-i-SEN-sis
martinicensis, martinicense
From Martinique, Lesser Antilles, as in *Trimezia martinicensis*

mas
MAS
—
masculus
MASK-yoo-lus
mascula, masculum
With masculine qualities, male, as in *Cornus mas*

Masdevallia
mas-duh-VAY-lee-uh
Named after José Masdeval (?–1801), Spanish physician and botanist (*Orchidaceae*)

Mathiasella
mu-thy-us-EL-uh
Named after Mildred Esther Mathias (1906–95), American botanist (*Apiaceae*)

Matricaria
ma-tri-KARE-ee-uh
From Latin *matrix*, meaning "womb," and *aria*, meaning "pertaining to," a reference to medicinal properties (*Asteraceae*)

matronalis
mah-tro-NAH-lis
matronalis, matronale
Relating to March 1, the Roman Matronalia, festival of motherhood and Juno as goddess of childbirth, as in *Hesperis matronalis*

Matteuccia
ma-TOO-chee-uh
Named after Carlo Matteucci (1811–68), Italian physicist (*Onocleaceae*)

Matthiola [2]
mat-ee-OH-luh
Named after Pietro Andrea Matthioli (1500–77), Italian physician and botanist (*Brassicaceae*)

Maurandya
maw-RAN-dee-uh
Named after Catherina Pancracia Maurandy (eighteenth century), Spanish botanist (*Plantaginaceae*)

mauritanicus
maw-rih-TAWN-ih-kus
mauritanica, mauritanicum
Connected with North Africa, especially Morocco, as in *Lavatera mauritanica*

mauritianus
maw-rih-tee-AH-nus
mauritiana, mauritianum
Connected with Mauritius, Indian Ocean, as in *Croton mauritianus*

Maxillaria
maks-i-LAIR-ee-uh
From Latin *maxilla*, meaning "jawbone," an allusion to the yawning flowers and the column base, which may resemble a jaw (*Orchidaceae*)

maxillaris
maks-ILL-ah-ris
maxillaris, maxillare
Relating to the jaw, as in *Zygopetalum maxillare*

maximus
MAKS-ih-mus
maxima, maximum
Largest, as in *Rudbeckia maxima*

Maytenus
MAY-ten-us
From Mapuche (Chile) name *mayten* for *M. boaria* (*Celastraceae*)

Mazus

MAY-zus

From Greek *mazos*, meaning "nipple," because the petals have noticeable swellings (*Phrymaceae*)

Meconopsis [3]

mek-uh-NOP-sis

From Greek *mekon*, meaning "poppy," and *opsis*, meaning "resemblance" (*Papaveraceae*)

Medeola

me-dee-OH-luh

Named after Medea, the sorceress in Greek mythology who helped Jason find the golden fleece (*Liliaceae*)

Medicago

me dee-KAH-go

From Greek *medikos*, meaning "lucerne" (aka alfalfa, *M. sativa*), a plant said to come from Media, now in northwestern Iran (*Fabaceae*)

medicus

MED-ih-kus

medica, medicum

Medicinal, as in *Citrus medica*

Medinilla

med-uh-NILL-uh

Named after José de Medinilla y Pineda (dates unknown), twice Spanish governor of the Mariana Islands (1812–22, 1826–31), (*Melastomataceae*)

mediopictus

MED-ee-o-pic-tus

mediopicta, mediopictum

With a stripe or color running down the middle, as in *Calathea mediopicta*

mediterraneus

med-ih-ter-RAY-nee-us

mediterranea, mediterraneum

Either from a land-locked region, or from the Mediterranean, as in *Minuartia mediterranea*

medius

MEED-ee-us

media, medium

Intermediate; middle, as in *Mahonia × media*

medullaris

med-yoo-LAH-ris

medullaris, medullare

—

medullus

med-DUL-us

medulla, medullum

Pithy, as in *Cyathea medullaris*

mega-

Used in compound words to denote big

megacanthus

meg-uh-KAN-thus

megacantha, megacanthum

With big spines, as in *Opuntia megacantha*

megacarpus

meg-uh-CAR-pus

megacarpa, megacarpum

With big fruit, as in *Ceanothus megacarpus*

megalanthus

meg-uh-LAN-thus

megalantha, megalanthum

With big flowers, as in *Potentilla megalantha*

megalophyllus

meg-uh-luh-FIL-us

megalophylla, megalophyllum

With big leaves, as in *Ampelopsis megalophylla*

megapotamicus

meg-uh-poh-TAM-ih-kus

megapotamica, megapotamicum

Connected with a big river, for instance the Amazon or Rio Grande, as in *Abutilon megapotamicum*

Megaskepasma

meg-uh-skuh-PAZ-muh

From Greek *megas*, meaning "great," and *skepasma*, meaning "covering," because the flowers are partly concealed under large, colorful bracts (*Acanthaceae*)

megaspermus

meg-uh-SPER-mus

megasperma, megaspermum

With big seed, as in *Callerya megasperma*

megastigma

meg-uh-STIG-ma

With a big stigma, as in *Boronia megastigma*

3

Meconopsis delavayi

Melaleuca [1]

me-luh-LOO-kuh

From Greek *melas*, meaning "black," and *leukos*, meaning "white"; early specimens had white stems blackened by fire (*Myrtaceae*)

melanocaulon

mel-an-oh-KAW-lon

With black stems, as in *Blechnum melanocaulon*

melanocentrus

mel-an-oh-KEN-trus

melanocentra, melanocentrum

With a black center, as in *Saxifraga melanocentra*

melanococcus

mel-an-oh-KOK-us

melanococca, melanococcum

With black berries, as in *Elaeis melanococcus*

melanoxylon

mel-an-oh-ZY-lon

With black wood, as in *Acacia melanoxylon*

Melastoma

mel-uh-STO-muh

From Greek *melas*, meaning "black," and *stoma*, meaning "mouth," because eating the fruit stains the tongue dark blue or black (*Melastomataceae*)

meleagris [3]

mel-EE-uh-gris

meleagris, meleagre

With spots like a guinea fowl, as in *Fritillaria meleagris*

Melia

ME-lee-uh

The classical Greek name for an ash tree (*Fraxinus*), which has similar leaves to *M. azedarach* (*Meliaceae*)

Melianthus

me-lee-ANTH-us

From Greek *mel*, meaning "honey," and *anthos*, meaning "flower," due to the copious nectar (*Francoaceae*)

Melica

MEL-i-kuh

From Greek *melike*, meaning "grass," or possibly derived from Latin *herba medica*, meaning "grass from Media," now in northeastern Iran (*Poaceae*)

Melicope

mel-i-KOH-pee

From Greek *mel*, meaning "honey," and *kope*, meaning "split," because the floral glands are notched (*Rutaceae*)

Melicytus [2]

mel-i-SY-tus

From Greek *mel*, meaning "honey," and *kytos*, meaning "container," because the nectar seeps from a hollow nectary (*Violaceae*)

Melilotus

mel-i-LOH-tus

From Greek *mel*, meaning "honey," and *Lotus*, a related genus, because this plant has sweet-scented foliage (*Fabaceae*)

1

Melaleuca cajuputi

2

Melicytus crassifolius

Melinis

mel-I-nis

From Greek *mel*, meaning "honey," because some species smell of molasses, or from *melas*, meaning "black," the color of the seed (*Poaceae*)

Meliosma

me-lee-OZ-muh

From Greek *mel*, meaning "honey," and *osme*, meaning "fragrance," the scent of the flowers (*Sabiaceae*)

Melissa

muh-LISS-uh

From Greek *melissa* or *melitta*, meaning "honeybee," which are strongly attracted to the flowers (*Lamiaceae*)

Melittis

MEH-lit-iss

From Greek *melissa* or *melitta*, meaning "honeybee," which are strongly attracted to the flowers (*Lamiaceae*)

melliferus

mel-IH-fer-us

mellifera, melliferum

Producing honey, as in *Euphorbia mellifera*

Melliodendron

mel-ee-oh-DEN-dron

From Greek *mel*, meaning "honey," and *dendron*, meaning "tree," suggesting fragrant flowers (*Styracaceae*)

melliodorus

mel-ee-uh-do-rus

melliodora, melliodorum

With the scent of honey, as in *Eucalyptus melliodora*

mellitus

mel-IT-tus

mellita, mellitum

Sweet, like honey, as in *Iris mellita*

Melocactus

mel-oh-KAK-tus

From Greek *melon*, meaning "apple," plus cactus, alluding to the spherical shape of young plants; sometimes taken to derive from English "melon" (*Cactaceae*)

meloformis

mel-OH-for-mis

meloformis, meloforme

Shaped like a melon, as in *Euphorbia meloforme*

Fritillaria meleagris

membranaceus

mem-bran-AY-see-us

membranacea, membranaceum

Like skin or membrane, as in *Scadoxus membranaceus*

meniscifolius

men-is-ih-FOH-lee-us

meniscifolia, meniscifolium

With leaves that have a crescent shape, as in *Serpocaulon meniscifolium*

Menispermum

men-i-SPER-mum

From Greek *mene*, meaning "moon," and *sperma*, meaning "seed," because the seed is crescentlike (*Menispermaceae*)

Mentha [4]

MEN-thuh

From the classical Latin name for mint; also a nymph in Greek mythology, who was transformed into a mint plant (*Lamiaceae*)

Mentzelia

ment-ZEE-lee-uh

Named after Christian Mentzel (1622–1701), German botanist and physician (*Loasaceae*)

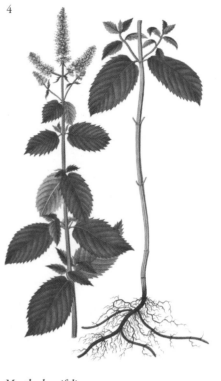

Mentha longifolia

Menyanthes

men-ee-ANTH-eez

From Greek *menanthos*, the name of an aquatic plant, probably the related *Nymphoides peltata*, although several other etymologies have been proposed (*Menyanthaceae*)

menziesii

menz-ESS-ee-eye

Named after Archibald Menzies (1754–1842), British naval surgeon and botanist, as in *Pseudotsuga menziesii*

Mercurialis

mer-kure-ee-AH-lis

Named after the Roman god Mercury, who is said to have discovered the plant (*Euphorbiaceae*)

meridianus

mer-id-ee-AH-nus

meridiana, meridianum

—

meridionalis

mer-id-ee-oh-NAH-lis

meridionalis, meridionale

Flowering at noon, as in *Primula × meridiana*

Metrosideros excelsa

Michauxia campanuloides

Mertensia
mer-TEN-see-uh
Named after Frank Carl Mertens (1764–1831), German botanist (*Boraginaceae*)

Mesembryanthemum
mez-em-bree-ANTH-uh-mum
From Greek *mesembria*, meaning "midday," and *anthemon*, meaning "flower," because the blooms open around noon (*Aizoaceae*)

Mespilus
MES-pi-lus
From the classical Latin name for medlar, *M. germanica*; possibly derived from Greek *mesos*, meaning "half," and *pilos*, meaning "ball," an allusion to the shape of the fruit (*Rosaceae*)

metallicus
meh-TAL-ih-kus
metallica, metallicum
Metallic, as in *Begonia metallica*

Metasequoia
met-uh-seh-KWOY-uh
From Greek *metra*, meaning "after" or "changed," plus *Sequoia*, a related genus (*Cupressaceae*)

Metrosideros [1]
met-roh-SID-ur-os
From Greek *metra*, meaning "heart" or "center," and *sideron*, meaning "iron," because the heartwood is robust (*Myrtaceae*)

mexicanus
meks-sih-KAH-nus
mexicana, mexicanum
Connected with Mexico, as in *Agastache mexicana*

Michauxia [2]
mich-OH-zee-uh
Named after André Michaux (1746–1802), French botanist (*Campanulaceae*)

michauxioides
mich-OH-zee-uh-deez
Resembling *Michauxia*, as in *Campanula michauxioides*

micracanthus
mik-ra-KAN-thus
micracantha, micracanthum
With small thorns, as in *Euphorbia micracantha*

micranthus
my-KRAN-thus
micrantha, micranthum
With small flowers, as in *Heuchera micrantha*

micro-
Used in compound words to denote small

Microbiota
my-kro-by-OH-tuh
From Greek *mikros*, meaning "small," plus *Biota*, a related genus [now *Platycladus*] (*Cupressaceae*)

Microcachrys
my-kro-KAK-ris
From Greek *mikros*, meaning "small," and *cachrus*, meaning "catkin," referring to the tiny pollen cones (*Podocarpaceae*)

microcarpus
my-kro-KAR-pus
microcarpa, microcarpum
With small fruit, as in × *Citrofortunella microcarpa*

microcephalus
my-kro-SEF-uh-lus
microcephala, microcephalum
With a small head, as in *Persicaria microcephala*

microdasys
my-kro-DAS-is
Small and shaggy, as in *Opuntia microdasys*

microdon
my-kro-DON
With small teeth, as in *Asplenium* × *microdon*

microglossus
my-kro-GLOS-us
microglossa, microglossum
With a small tongue, as in *Ruscus* × *microglossum*

micropetalus
my-kro-PET-uh-lus
micropetala, micropetalum
With small petals, as in *Cuphea micropetala*

microphyllus
my-kro-FIL-us
microphylla, microphyllum
With small leaves, as in *Sophora microphylla*

micropterus
my-krop-TER-us
microptera, micropterum
With small wings, as in *Promenaea microptera*

Promenaea microptera

microsepalus

my-kro-SEP-a-lus

microsepala, microsepalum

With small sepals, as in *Pentadenia microsepala*

Mikania

my-KAY-nee-uh

Named after Josef Gottfried Mikan (1743–1814), Czech botanist (*Asteraceae*)

miliaceus

mil-ee-AY-see-us

miliacea, miliaceum

Relating to millet, as in *Panicum miliaceum*

militaris

mil-ih-TAH-ris

militaris, militare

Relating to soldiers; like a soldier, as in *Orchis militaris*

Milium

MIL-ee-um

From Greek *meline*, meaning "millet" (*Poaceae*)

millefoliatus

mil-le-foh-lee-AH-tus

millefoliata, millefoliatum

—

millefolius

mil-le-FOH-lee-us

millefolia, millefolium

With many leaves (literally a thousand leaves), as in *Achillea millefolium*

Millettia

mil-ETT-ee-uh

Named after Charles Millett (1792–1873) of the East India Company (*Fabaceae*)

Miltonia

mil-TOH-nee-uh

Named after Charles William Wentworth-Fitzwilliam (1786–1857), formerly Viscount Milton, British politician (*Orchidaceae*)

Miltoniopsis

mil-toh-nee-OP-sis

Resembling the related genus *Miltonia* (*Orchidaceae*)

Mimosa [1]

mim-OH-suh

From Greek *mimos*, an "imitator," because the sensitive plant (*M. pudica*) appears to mimic the movements of animals (*Fabaceae*)

mimosoides

mim-yoo-SOY-deez

Resembling *Mimosa*, as in *Caesalpinia mimosoides*

Mimulus

MIM-ew-lus

From Greek *mimos*, an "imitator," or *mimo*, a "monkey," because the flowers of some species (especially *M. guttatus*) resemble a face (*Phrymaceae*)

miniatus

min-ee-AH-tus

miniata, miniatum

Cinnabar-red, as in *Clivia miniata*

minimus

MIN-eh-mus

minima, minimum

Smallest, as in *Myosurus minimus*

minor

MY-nor

minor, minus

Smaller, as in *Vinca minor*

Minuartia

min-ew-ART-ee-uh

Named after Joan Minuart i Parets (1693–1768), Spanish/Catalan botanist and pharmacist (*Caryophyllaceae*)

minutiflorus

min-yoo-tih-FLOR-us

minutiflora, minutiflorum

With minute flowers, as in *Narcissus minutiflorus*

1

Mimosa pudica

minutifolius

min-yoo-tih-FOH-lee-us

minutifolia, minutifolium

With minute leaves, as in *Rosa minutifolia*

minutissimus

min-yoo-TEE-sih-mus

minutissima, minutissimum

The most minute, as in *Primula minutissima*

minutus

min-YOO-tus

minuta, minutum

Especially small, as in *Tagetes minuta*

Mirabilis

mi-RAB-il-is

From Latin *mirabilis*, meaning "wonderful" (*Nyctaginaceae*)

Miscanthus

mis-KAN-thus

From Greek *miskos*, meaning "stalk," and *anthos*, meaning "flower," because the spikelets are stalked (*Poaceae*)

missouriensis

miss-oor-ee-EN-sis

missouriensis, missouriense

From Missouri, as in *Iris missouriensis*

Mitchella

mi-CHEL-uh

Named after John Mitchell (1711–1768), American botanist and cartographer (*Rubiaceae*)

Mitella

my-TEL-uh

From Greek *mitra*, meaning "cap," and the diminutive *ella*, because the fruit resembles a small hat (*Saxifragaceae*)

mitis

MIT-is

mitis, mite

Mild; gentle; without spines, as in *Caryota mitis*

Mitraria

my-TRAIR-ee-uh

From Greek *mitra*, meaning "cap," in reference to the hat-shaped bracts enclosing the calyx (*Gesneriaceae*)

mitratus

my-TRAH-tus

mitrata, mitratum

With a turban or miter, as in *Mitrophyllum mitratum*

mitriformis

my-tri-FOR-mis

mitriformis, mitriforme

Like a cap, as in *Aloe mitriformis*

2

Hamamelis mollis

mixtus

MIKS-tus

mixta, mixtum

Mixed, as in *Potentilla × mixta*

modestus

mo-DES-tus

modesta, modestum

Modest, as in *Aglaonema modestum*

moesiacus

mee-shee-AH-kus

moesiaca, moesiacum

Connected with Moesia, the Balkans, as in *Campanula moesiaca*

moldavicus

mol-DAV-ih-kus

moldavica, moldavicum

From Moldavia, eastern Europe, as in *Dracocephalum moldavica*

Molinia

muh-LIN-ee-uh

Named after Juan Ignacio Molina (1740–1829), Chilean cleric and botanist (*Poaceae*)

mollis [2]

MO-lis

mollis, molle

Soft; with soft hairs, as in *Alchemilla mollis*

mollissimus

maw-LISS-ih-mus

mollissima, mollissimum

Very soft, as in *Passiflora mollissima*

Moltkia

MOLT-kee-uh

Named after Joachim Gadske Moltke
(1746–1818), Danish politician
(*Boraginaceae*)

moluccanus

mol-oo-KAH-nus

moluccana, moluccanum

Connected with the Moluccas or Spice
Islands, Indonesia, as in *Pittosporum
moluccanum*

Moluccella

mol-oo-KEL-uh

Named after the Molucca Islands of
Indonesia, in the mistaken belief that this
was their point of origin; actually distributed
from Mediterranean to northwest India
(*Lamiaceae*)

monacanthus

mon-ah-KAN-thus

monacantha, monacanthum

With one spine, as in *Rhipsalis monacantha*

monadelphus

mon-ah-DEL-fus

monadelpha, monadelphum

With filaments united, as in *Dianthus
monadelphus*

monandrus

mon-AN-drus

monandra, monandrum

With one stamen, as in *Bauhinia monandra*

Monanthes

mon-ANTH-eez

From Greek *monos*, meaning "single," and
anthos, meaning "flower," because unlike
related genera, flowers are typically solitary,
not in inflorescences (*Crassulaceae*)

Monarda [1]

mon-AR-duh

Named after Nicolás Bautista Monardes
(1493–1588), Spanish physician and
botanist (*Lamiaceae*)

Monardella

mon-ar-DEL-uh

Resembling the related genus *Monarda*,
although smaller; note the diminutive *ella*
(*Lamiaceae*)

monensis

mon-EN-sis

monensis, monense

From Mona, either the Isle of Man or
Anglesey, as in *Coincya monensis*

mongolicus

mon-GOL-ih-kus

mongolica, mongolicum

Connected with Mongolia, as in *Quercus
mongolica*

moniliferus

mon-ih-LIH-fer-us

monilifera, moniliferum

With a necklace, as in *Chrysanthemoides
monilifera*

moniliformis

mon-il-lee-FOR-mis

moniliformis, moniliforme

Like a necklace; with structures resembling
strings of beads, as in *Melpomene
moniliformis*

mono-

Used in compound words to denote single

monogynus

mon-OH-gy-nus

monogyna, monogynum

With one pistil, as in *Crataegus monogyna*

monopetalus

mon-OH-PET-uh-lus

monopetala, monopetalum

With a single petal, as in *Limoniastrum
monopetalum*

monophyllus

mon-oh-FIL-us

monophylla, monophyllum

With one leaf, as in *Pinus monophylla*

monopyrenus

mon-OH-py-ree-nus

monopyrena, monopyrenum

With a single stone, as in *Cotoneaster
monopyrenus*

monostachyus

mon-oh-STAK-ee-us

monostachya, monostachyum

With one spike, as in *Guzmania
monostachya*

1

Monarda didyma

Clematis montana

monspessulanus

monz-pess-yoo-LAH-nus

monspessulana, monspessulanum

Connected with Montpellier, France, as in *Acer monspessulanum*

Monstera

mon-STEER-uh

From Latin *monstrosus*, meaning "abnormal," because the leaves of many species are perforated (*Araceae*)

monstrosus

mon-STROH-sus

monstrosa, monstrosum

Abnormal, as in *Gypsophila* × *monstrosa*

montanus [2]

MON-tah-nus

montana, montanum

Relating to mountains, as in *Clematis montana*

montensis

mont-EN-sis

montensis, montense

—

monticola

mon-TIH-koh-luh

Growing on mountains, as in *Halesia monticola*

Montia

MON-tee-uh

Named after Giuseppe Monti (1682–1760), Italian botanist and chemist (*Montiaceae*)

montigenus

mon-TEE-gen-us

montigena, montigenum

Born of the mountains, as in *Picea montigena*

Moraea

MORE-ee-uh

Originally named *Morea* after Robert More (1703–80), British botanist, Linnaeus had a change of heart and rechristened it *Moraea*, after Sara Elisabeth Moraea (1716–1806), his wife (*Iridaceae*)

Morella

mo-REL-uh

Resembling the mulberry, *Morus*, but smaller; note the diminutive *ella* (*Myricaceae*)

morifolius

mor-ee-FOH-lee-us

morifolia, morifolium

With leaves like the mulberry (*Morus*), as in *Passiflora morifolia*

Morina

mo-REE-nuh

Named after Louis Morin de Saint-Victor (1635–1715), French botanist and physician (*Caprifoliaceae*)

Morus

MO-russ

From Latin *morum*, the vernacular name for *M. nigra* (*Moraceae*)

moschatus

MOSS-kuh-tus

moschata, moschatum

Musky, as in *Malva moschata*

mucosus

moo-KOZ-us

mucosa, mucosum

Slimy, as in *Rollinia mucosa*

mucronatus

muh-kron-AH-tus

mucronata, mucronatum

With a point, as in *Gaultheria mucronata*

mucronulatus

muh-kron-yoo-LAH-tus

mucronulata, mucronulatum

With a sharp, hard point, as in *Rhododendron mucronulatum*

Muehlenbeckia

moo-lun-BEK-ee-uh

Named after Henri Gustav Mühlenbeck (1798–1845), French physician and botanist (*Polygonaceae*)

Muhlenbergia

moo-lun-BERG-ee-uh

Named after Gotthilf Heinrich Ernst Muhlenberg (1753–1815), American cleric and botanist (*Poaceae*)

Mukdenia

muk-DEEN-ee-uh

Named for Mukden (now Shenyang), the capital of China's Liaoning Province (*Saxifragaceae*)

multi-

Used in compound words to denote many

multibracteatus
mul-tee-brak-tee-AH-tus
multibracteata, multibracteatum
With many bracts, as in *Rosa multibracteata*

multicaulis
mul-tee-KAW-lis
multicaulis, multicaule
With many stems, as in *Salvia multicaulis*

multiceps
MUL-tee-seps
With many heads, as in *Gaillardia multiceps*

multicolor
mul-tee-kol-or
Multicolor, as in *Echeveria multicolor*

multicostatus
mul-tee-koh-STAH-tus
multicostata, multicostatum
With many ribs, as in *Echinofossulocactus multicostatus*

multifidus
mul-TIF-id-us
multifida, multifidum
With many divisions, usually of leaves with many tears, as in *Helleborus multifidus*

multiflorus
mul-tee-FLOR-us
multiflora, multiforum
With many flowers, as in *Cytisus multiflorus*

multilineatus
mul-tee-lin-ee-AH-tus
multilineata, multilineatum
With many lines, as in *Hakea multilineata*

multinervis
mul-tee-NER-vis
multinervis, multinerve
With many nerves, as in *Quercus multinervis*

multiplex
MUL-tih-pleks
With many folds, as in *Bambusa multiplex*

multiradiatus
mul-tee-rad-ee-AH-tus
multiradiata, multiradiatum
With many rays, as in *Pelargonium multiradiatum*

multisectus
mul-tee-SEK-tus
multisecta, multisectum
With many cuts, as in *Geranium multisectum*

mundulus
mun-DYOO-lus
mundula, mundulum
Trim; neat, as in *Gaultheria mundula*

muralis [1]
mur-AH-lis
muralis, murale
Growing on walls, as in *Cymbalaria muralis*

muricatus
mur-ee-KAH-tus
muricata, muricatum
With rough and hard points, as in *Solanum muricatum*

Murraya
MUR-ee-uh
Named after Johan Andreas Murray (1740–91), Swedish physician and botanist (*Rutaceae*)

Musa
MOO-suh
From Arabic *mouz*, the vernacular name for banana (*Musaceae*)

musaicus
muh-ZAY-ih-kus
musaica, musaicum
Like a mosaic, as in *Guzmania musaica*

Muscari
mus-KAR-ee
From Greek *moschos*, meaning "musk," referring to the scent of the flowers, via the Turkish name (*Asparagaceae*)

muscipula
musk-IP-yoo-luh
Catches flies, as in *Dionaea muscipula*

muscivorus
mus-SEE-ver-us
muscivora, muscivorum
Appearing to eat flies, as in *Helicodiceros muscivorus*

muscoides
mus-COY-deez
Resembling moss, as in *Saxifraga muscoides*

muscosus
mus-KOH-sus
muscosa, muscosum
Like moss, as in *Selaginella muscosa*

Musella
moo-ZEL-uh
Resembling the related genus *Musa* but smaller; note the diminutive *ella* (*Musaceae*)

Mussaenda
moos-ay-EN-duh
From Sinhalese (Sri Lanka) *mussenda*, the vernacular name (*Rubiaceae*)

mutabilis [2]
mew-TAH-bih-lis
mutabilis, mutabile
Changeable, particularly relating to color, as in *Hibiscus mutabilis*

mutatus
mew-TAH-tus
mutata, mutatum
Changed, as in *Saxifraga mutata*

muticus
MEW-tih-kus
mutica, muticum
Blunt, as in *Pycnanthemum muticum*

mutilatus
mew-til-AH-tus
mutilata, mutilatum
Divided as though by tearing, as in *Peperomia mutilata*

Mutisia
mew-TIZ-ee-uh
Named after José Celestino Mutis (1732–1808), Spanish cleric, botanist and mathematician (*Asteraceae*)

Myoporum
my-oh-POR-um
From Greek *myein*, meaning "to close," and *poros*, meaning "pore," a reference to the arid habit of many species, where leaf pores are closed to save water (*Scrophulariaceae*)

Myosotis [3]
my-oh-SOH-tis
From Greek *mus*, meaning "mouse," and *otis*, meaning "ear," to which the leaves are said to resemble (*Boraginaceae*)

myri-
Used in compound words to denote very many

Cymbalaria muralis

Hibiscus mutabilis

Myosotis scorpioides

Myrica
MI-ri-kuh
From Greek *myrike*, the vernacular name for tamarisk, *Tamarix* (*Myricaceae*)

myriacanthus
mir-ee-uh-KAN-thus
myriacantha, myriacanthum
With many thorns, as in *Aloe myriacantha*

myriocarpus
mir-ee-oh-KAR-pus
myriocarpa, myriocarpum
With many fruits, as in *Schefflera myriocarpa*

Myriophyllum
mi-ree-oh-FIL-um
From Greek *myrios*, meaning "many," and *phyllon*, meaning "leaf," because the finely divided foliage appears plentiful (*Haloragaceae*)

myriophyllus
mir-ee-oh-FIL-us
myriophylla, myriophyllum
With many leaves, as in *Acaena myriophylla*

myriostigma
mir-ee-oh-STIG-muh
With many spots, as in *Astrophytum myriostigma*

Myrmecodia
mir-meh-KOH-dee-um
From Greek *myrmekodes*, meaning "full of ants," because the swollen stem provides nesting space for ants (*Rubiaceae*)

myrmecophilus
mir-meh-koh-FIL-us
myrmecophila, myrmecophilum
Ant-loving, as in *Aeschynanthus myrmecophilus*

Myrrhis
MUR-iss
From Greek *murrha*, meaning "myrrh-scented"; also the name of this plant and the resin extracted from trees of the genus *Commiphora*, the true myrrh (*Apiaceae*)

Myrsine
mir-SY-nee
From Greek *myrsine*, meaning "myrtle" (*Primulaceae*)

myrsinifolius
mir-sin-ee-FOH-lee-us
myrsinifolia, myrsinifolium
With leaves like *Myrsine*, often referring to myrtle for which this is an ancient Greek name, as in *Salix myrsinifolia*

myrsinites
mir-SIN-ih-teez
—

myrsinoides
mir-SIN-OY-deez
Resembling *Myrsine*, as in *Gaultheria myrsinoides*

myrtifolius
mir-tih-FOH-lee-us
myrtifolia, myrtifolium
With leaves like *Myrsine*, as in *Leptospermum myrtifolium*

Myrtillocactus
mir-til-oh-KAK-tus
A cactus whose fruit has been likened to those of true myrtle (*Myrtus communis*) or bilberry (*Vaccinium myrtillus*), (*Cactaceae*)

Myrtus
MIR-tus
From Greek *myrtos*, the vernacular name for *M. communis* (*Myrtaceae*)

Nandina
nan-DEEN-nuh
From Japanese *nanten*, the vernacular name for *N. domestica* (*Berberidaceae*)

nanellus
nan-EL-lus
nanella, nanellum
Very dwarf, as in *Lathyrus odoratus* var. *nanellus*

nankingensis
nan-king-EN-sis
nankingensis, nankingense
From Nanking (Nanjing), China, as in *Chrysanthemum nankingense*

nanus
NAH-nus
nana, nanum
Dwarf, as in *Betula nana*

napaulensis
nap-awl-EN-sis
napaulensis, napaulense
From Nepal, as in *Meconopsis napaulensis*

napellus
nap-ELL-us
napella, napellum
Like a little turnip, referring to the roots, as in *Aconitum napellus*

napifolius
nap-ih-FOH-lee-us
napifolia, napifolium
With leaves shaped like a turnip (*Brassica rapa*), that is,. a flattened sphere, as in *Salvia napifolia*

narbonensis
nar-bone-EN-sis
narbonensis, narbonense
From Narbonne, France, as in *Linum narbonense*

narcissiflorus
nar-sis-si-FLOR-us
narcissiflora, narcissiflorum
With flowers like daffodil (*Narcissus*), as in *Iris narcissiflora*

GIFT OF THE GODS

Greek and Roman mythology is alive with tales of heroes and monsters, gods, and goddesses. Their importance to these ancient cultures is reflected in the many genera that are named after mythological characters or places, including *Centaurea*, *Dryas*, *Heracleum*, *Hyacinthus*, and *Pieris*. The tale of Narcissus is one of the best known in Greek mythology, spawning the term narcissism, but it is unclear whether the flower that sprang up after his death was a daffodil. Some sources suggest that the flower gets its name from Greek *narkao*, meaning "I grow numb" (also the origin of the word "narcotic"), alluding to the intoxicating fragrance.

Narcissus
nar-SIS-us
Named after Narcissus of Greek mythology, a hunter who saw his reflection in a forest pool and fell in love with it, but upon realizing that the love could not be reciprocated, melted away and turned into a flower (*Amaryllidaceae*)

Nassella
nas-EL-uh
From Latin *nassa*, meaning "basket with narrow neck," a reference to the overlapping lemmas (*Poaceae*)

Nasturtium
nuh-STUR-shum
From Latin *nasus*, meaning "nose," and *tortus*, meaning "twisted," alluding to the strong scent of the foliage; not to be confused with *Tropaeolum* (*Brassicaceae*)

natalensis
nuh-tal-EN-sis
natalensis, natalense
From Natal, South Africa, as in *Tulbaghia natalensis*

natans
NAT-anz
Floating, as in *Trapa natans*

nauseosus
naw-see-OH-sus
nauseosa, nauseosum
Causing nausea, as in *Chrysothamnus nauseosus*

navicularis
nav-ik-yoo-LAH-ris
navicularis, naviculare
Shaped like a boat, as in *Callisia navicularis*

neapolitanus
nee-uh-pol-ih-TAH-nus
neapolitana, neapolitanum
Connected with Naples, Italy, as in *Allium neapolitanum*

nebulosus
neb-yoo-LOH-sus
nebulosa, nebulosum
Like a cloud, as in *Aglaonema nebulosum*

Nectaroscordum
nek-tuh-roh-SKOR-dum
From Greek *nektar*, meaning "drink of the gods," and *skordion*, meaning "garlic," a reference to the heavy nectar production in this garlic relative (*Amaryllidaceae*)

neglectus
nay-GLEK-tus
neglecta, neglectum
Previously neglected, as in *Muscari neglectum*

Neillia
NEE-lee-uh
Named after Patrick Neill (1776–1851), Scottish botanist (*Rosaceae*)

nelumbifolius
nel-um-bee-FOH-lee-us
nelumbifolia, nelumbifolium
With leaves like lotus (*Nelumbo*), as in *Ligularia nelumbifolia*

i

Nelumbo [1]

ne-LUM-boh

From Sinhalese (Sri Lanka) *nelum*, the vernacular name for lotus flower, from *N. nucifera* (*Nelumbonaceae*)

Nematanthus

nem-uh-TAN-thus

From Greek *nema*, meaning "thread," and *anthos*, meaning "flower," because the blooms of some species hang from threadlike pedicels (*Gesneriaceae*)

Nemesia

nuh-ME-shuh

From Greek *nemesion*, the vernacular name for a similar plant (*Scrophulariaceae*)

Nemophila

nem-OF-il-uh

From Greek *nemos*, meaning "woodland glade," and *phileo*, meaning "to love," a reference to the habitat of some species (*Boraginaceae*)

nemoralis

nem-or-RAH-lis

nemoralis, nemorale

—

nemorosus

nem-or-OH-sus

nemorosa, nemorosum

Of woodland, as in *Anemone nemorosa*

Neolitsea

ne-oh-LIT-see-uh

From Greek *neos*, meaning "new," plus the related genus *Litsea* (*Lauraceae*)

Neolloydia

ne-oh-LOY-dee-uh

Named after Francis Ernest Lloyd (1868–1947), American botanist, plus Greek *neos*, meaning "new," to differentiate the genus from *Lloydia* (*Cactaceae*)

Neomarica

ne-oh-MAR-i-kuh

From Greek *neos*, meaning "new," plus the related genus *Marica*, an illegitimate name; species of *Marica* are now distributed among nine other genera, including this one (*Iridaceae*)

Neoregelia

ne-oh-reg-EE-lee-uh

Named after Eduard August von Regel (1815–92), German botanist, plus Greek *neos*, meaning "new," to differentiate the genus from *Regelia* (*Bromeliaceae*)

nepalensis

nep-al-EN-sis

nepalensis, nepalense

—

nepaulensis

nep-al-EN-sis

nepaulensis, nepaulense

From Nepal, as in *Hedera nepalensis*

Nepenthes

nuh-PEN-theez

From Greek *ne*, meaning "not," and *penthes*, meaning "mourning"; in Greek mythology, the potion *nepenthes pharmakon* was given to Helen of Troy to dispel her grief (*Nepenthaceae*)

Nepeta

ne-PEE-tuh

From Latin *nepeta*, the vernacular name for an aromatic plant, possibly derived from Nepete (now Nepi), a city in central Italy (*Lamiaceae*)

nepetoides

nep-et-OY-deez

Resembling catmint (*Nepeta*), as in *Agastache nepetoides*

Nephrolepis

nef-roh-LEE-pis

From Greek *nephros*, meaning "kidney," and *lepis*, meaning "scale," alluding to the shape of the membrane over the sporangia (*Polypodiaceae*)

neriifolius

ner-ih-FOH-lee-us

neriifolia, neriifolium

With leaves like oleander (*Nerium*), as in *Podocarpus neriifolius*

Nerine [2]

nuh-RY-nee

Named after a sea nymph from Greek mythology (*Amaryllidaceae*)

Nerium

NEER-ee-um

From Greek *nerion*, the vernacular name for *N. oleander* (*Apocynaceae*)

Nelumbo nucifera

Nertera

NER-tuh-ruh

From Greek *nerteros*, meaning "low down," a reference to the dainty, creeping habit (*Rubiaceae*)

nervis

NERV-is

nervis, nerve

—

nervosus

ner-VOH-sus

nervosa, nervosum

With visible nerves, as in *Astelia nervosa*

nicaeensis

ny-see-EN-sis

nicaeensis, nicaeense

From Nice, France, as in *Acis nicaeensis*

Nicandra [3]

NY-kan-druh

Named after Nicander, a Greek poet from Colophon in modern-day Turkey (*Solanaceae*)

Nicotiana

ni-koh-tee-AH-nuh

Named after Jean Nicot (1530–1604), French diplomat who introduced tobacco to France (*Solanaceae*)

nictitans

NIC-tih-tanz

Blinking; moving, as in *Chamaecrista nictitans*

Nerine humilis

Nicandra physalodes

Galanthus nivalis

Nidularium

ny-dew-LAIR-ee-um

From Greek *nidus*, meaning "nest," and *aria*, meaning "pertaining to"; the leaves are arranged in a nestlike rosette (*Bromeliaceae*)

nidus

NID-us

Like a nest, as in *Asplenium nidus*

Nierembergia

near-um-BERG-ee-uh

Named after Juan Eusebio Nieremberg (1595–1658), Spanish cleric (*Solanaceae*)

Nigella

ny-JEL-uh

From Latin *niger*, meaning "black," and diminutive *ella*, referring to the small black seed (*Ranunculaceae*)

niger

NY-ger

nigra, nigrum

Black, as in *Phyllostachys nigra*

nigratus

ny-GRAH-tus

nigrata, nigratum

Blackened, blackish, as in *Oncidium nigratum*

nigrescens

ny-GRESS-enz

Turning black, as in *Silene nigrescens*

nigricans

ny-GRIH-kanz

Blackish, as in *Salix nigricans*

nikoensis

nik-o-EN-sis

nikoensis, nikoense

From Nike, Japan, as in *Adenophora nikoensis*

niloticus

nil-OH-tih-kus

nilotica, niloticum

Connected with the Nile Valley, as in *Salvia nilotica*

Nipponanthemum

nip-on-ANTH-uh-mum

From Japanese *Nippon*, meaning "Japan," plus Greek *anthemon*, meaning "flower" (*Asteraceae*)

nipponicus

nip-PON-ih-kus

nipponica, nipponicum

Connected with Japan (Nippon), as in *Phyllodoce nipponica*

nitens

NI-tenz

—

nitidus

NI-ti-dus

nitida, nitidum

Shining, as in *Lonicera nitida*

nivalis [4]

niv-VAH-lis

nivalis, nivale

—

niveus

NIV-ee-us

nivea, niveum

—

nivosus

niv-OH-sus

nivosa, nivosum

As white as snow, or growing near snow, as in *Galanthus nivalis*

nobilis

NO-bil-is

nobilis, nobile

Noble; renowned, as in *Laurus nobilis*

noctiflorus

nok-tee-FLOR-us

noctiflora, noctiflorum

—

nocturnus

NOK-ter-nus

nocturna, nocturnum

Flowering at night, as in *Silene noctiflora*

nodiflorus

no-dee-FLOR-us

nodiflora, nodiflorum

Flowering at the nodes, as in *Eleutherococcus nodiflorus*

A B C D E F G H I J K L M N O P Q R S T U V W X Y Z i

1

Nolana spathulata

2

Hyacinthoides non-scripta

3

Symphyotrichum novi-belgii

nodosus
nod-OH-sus
nodosa, nodosum
With conspicuous joints or nodes, as in *Geranium nodosum*

nodulosus
no-du-LOH-sus
nodulosa, nodulosum
With small nodes, as in *Echeveria nodulosa*

Nolana [1]
noh-LAH-nuh
From Latin *nola*, meaning "small bell," like the flowers (*Solanaceae*)

noli-tangere
NO-lee TAN-ger-ee
"Touch not" (because the seed pods burst), as in *Impatiens noli-tangere*

Nomocharis
noh-MOK-ur-is
From Greek *nomos*, meaning "meadow," and *charis*, meaning "grace," graceful plants from meadows (*Liliaceae*)

non-scriptus [2]
non-SKRIP-tus
non-scripta, non-scriptum
Without any markings, as in *Hyacinthoides non-scripta*

norvegicus
nor-VEG-ih-kus
norvegica, norvegicum
Connected with Norway, as in *Arenaria norvegica*

notatus
no-TAH-tus
notata, notatum
With spots or marks, as in *Glyceria notata*

Nothofagus
noh-toh-FAY-gus
From Greek *nothos*, meaning "false," and *Fagus*, another genus (*Nothofagaceae*)

Notholirion
noh-toh-LIR-ee-on
From Greek *nothos*, meaning "false," and *lirion*, meaning "lily" (*Liliaceae*)

Notholithocarpus
noh-tho-lith-oh-KAR-pus
From Greek *nothos*, meaning "false," and *Lithocarpus*, a related genus (*Fagaceae*)

Nothoscordum
noh-toh-SKOR-dum
From Greek *nothos*, meaning "false," and *skordion*, meaning "garlic" (*Amaryllidaceae*)

novae-angliae
NO-vay ANG-lee-a
Connected with New England, as in *Aster novae-angliae*

novae-zelandiae
NO-vay zee-LAN-dee-ay
Connected with New Zealand, as in *Acaena novae-zelandiae*

novi-
Used in compound words to denote new

novi-belgii [3]
NO-vee BEL-jee-eye
Connected with New York, as in *Aster novi-belgii*

nubicola
new-BIH-koh-luh
Growing up in the clouds, as in *Salvia nubicola*

nubigenus
new-bee-GEE-nus
nubigena, nubigenum
Born up in the clouds, as in *Kniphofia nubigena*

nucifer
NEW-siff-er
nucifera, nuciferum
Producing nuts, as in *Cocos nucifera*

nudatus

new-DAH-tus

nudata, nudatum

–

nudus

NEW-dus

nuda, nudum

Bare; naked, as in *Nepeta nuda*

nudicaulis

new-dee-KAW-lis

nudicaulis, nudicaule

With bare stems, as in *Papaver nudicaule*

nudiflorus

new-dee-FLOR-us

nudiflora, nudiflorum

With flowers that appear before the leaves, as in *Jasminum nudiflorum*

numidicus

nu-MID-ih-kus

numidica, numidicum

Connected with Algeria, as in *Abies numidica*

nummularius

num-ew-LAH-ree-us

nummularia, nummularium

Like coins, as in *Lysimachia nummularia*

Nuphar

NEW-far

From Arabic or Persian *nenufar*, meaning water lily (*Nymphaeaceae*)

nutans

NUT-anz

Nodding, as in *Billbergia nutans*

nyctagineus

nyk-ta-JEE-nee-us

nyctaginea, nyctagineum

Flowering at night, as in *Mirabilis nyctaginea*

Nymphaea

nim-FAY-uh

From Greek *nymphaia*, meaning "water lily," derived from the nymphs of Greek mythology (*Nymphaeaceae*)

Nymphoides (also nymphoides)

nim-FOY-deez

Resembling the genus *Nymphaea* (*Menyanthaceae*)

Nyssa

NIS-uh

The classical Latin name for a water nymph, alluding to the aquatic habitat of *N. aquatica* (*Nyssaceae*)

Nymphaea

Water lilies are aquatic perennials with floating leaves and elaborate flowers. This description, although accurate, fails to convey the romance of these plants, a romance that inspired French Impressionist Claude Monet to paint them so frequently. They are important to biologists, too, as one of the first families of flowering plants to diverge. Their beautiful blooms reveal characteristics considered by some as primitive, such as having numerous petals (most flowers have three, four, or five petals) and stamens where the anther and filament can be difficult to tell apart. Observing these characteristics can help us understand how flowers first evolved.

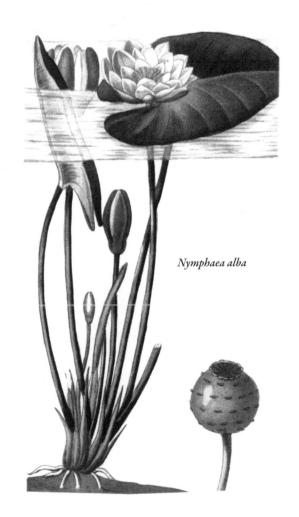

Nymphaea alba

obconicus
ob-KON-ih-kus
obconica, obconicum
In the shape of an inverted cone, as in
Primula obconica

obesus
oh-BEE-sus
obesa, obesum
Fat, as in *Euphorbia obesa*

oblatus
ob-LAH-tus
oblata, oblatum
With flattened ends, as in *Syringa oblata*

obliquus
oh-BLIK-wus
obliqua, obliquum
Lopsided, as in *Nothofagus obliqua*

oblongatus
ob-long-GAH-tus
oblongata, oblongatum
—

oblongus
ob-LONG-us
oblonga, oblongum
Oblong, as in *Passiflora oblongata*

oblongifolius
ob-long-ih-FOH-lee-us
oblongifolia, oblongifolium
With oblong leaves, as in *Asplenium
oblongifolium*

obovatus
ob-oh-VAH-tus
obovata, obovatum
In the shape of an inverted egg, as in *Paeonia
obovata*

Obregonia
oh-bre-GOH-nee-uh
Named after Álvaro Obregón Salido
(1880–1928), 39th President of Mexico
(*Cactaceae*)

obscurus
ob-SKEW-rus
obscura, obscurum
Not clear or certain, as in *Digitalis obscura*

obtectus
ob-TEK-tus
obtecta, obtectum
Covered; protected, as in *Cordyline obtecta*

obtusatus
ob-tew-SAH-tus
obtusata, obtusatum
Blunt, as in *Asplenium obtusatum*

obtusifolius
ob-too-sih-FOH-lee-us
obtusifolia, obtusifolium
With blunt leaves, as in *Peperomia obtusifolia*

obtusus
ob-TOO-sus
obtusa, obtusum
Blunt, as in *Chamaecyparis obtusa*

obvallatus
ob-val-LAH-tus
obvallata, obvallatum
Enclosed, within a wall, as in *Saussurea
obvallata*

occidentalis
ok-sih-den-TAH-lis
occidentalis, occidentale
Relating to the west, as in *Thuja occidentalis*

occultus
ock-ULL-tus
occulta, occultum
Hidden, as in *Huernia occulta*

ocellatus
ock-ell-AH-tus
ocellata, ocellatum
With an eye; with a spot surrounding
a smaller spot of a different color, as in
Convolvulus ocellatus

Ochagavia
ok-uh-GAH-vee-uh
Named after Sylvestris Ochagavia
(nineteenth century), Chilean education
minister (*Bromeliaceae*)

Ochna
OK-nuh
From Greek *ochne*, meaning "pear tree," in
reference to the shape of the fruit or possibly
the leaves (*Ochnaceae*)

ochraceus
oh-KRA-see-us
ochracea, ochraceum
An ocher color, as in *Hebe ochracea*

ochroleucus
ock-roh-LEW-kus
ochroleuca, ochroleucum
Yellowish white, as in *Crocus ochroleucus*

Ocimum
OS-i-mum
From Greek *okimom*, the vernacular name
for an aromatic plant (*Lamiaceae*)

oct-
Used in compound words to denote eight

octandrus
ock-TAN-drus
octandra, octandrum
With eight stamens, as in *Phytolacca
octandra*

octopetalus
ock-toh-PET-uh-lus
octopetala, octopetalum
With eight petals, as in *Dryas octopetala*

oculatus
ock-yoo-LAH-tus
oculata, oculatum
With an eye, as in *Haworthia oculata*

oculiroseus
ock-yoo-lee-ROH-sus
oculirosea, oculiroseum
With an eye in a rose color, as in *Hibiscus
palustris* f. *oculiroseus*

ocymoides
ok-kye-MOY-deez
Resembling basil (*Ocimum*), as in *Halimium
ocymoides*

Odontoglossum
oh-don-toh-GLOS-um
From Greek *odontos*, meaning "tooth," and
glossos, meaning "tongue," because the floral
lip has toothlike projections (*Orchidaceae*)

odoratissimus
oh-dor-uh-TISS-ih-mus
odoratissima, odoratissimum
With a particularly fragrant scent, as in
Viburnum odoratissimum

odoratus

oh-dor-AH-tus

odorata, odoratum

—

odoriferus

oh-dor-IH-fer-us

odorifera, odoriferum

—

odorus

oh-DOR-us

odora, odorum

With a fragrant scent, as in *Lathyrus odoratus*

Oemleria

oom-LEER-ee-uh

Named after Augustus Gottlieb Oemler (1773–1852), German naturalist (*Rosaceae*)

Oenanthe

ee-NAN-thee

From Greek *oinos*, meaning "wine," and *anthos*, meaning "flower," because the blooms have a vinous scent (*Apiaceae*)

Oenothera

ee-NOTH-ur-uh

Etymology uncertain, although perhaps from Greek *oinotheras*, a plant added to wine to induce sleep or improve the taste of wine; alternatively from Latin *onotheras*, a toxic plant given to animals so they can be caught (*Onagraceae*)

officinalis

oh-fiss-ih-NAH-lis

officinalis, officinale

Sold in stores, hence denoting a useful plant (vegetable, culinary, or medicinal herb), as in *Rosmarinus officinalis*

officinarum

off-ik-IN-ar-um

From a store, usually an apothecary, as in *Mandragora officinarum*

olbius

OL-bee-us

olbia, olbium

Connected with the Îles d'Hyères, France, as in *Lavatera olbia*

Olea [1]

OH-lee-uh

From classical Latin name for olive, *O. europaea* (*Oleaceae*)

Olea europaea

Olearia

oh-LEAR-ee-uh

Named after the German scholar Adam Ölschläger (ca. 1603–71)—Adam Olearius in Latin (*Asteraceae*)

oleiferus

oh-lee-IH-fer-us

oleifera, oleiferum

Producing oil, as in *Elaeis oleifera*

oleifolius

oh-lee-ih-FOH-lee-us

oleifolia, oleifolium

With leaves like olive (*Olea*), as in *Lithodora oleifolia*

oleoides

oh-lee-OY-deez

Resembling olive (*Olea*), as in *Daphne oleoides*

oleraceus

awl-lur-RAY-see-us

oleracea, oleraceum

Used as a vegetable, as in *Spinacia oleracea*

oliganthus

ol-ig-AN-thus

oligantha, oliganthum

With few flowers, as in *Ceanothus oliganthus*

oligocarpus

ol-ig-oh-KAR-pus

oligocarpa, oligocarpum

With few fruit, as in *Cayratia oligocarpa*

oligophyllus

ol-ig-oh-FIL-us

oligophylla, oligophyllum

With few leaves, as in *Senna oligophylla*

oligospermus

ol-ig-oh-SPERM-us

oligosperma, oligospermum

With few seed, as in *Draba oligosperma*

olitorius

ol-ih-TOR-ee-us

olitoria, olitorium

Relating to culinary herbs, as in *Corchorus olitorius*

olivaceus

oh-lee-VAY-see-us

olivacea, olivaceum

Olive; green-brown, as in *Lithops olivacea*

Olsynium

ol-SY-nee-um

From Greek *ol*, meaning "little," and *syn*, meaning "joined," because the stamens are only partly fused together (*Iridaceae*)

Oncidium leucochilum

olympicus
oh-LIM-pih-kus
olympica, olympicum
Connected with Mount Olympus, Greece, as in *Hypericum olympicum*

Omphalodes
om-fah-LOH-deez
From Greek *omphalos*, meaning "navel," because the fruit is dimpled like the human navel (*Boraginaceae*)

Omphalogramma
om-fah-loh-GRAM-uh
From Greek *omphalos*, meaning "navel," and *gramme*, meaning "line," perhaps an allusion to lines of dimples on the seed? (*Primulaceae*)

Oncidium [2]
on-SID-ee-um
From Greek *oncos*, meaning "swelling" or "tumor," referring to calluses on the floral lip (*Orchidaceae*)

Onoclea
on-OK-lee-uh
From Greek *onos*, meaning "vessel," and *kleiein*, meaning "to close," because the leaf edge is folded over to enclose the sporangia (*Onocleaceae*)

Ononis
o-NOH-nis
From Greek *onos*, meaning "donkey," and *oninemi*, meaning "useful," perhaps because only a donkey can safely eat from spiny *O. spinosa* (*Fabaceae*)

Onopordum
o-noh-POR-dum
From Greek *onos*, meaning "donkey," and *perdo*, meaning "to break wind," although the reasons for this name are obscure (*Asteraceae*)

Onosma
uh-NOZ-muh
From Greek *onos*, meaning "donkey," and *osme*, meaning "fragrance," either because some part of the plant smells or because donkeys consume it (*Boraginaceae*)

Onychium
on-IK-ee-um
From Greek *onychion*, meaning "little claw," referring to the shape of the leaflets (*Pteridaceae*)

opacus
oh-PAH-kus
opaca, opacum
Dark; dull; shaded, as in *Crataegus opaca*

operculatus
oh-per-koo-LAH-tus
operculata, operculatum
With a cover or lid, as in *Luffa operculata*

ophioglossifolius
oh-fee-oh-gloss-ih-FOH-lee-us
ophioglossifolia, ophioglossifolium
With leaves like adder's tongue fern (*Ophioglossum*), as in *Ranunculus ophioglossifolius*

Ophiopogon
oh-fee-oh-POH-gon
From Greek *ophis*, meaning "snake," and *pogon*, meaning "beard," a translation of the Japanese name "snake's beard," perhaps referring to the tufted habit (*Asparagaceae*)

Ophrys
OF-ris
From Greek *ophyras*, meaning "eyebrow," because several species have a fringe of hairs on the edge of the flower lip (*Orchidaceae*)

Ophthalmophyllum
of-thal-moh-FIL-um
From Greek *ophthalmos*, meaning "eye," and *phyllon*, meaning "leaf," because each leaf has a lenslike tip (*Aizoaceae*)

Oplismenus
op-lis-MEN-us
From Greek *oplisma*, meaning "armament," and *menos*, meaning "courage," because the flower spikelets have spearlike awns (*Poaceae*)

oppositifolius
op-po-sih-tih-FOH-lee-us
oppositifolia, oppositifolium
With leaves that grow opposite each other from the stem, as in *Chiastophyllum oppositifolium*

Opuntia
oh-POON-tee-uh
From Greek, the name for a spiny plant found near the ancient Greek city of Opus (*Cactaceae*)

Orbea
OR-bee-uh
From Latin *orbis*, meaning "ring," because the flowers have a prominent ringlike annulus in the center (*Apocynaceae*)

orbicularis
or-bik-yoo-LAH-ris
orbicularis, orbiculare
—

orbiculatus
or-bee-kul-AH-tus
orbiculata, orbiculatum
In the shape of a disk; flat and round, as in *Cotyledon orbiculata*

orchideus
or-KI-de-us
orchidea, orchideum
—

orchioides
or-ki-OY-deez
Like an orchid (*Orchis*), as in *Veronica orchidea*

orchidiflorus
or-kee-dee-FLOR-us
orchidiflora, orchidiflorum
With flowers like an orchid (*Orchis*), as in *Gladiolus orchidiflorus*

Orchis
OR-kis
From Greek *orkhis*, meaning "testicle," alluding to the shape of the tubers (*Orchidaceae*)

oreganus
or-reh-GAH-nus
oregana, oreganum
Connected with Oregon, as in *Sidalcea oregana*

Oreocereus
or-ee-oh-SER-ee-us
From Greek *oros*, meaning "mountain," plus *Cereus*, a related genus; these cacti hail from the Andes (*Cactaceae*)

oreophilus
or-ee-O-fil-us
oreophila, oreophilum
Loving mountains, as in *Sarracenia oreophila*

oresbius
or-ES-bee-us
oresbia, oresbium
Growing on mountains, as in *Castilleja oresbia*

orientalis
or-ee-en-TAH-lis
orientalis, orientale
Relating to Asia, also referred to as the Orient; eastern, as in *Thuja orientalis*

origanifolius
or-ih-gan-ih-FOH-lee-us
origanifolia, origanifolium
With leaves like marjoram (*Origanum*), as in *Chaenorhinum origanifolium*

origanoides
or-ig-an-OY-deez
Resembling marjoram (*Origanum*), as in *Dracocephalum origanoides*

Origanum
o-ri-GAH-num
From Greek *origanon*, an acrid herb; derived from *oros*, meaning "mountain," and *ganos*, meaning "beauty' (*Lamiaceae*)

Orixa
o-RIKS-uh
From Japanese vernacular name for *O. japonica* (*Rutaceae*)

Orlaya
or-LAY-uh
Named after Johann Orlay (ca. 1770–1827), Hungarian physician and botanist (*Apiaceae*)

ornans
OR-nanz
—

ornatus
or-NA-tus
ornata, ornatum
Ornamental; showy, as in *Musa ornata*

ornatissimus
or-nuh-TISS-ih-mus
ornatissima, ornatissimum
Especially showy, as in *Bulbophyllum ornatissimum*

Ornithogalum [1]
or-ni-THO-gal-um
From Greek *ornithos*, meaning "bird," and *gala*, meaning "milk," on account of the white flowers of some species, although Romans are said to have used the term "bird's milk" to refer to something wonderful or rare, as with "hen's teeth" (*Asparagaceae*)

ornithopodus
or-nith-OP-oh-dus
ornithopoda, ornithopodum
—

ornithopus
or-nith-OP-pus
Like a bird's foot, as in *Carex ornithopoda*

Orobanche
or-oh-BAN-kee
From Greek *orobos*, meaning a type of vetch, plus *ankho*, meaning "to strangle," referring to the parasitic habit (*Orobanchaceae*)

Orontium
o-RON-tee-um
From Greek name for a plant that grew by/in the Orontes River, which flows through Lebanon, Syria, and Turkey; this aquatic plant is native to the eastern United States (*Araceae*)

Orostachys
or-oh-STAK-is
From Greek *oros*, meaning "mountain," and *stachys*, meaning "ear of grain," because these are alpine plants with wheatlike flower spikes (*Crassulaceae*)

Oroya
uh-ROY-uh
Named after La Oroya, a city in central Peru, where the plant was first discovered (*Cactaceae*)

Ortegocactus
or-TAY-goh-kak-tus
Named after the Ortega family of San José Lachiguirí in Oaxaca, Mexico, who helped discover this cactus (*Cactaceae*)

ortho-
Used in compound words to denote straight or upright

1

Ornithogalum pyrenaicum

2

Oryza sativa

orthobotrys
or-THO-bot-ris
With upright clusters, as in *Berberis orthobotrys*

orthocarpus
or-tho-KAR-pus
orthocarpa, orthocarpum
With upright fruit, as in *Malus orthocarpa*

orthoglossus
or-tho-GLOSS-us
orthoglossa, orthoglossum
With a straight tongue, as in *Bulbophyllum orthoglossum*

Orthophytum
or-tho-FY-tum
From Greek *orthos*, meaning "straight," and *phyton*, meaning "plant," because the tall, straight inflorescences bear leaves (*Bromeliaceae*)

orthosepalus
or-tho-SEP-a-lus
orthosepala, orthosepalum
With straight sepals, as in *Rubus orthosepalus*

Oryza [2]
or-EYE-zuh
From Greek *oryza*, meaning "rice," both words deriving from Arabic *eruz* (*Poaceae*)

Osmanthus
oz-MAN-thus
From Greek *osme*, meaning "fragrant," and *anthos*, meaning "flower' (*Oleaceae*)

3

Osmunda regalis

Osmunda [3]
oz-MOON-duh
Many possible etymologies have been suggested for this Linnean name, which he did not explain. Osmunder was the Saxon name for the Norse god Thor, while Osmund the Waterman, also from Saxon mythology, hid his family in a clump of ferns; bog iron, an iron deposit found in marshland where these ferns grew, was reduced in a furnace called an osmund (*Osmundaceae*)

Osmundastrum
oz-moon-DAS-trum
Referring to *Osmunda*, a related genus (*Osmundaceae*)

Osteomeles
oss-tee-oh-ME-leez
From Greek *osteon*, meaning "bone," and *melon*, meaning "apple," because the seed within the applelike fruit has a hard outer coat (*Rosaceae*)

Osteospermum
oss-tee-oh-SPER-mum
From Greek *osteon*, meaning "bone," and *sperma*, meaning "seed," because the small, seedlike fruit is hard (*Asteraceae*)

Ostrowskia
oss-TROH-skee-uh
Named after Michael Nicolajewitsch von Ostrowski (dates unknown), Russian Imperial government minister (*Campanulaceae*)

Ostrya
oss-TRY-uh
From Greek *ostryos*, meaning "scale," because each seed is enclosed within a scalelike bract (*Betulaceae*)

Ostryopsis
oss-try-OP-sis
Resembling the related genus *Ostrya* (*Betulaceae*)

Othonna
oth-ON-uh
From Greek *othonne*, meaning "linen," because some species have soft hairs on their leaves (*Asteraceae*)

ovalis
oh-VAH-lis
ovalis, ovale
Oval, as in *Amelanchier ovalis*

ovatus
oh-VAH-tus
ovata, ovatum
Shaped like an egg; ovate, as in *Lagurus ovatus*

ovinus
oh-VIN-us
ovina, ovinum
Relating to sheep or sheep feed, as in *Festuca ovina*

Oxalis
oks-AH-lis
From Greek *oxys*, meaning "acid," referring to the sharp taste of the leaves caused by oxalic acid (*Oxalidaceae*)

oxyacanthus
oks-ee-a-KAN-thus
oxyacantha, oxyacanthum
With sharp spines, as in *Asparagus oxyacanthus*

Oxydendrum
oks-ee-DEN-drum
From Greek *oxys*, meaning "acid," and *dendron*, meaning "tree," because the leaves taste sharp when eaten (*Ericaceae*)

oxygonus
ok-SY-goh-nus
oxygona, oxygonum
With sharp angles, as in *Echinopsis oxygona*

oxyphilus
oks-ee-FIL-us
oxyphila, oxyphilum
Growing in acid soil, as in *Allium oxyphilum*

oxyphyllus
oks-ee-FIL-us
oxyphylla, oxyphyllum
With sharp, pointed leaves, as in *Euonymus oxyphyllus*

Ozothamnus
oh-zoh-THAM-nus
From Greek *ozo*, meaning "to smell," and *thamnus*, meaning "shrub," due to aromatic foliage (*Asteraceae*)

pachy-
Used in compound words to denote thick

pachycarpus
pak-ee-KAR-pus
pachycarpa, pachycarpum
With a thick pericarp, as in *Angelica pachycarpa*

Pachycereus
pak-ee-SER-ee-us
From Greek *pachys*, meaning "thick," plus *Cereus*, a related genus (*Cactaceae*)

Pachycormus
pak-ee-KOR-mus
From Greek *pachys*, meaning "thick," and *kormos*, meaning "stump," because this shrub has a thickened stem (*Anacardiaceae*)

Pachycymbium
pak-ee-SIM-bee-um
From Greek *pachys*, meaning "thick," and *kumbivon*, meaning "small cup," alluding to the flower shape (*Apocynaceae*)

Pachyphragma
pak-ee-FRAG-muh
From Greek *pachys*, meaning "thick," and *phragma*, meaning "fence," because the chambers within the fruit is divided by a thick wall (*Brassicaceae*)

pachyphyllus
pak-ih-FIL-us
pachyphylla, pachyphyllum
Thick-leaved, as in *Callistemon pachyphyllus*

Pachyphytum
pak-ee-FY-tum
From Greek *pachys*, meaning "thick," and *phyton*, meaning "plant," alluding to the thick, succulent leaves and stems (*Crassulaceae*)

Pachypodium [1]
pak-ee-POH-dee-um
From Greek *pachys*, meaning "thick," and *podion*, meaning "foot," because these succulents have swollen trunks (*Apocynaceae*)

pachypodus
pak-ih-POD-us
pachypoda, pachypodum
With a fat stem, as in *Actaea pachypoda*

pachypterus
pak-IP-ter-us
pachyptera, pachypterum
With thick wings, as in *Rhipsalis pachyptera*

Pachysandra
pak-ee-SAN-druh
From Greek *pachys*, meaning "thick," and *andros*, meaning "male," referencing the thick, white stamen filaments (*Buxaceae*)

pachysanthus
pak-ee-SAN-thus
pachysantha, pachysanthum
With thick flowers, as in *Rhododendron pachysanthum*

Pachystachys
pak-ee-STAK-iss
From Greek *pachys*, meaning "thick," and *stachys*, meaning "ear of grain," because the flowers are arranged in sturdy, upright inflorescences (*Acanthaceae*)

Pachystegia
pak-ee-STEE-jee-uh
From Greek *pachys*, meaning "thick," and *stegos*, meaning "cover," alluding to the thick, woolly coating on the stems and leaf undersides (*Asteraceae*)

pacificus
pa-SIF-ih-kus
pacifica, pacificum
Connected with the Pacific Ocean, as in *Chrysanthemum pacificum*

Packera
PAK-er-uh
Named after John G. Packer (1929-), Canadian botanist (*Asteraceae*)

1

Pachypodium succulentum

1

Paeonia suffruticosa

padus
PAD-us
Ancient Greek name for a kind of wild cherry, as in *Prunus padus*

Paeonia [1]
pee-OH-nee-uh
Named after Paeon, physician to the gods in Greek mythology, alluding to the medicinal properties of peony (*Paeoniaceae*)

paganus
PAG-ah-nus
pagana, paganum
From wild or country regions, as in *Rubus paganus*

palaestinus
pal-ess-TEEN-us
palaestina, palaestinum
Connected with Palestine, as in *Iris palaestina*

Paliurus
pal-ee-EW-rus
From Greek *paliouros*, the classical name for *P. spina-christi* or the related *Ziziphus spina-christi* (*Rhamnaceae*)

pallens
PAL-lenz
—

pallidus
PAL-lid-dus
pallida, pallidum
Pale, as in *Tradescantia pallida*

pallescens
pa-LESS-enz
Somewhat pale, as in *Sorbus pallescens*

pallidiflorus
pal-id-uh-FLOR-us
pallidiflora, pallidiflorum
With pale flowers, as in *Eucomis pallidiflora*

palmaris
pal-MAH-ris
palmaris, palmare
A hand's breadth wide, as in *Limonium palmare*

palmatus
pahl-MAH-tus
palmata, palmatum
Palmate, as in *Acer palmatum*

palmensis
pal-MEN-sis
palmensis, palmense
From Las Palmas, Canary Islands, as in *Aichryson palmense*

palmeri
PALM-er-ee
Named after Ernest Jesse Palmer (1875–1962), British explorer and plant collector in the United States, as in *Agave palmeri*

palmetto
pahl-MET-oh
A small palm, as in *Sabal palmetto*

palmifolius
palm-ih-FOH-lee-us
palmifolia, palmifolium
With palmlike leaves, as in *Sisyrinchium palmifolium*

paludosus
pal-oo-DOH-sus
paludosa, paludosum
—

palustris
pal-US-tris
palustris, palustre
Of marshland, as in *Quercus palustris*

Pamianthe
pam-ee-ANTH-ee
Named after Albert Samuel Pam (1875–1955), British soldier and banker (*Amaryllidaceae*)

Panax
PA-nax
From Greek *panakes*, meaning "all healing" or "panacea," for its medicinal properties (*Araliaceae*)

Pancratium
pan-KRAT-ee-um
From Greek *pan*, meaning "all," and *kratos*, meaning "strength," because many of these bulbous plants tolerate extreme drought (*Amaryllidaceae*)

pandanifolius
pan-dan-uh-FOH-lee-us
pandanifolia, pandanifolium
With leaves like *Pandanus*, as in *Eryngium pandanifolium*

Pandanus
pan-DAN-us
From Malay *pandan*, the name for the edible *P. amaryllifolius* (*Pandanaceae*)

Pandorea
pan-DOR-ee-uh
Named after Pandora of Greek mythology; the fruit is said to resemble her famous jar, often mistranslated as a box (*Apocynaceae*)

panduratus
pand-yoor-RAH-tus
pandurata, panduratum
Shaped like a fiddle, as in *Coelogyne pandurata*

paniculatus
pan-ick-yoo-LAH-tus
paniculata, paniculatum
With flowers arranged in panicles, as in *Koelreuteria paniculata*

Panicum
PAN-ik-um
From Latin *panicum*, the classical name for millet (*Poaceae*)

pannonicus
pa-NO-nih-kus
pannonica, pannonicum
Of Pannonia, a Roman region, as in *Lathyrus pannonicus*

pannosus
pan-OH-sus
pannosa, pannosum
Tattered, as in *Helianthemum pannosum*

Papaver [2]
puh-PAH-ver
From Latin *papaver*, the classical name for poppy (*Papaveraceae*)

Paphiopedilum
paf-ee-oh-PED-uh-lum
Named after the city of Paphos in Cyprus, the site of the temple of Aphrodite, plus the Greek *pedilon*, meaning "slipper"—see *Cypripedium* (*Orchidaceae*)

papilio
pap-ILL-ee-oh
A butterfly, as in *Hippeastrum papilio*

papilionaceus
pap-il-ee-on-uh-SEE-us
papilionacea, papilionaceum
Like a butterfly, as in *Pelargonium papilionaceum*

papyraceus
pap-ih-REE-see-us
papyracea, papyraceum
Like paper, as in *Narcissus papyraceus*

papyrifer
pap-IH-riff-er
—

papyriferus
pap-ih-RIH-fer-us
papyrifera, papyriferum
Producing paper, as in *Tetrapanax papyrifer*

papyrus
pa-PY-rus
Ancient Greek word for paper, as in *Cyperus papyrus*

Paradisea
par-uh-DIZ-ee-uh
Named after Giovanni Paradisi (1760–1820), Italian botanist and politician (*Asparagaceae*)

paradisi
par-uh-DEE-see
—

paradisiacus
par-uh-DEE-see-cus
paradisiaca, paradisiacum
From a park or garden, as in *Citrus × paradisi*

paradoxus
par-uh-DOKS-us
paradoxa, paradoxum
Unexpected or paradoxical, as in *Acacia paradoxa*

paraguayensis
par-uh-gway-EN-sis
paraguayensis, paraguayense
From Paraguay, as in *Ilex paraguayensis*

Paraquilegia
par-ak-wi-LEE-juh
From Greek *para*, meaning "near," plus *Aquilegia*, a related genus (*Ranunculaceae*)

Paraserianthes
par-uh-ser-ee-ANTH-eez
From Greek *para*, meaning "near," plus *Serianthes*, a related genus (*Fabaceae*)

parasiticus
par-uh-SIT-ih-kus
parasitica, parasiticum
Parasitic, as in *Agalmyla parasitica*

2

Papaver somniferum

Parthenocissus

As its name suggests, Virginia creeper (*Parthenocissus quinquefolia*) is native to Virginia and much of eastern North America. The genus name originates in the French common name *vigne-vierge*, or "virgin vine," which was translated into Latinized Greek as *Parthenocissus* (or "virgin ivy"). The virgin is, of course, the Virgin Queen, Elizabeth I, after whom the original colony was named. The Greek *kissos*, meaning "ivy," has been used in Latin names for several other vines, including *Cissus*, *Ampelocissus*, and *Rhoicissus*. However, all are members of the grape family (*Vitaceae*), while true ivy (*Hedera*) belongs elsewhere (*Araliaceae*).

Parthenocissus tricuspidata

pardalinus
par-da-LEE-nus
pardalina, pardalinum
—

pardinus
par-DEE-nus
pardina, pardinum
With spots like a leopard, as in *Hippeastrum pardinum*

pari-
Used in compound words to denote equal

Parietaria
par-ee-uh-TAIR-ee-uh
From Latin *paries*, meaning "wall," a common habitat for upright pellitory, *P. officinalis* (*Urticaceae*)

Paris
PA-ris
From Latin *par*, meaning "equal," alluding to *P. quadrifolia*, which typically has four symmetrically arranged leaves (*Melanthiaceae*)

Parkinsonia
par-kin-SOWN-ee-uh
Named after John Parkinson (1567–1650), British botanist and apothecary (*Fabaceae*)

Parnassia
par-NASS-ee-uh
Named after Mount Parnassus in Greece, which was sacred to the gods Dionysus and Apollo and home to the Muses (*Celastraceae*)

parnassicus
par-NASS-ih-kus
parnassica, parnassicum
Connected with Mount Parnassus, Greece, as in *Thymus parnassicus*

parnassifolius
par-nass-ih-FOH-lee-us
parnassifolia, parnassifolium
With leaves like the grass of Parnassus (*Parnassia*), as in *Saxifraga parnassifolia*

Parochetus
par-oh-KEE-tus
From Greek *para*, meaning "near," and *ochetos*, meaning "stream," a reference to habitat (*Fabaceae*)

Parodia
puh-ROH-dee-uh
Named after Domingo Parodi (1823–90), Italian pharmacist and botanist (*Cactaceae*)

Paronychia
par-oh-NY-kee-uh
From Greek *para*, meaning "near," and *onyx*, meaning "nail," because the plant was used to treat paronychia, or inflammation around the fingernails (*Caryophyllaceae*)

Parrotia
puh-ROT-ee-uh
Named after Johann Jacob Friedrich Wilhelm Parrot (1791–1841), Baltic German physician and explorer (*Hamamelidaceae*)

Parrotiopsis
puh-roh-tee-OP-sis
Resembling the related genus *Parrotia* (*Hamamelidaceae*)

parryae
PAR-ee-eye
—

parryi
PAIR-ree
Named after Dr. Charles Christopher Parry (1823–90), British-born botanist and plant collector. The form *parryae* commemorates his wife, Emily Richmond Parry (1821–1915), as in *Linanthus parryae*

Passiflora

Passionflowers (*Passiflora*) were named by early Catholic missionaries in South America. To them, the intricate blooms were highly symbolic, with the ten petals representing the apostles (minus Peter, who denied Christ, and Judas, who betrayed him), the corona was the crown of thorns, the five stamens were the five wounds, and the three stigmas, the nails. Only a handful of species are suitable for cultivation outside in frost-prone regions, but the best of these, the blue passionflower (*P. caerulea*), clearly demonstrates this religious imagery. Edible passionfruit also come from *Passiflora*, from *P. edulis* and a few other species.

Passiflora caerulea

Parthenium (also parthenium)
par-THEE-nee-um
From Greek *parthenos*, meaning "virgin," or *parthenion*, the name of an unidentified plant; the etymology is unclear (*Asteraceae*)

Parthenocissus
parth-en-oh-SIS-us
From Greek *parthenos*, meaning "virgin," and *kissos*, meaning "ivy," because Virginia creeper, *P. quinquefolia*, is an ivylike plant from Virginia (*Vitaceae*)

partitus
par-TY-tus
partita, partitum
Parted, as in *Hibiscus partitus*

parvi-
Used in compound words to denote small

parviflorus
par-vih-FLOR-us
parviflora, parviflorum
With small flowers, as in *Aesculus parviflora*

parvifolius
par-vih-FOH-lee-us
parvifolia, parvifolium
With small leaves, as in *Eucalyptus parvifolia*

parvus
PAR-vus
parva, parvum
Small, as in *Lilium parvum*

Passiflora
pas-i-FLAW-ruh
From Latin *passio*, meaning "passion," and *flos*, meaning "flower," because the parts of the flower were thought to symbolize the Crucifixion of Christ (*Passifloraceae*)

Pastinaca
pas-tin-AH-kuh
From Latin *pastinaca*, vernacular name for parsnips and carrots (*Apiaceae*)

patagonicus
pat-uh-GOH-nih-kus
patagonica, patagonicum
Connected with Patagonia, as in *Sisyrinchium patagonicum*

patavinus
pat-uh-VIN-us
patavina, patavinum
Connected with Padua (previously Patavium), Italy, as in *Haplophyllum patavinum*

patens
PAT-enz
—

patulus
PAT-yoo-lus
patula, patulum
With a spreading habit, as in *Salvia patens*

Patersonia
pat-ur-SOH-nee-uh
Named after William Paterson (1755–1810), Scottish soldier and botanist (*Iridaceae*)

Patrinia
puh-TRIN-ee-uh
Named after Eugène Louis Melchior Patrin (1742–1815), French naturalist and mineralogist (*Caprifoliaceae*)

pauci-
Used in compound words to denote few

pauciflorus
PAW-si-flor-us
pauciflora, pauciflorum
With few flowers, as in *Corylopsis pauciflora*

paucifolius
paw-se-FOH-lee-us
paucifolia, paucifolium
With few leaves, as in *Scilla paucifolia*

paucinervis
paw-se-NER-vis
paucinervis, paucinerve
With few nerves, as in *Cornus paucinervis*

Paulownia
paw-LOH-nee-uh
Named after Anna Pavlovna (1795–1865), daughter of Czar Paul I of Russia and wife of King William II of the Netherlands (*Paulowniaceae*)

pauperculus
paw-PER-yoo-lus
paupercula, pauperculum
Poor, as in *Alstroemeria paupercula*

Pavetta
puh-VET-uh
From Sinhalese (Sri Lanka) *pawetta*, the vernacular name for *P. indica* (*Rubiaceae*)

pavia
PAH-vee-uh
Named after Peter Paaw (1564–1617), Dutch physician, as in *Aesculus pavia*

Pavonia
puh-VOH-nee-uh
Named after José Antonio Pavón Jiménez (1754–1840), Spanish botanist (*Malvaceae*)

pavoninus
pav-ON-ee-nus
pavonina, pavoninum
Peacock blue, as in *Anemone pavonina*

Paxistima
paks-i-STEEM-uh
From Greek *pachys*, meaning "thick," and stigma (*Celastraceae*)

pectinatus
pek-tin-AH-tus
pectinata, pectinatum
Like a comb, as in *Euryops pectinatus*

pectoralis
pek-TOR-ah-lis
pectoralis, pectorale
Of the chest, as in *Justicia pectoralis*

peculiaris
pe-kew-lee-AH-ris
peculiaris, peculiare
Peculiar or special, as in *Cheiridopsis peculiaris*

pedatifidus
ped-at-ee-FEE-dus
pedatifida, pedatifidum
Divided like a bird's foot, as in *Viola pedatifida*

pedatus
ped-AH-tus
pedata, pedatum
Shaped like a bird's foot, often in reference to the shape of a palmate leaf, as in *Adiantum pedatum*

pedemontanus
ped-ee-MON-tah-nus
pedemontana, pedemontanum
Connected with Piedmont, Italy, as in *Saxifraga pedemontana*

Pedilanthus
ped-i-LAN-thus
From Greek *pedilon*, meaning "slipper," and *anthos*, meaning "flower," because the flowers are sometimes enclosed within bracts shaped like a shoe (*Euphorbiaceae*)

Pediocactus
pee-dee-oh-KAK-tus
From Greek *pedion*, meaning "plain" (the geographical feature), plus cactus; a reference to preferred habitat (*Cactaceae*)

peduncularis
ped-unk-yoo-LAH-ris
peduncularis, pedunculare
—

pedunculatus
ped-unk-yoo-LA-tus
pedunculata, pedunculatum
With a flower stalk, as in *Lavandula pedunculata*

pedunculosus
ped-unk-yoo-LOH-sus
pedunculosa, pedunculosum
With many or particularly well-developed flower stems, as in *Ilex pedunculosa*

pekinensis
pee-keen-EN-sis
pekinensis, pekinense
From Peking (Beijing), China, as in *Euphorbia pekinensis*

Pelargonium
pel-ur-GOH-nee-um
From Greek *pelargos*, meaning "stork"; the fruit resembles a stork's bill (*Geraniaceae*)

pelegrina
pel-e-GREE-nuh
Local name for *Alstroemeria pelegrina*

Pellaea
PEL-ay-uh
From Greek *pellos*, meaning "dark," referring to the fronds (*Pteridaceae*)

Pellionia
pel-ee-OH-nee-uh
Named after Marie Joseph Alphonse Pellion (1796–1868), French naval officer (*Urticaceae*)

pellucidus
pel-LOO-sid-us
pellucida, pellucidum
Transparent; clear, as in *Conophytum pellucidum*

peloponnesiacus
pel-uh-pon-ee-see-AH-kus
peloponnesiaca, peloponnesiacum
Connected with the Peloponnese, Greece, as in *Colchicum peloponnesiacum*

Peltandra
pel-TAN-druh
From Greek *pelte*, meaning "small shield," and *andros*, meaning "male," because the male flowers are shaped like a shield (*Araceae*)

peltatus
pel-TAH-tus
peltata, peltatum
Shaped like a shield, as in *Darmera peltata*

Peltoboykinia
pel-toh-boy-KIN-ee-uh
From Greek *pelte*, meaning "small shield," and *Boykinia*, a related genus; this perennial with shieldlike leaves used to belong in *Boykinia* (*Saxifragaceae*)

Peltophorum
pel-toh-FOR-um
From Greek *pelte*, meaning "small shield," and *phorus*, meaning "bearing," because the stigma is shaped like a shield (*Fabaceae*)

Paulownia tomentosa

pelviformis

pel-vih-FORM-is

pelviformis, pelviforme

In the shape of a shallow cup, as in *Campanula pelviformis*

pendulinus

pend-yoo-LIN-us

pendulina, pendulinum

Hanging, as in *Salix × pendulina*

pendulus

PEND-yoo-lus

pendula, pendulum

Hanging, as in *Betula pendula*

Penstemon rupicola

CHEMICAL COMPOSITION

Plants are a major source of naturally occurring chemicals that are used as medicines, stimulants, flavorings, and in manufacturing. When chemicals are isolated, they are traditionally named after the plant from which they were extracted. So, nicotine is derived from *Nicotiana*, salicylic acid from *Salix*, atropine from *Atropa*, and ricin from *Ricinus*. Frenchman Jean Nicot was the French ambassador in Portugal when he first came across tobacco (*Nicotiana tabacum*). He took the plant back to France, where it became popular as a miracle cure for a number of maladies. Nicot's celebrity grew with that of tobacco, although the shine has somewhat worn off his namesake plant since.

penicillatus

pen-iss-sil-LAH-tus

penicillata, penicillatum

—

penicillius

pen-iss-SIL-ee-us

penicillia, penicillium

With a tuft of hair, as in *Parodia penicillata*

peninsularis

pen-in-sul-AH-ris

peninsularis, peninsulare

From peninsular regions, as in *Allium peninsulare*

Peniocereus

pen-ee-oh-SER-ee-us

From Greek *penios*, meaning "thread," and *Cereus*, a related genus; the stems are long, narrow, and sprawling (*Cactaceae*)

penna-marina

PEN-uh mar-EE-nuh

Sea feather, as in *Blechnum penna-marina*

pennatus

pen-AH-tus

pennata, pennatum

With feathers, pinnate, as in *Stipa pennata*

pennigerus

pen-NY-ger-us

pennigera, pennigerum

With leaves like feathers, as in *Thelypteris pennigera*

Pennisetum

pen-uh-SEE-tum

From Latin *penna*, meaning "feather," and *seta*, meaning "bristle," because the inflorescences may contain bristles covered in plumelike hairs (*Poaceae*)

pennsylvanicus

pen-sil-VAN-ih-kus

pennsylvanica, pennsylvanicum

—

pensylvanicus

pen-sil-VAN-ih-kus

pensylvanica, pensylvanicum

Connected with Pennsylvania, as in *Acer pensylvanicum*

pensilis

PEN-sil-is

pensilis, pensile

Hanging, as in *Glyptostrobus pensilis*

Penstemon [1]

pen-STEM-on, PEN-stem-on

From Greek *pente*, meaning "five," and *stemon*, meaning "stamen," because each flower has five stamens (*Plantaginaceae*)

penta-

Used in compound words to denote five

Pentaglottis

pen-tuh-GLOT-iss

From Greek *pente*, meaning "five," and *glottis*, meaning "tongue," referring to the five white scales in the flower center (*Boraginaceae*)

pentagonius

pen-ta-GON-ee-us

pentagonia, pentagonium

—

pentagonus

pen-ta-GON-us

pentagona, pentagonum

With five angles, as in *Rubus pentagonus*

pentagynus

pen-ta-GY-nus

pentagyna, pentagynum

With five pistils, as in *Crataegus pentagyna*

pentandrus

pen-TAN-drus

pentandra, pentandrum

With five stamens, as in *Ceiba pentandra*

pentapetaloides

pen-ta-pet-al-OY-deez

Appearing to possess five petals, as in *Convolvulus pentapetaloides*

pentaphyllus

pen-tuh-FIL-us

pentaphylla, pentaphyllum

With five leaves or leaflets, as in *Cardamine pentaphylla*

Pentas

PEN-tass

From Greek *pente*, meaning "five," because the petals and other flower parts are arranged in whorls of five, unlike many related genera where they are in whorls of four (*Rubiaceae*)

Peperomia

pep-uh-ROH-mee-uh

From Greek *peperi*, meaning "pepper," and *homoios*, meaning "similar," because some species resemble the related black pepper, *Piper nigrum* (*Piperaceae*)

pepo

PEP-oh

Latin word for a large melon or pumpkin, as in *Cucurbita pepo*

perbellus

per-BELL-us

perbella, perbellum

Very beautiful, as in *Mammillaria perbella*

peregrinus

per-uh-GREE-nus

peregrina, peregrinum

To wander, as in *Delphinium peregrinum*

perennis

per-EN-is

perennis, perenne

Perennial, as in *Bellis perennis*

Pereskia [2]

puh-RES-kee-uh

Named after Nicolas-Claude Fabri de Peiresc (1580–1637), French astronomer (*Cactaceae*)

perfoliatus

per-foh-lee-AH-tus

perfoliata, perfoliatum

With the leaf surrounding the stem, as in *Parahebe perfoliata*

perforatus

per-for-AH-tus

perforata, perforatum

With, or appearing to have, small holes, as in *Hypericum perforatum*

pergracilis

per-GRASS-il-is

pergracilis, pergracile

Especially slender, as in *Scleria pergracilis*

Pericallis

per-i-KAL-iss

From Greek *peri*, meaning "around," and *kallos*, meaning "beautiful' (*Asteraceae*)

Perilla

puh-RIL-uh

From Greek *pera*, meaning "pouch," plus diminutive *ella*, possibly referring to the shape of the calyx, although derivation unclear (*Lamiaceae*)

Periploca

per-i-PLOK-uh

From Greek *periplokos*, meaning "entwined," because this climber has twining stems (*Apocynaceae*)

Peritoma

per-i-TOH-muh

From Greek *peri*, meaning "around," and *tome*, meaning "to cut," referring to the fruit, which splits all the way round (*Cleomaceae*)

pernyi

PERN-yee-eye

Named after Paul Hubert Perny (1818–1907), French missionary and botanist, as in *Ilex pernyi*

Perovskia

per-OV-ski-uh

Named after Vasily Alekseevich Perovsky (1794–1857), Russian soldier (*Lamiaceae*)

Persea

PER-see-uh

The classical Greek name for a fruiting tree, possibly *Mimusops laurifolia*, as used by Dioscorides (*Lauraceae*)

Persicaria

per-si-KAIR-ee-uh

Named after the peach, *Prunus persica*, whose leaves resemble those of some *Persicaria* species (*Polygonaceae*)

2

Pereskia grandifolia

Euphorbia petiolaris

Petrocallis pyrenaica

persicifolius
per-sik-ih-FOH-lee-us
persicifolia, persicifolium
With leaves like the peach (*Prunus persica*), as in *Campanula persicifolia*

persicus
PER-sih-kus
persica, persicum
Connected with Persia (Iran), as in *Parrotia persica*

persistens
per-SIS-tenz
Persistent, as in *Elegia persistens*

persolutus
per-sol-YEW-tus
persoluta, persolutum
Very loose, as in *Erica persoluta*

perspicuus
PER-spic-kew-us
perspicua, perspicuum
Transparent, as in *Erica perspicua*

pertusus
per-TUS-us
pertusa, pertusum
Perforated; pierced, as in *Listrostachys pertusa*

perulatus
per-uh-LAH-tus
perulata, perulatum
With perules (bud scales), as in *Enkianthus perulatus*

peruvianus
per-u-vee-AH-nus
peruviana, peruvianum
Connected with Peru, as in *Scilla peruviana*

petaloideus
pet-a-LOY-dee-us
petaloidea, petaloideum
Like a petal, as in *Thalictrum petaloideum*

Petasites
pet-uh-SY-teez
From Greek *petasos*, meaning "broad-brimmed hat," alluding to the large parasol-like leaves (*Asteraceae*)

petiolaris [1]
pet-ee-OH-lah-ris
petiolaris, petiolare
—

petiolatus
pet-ee-oh-LAH-tus
petiolata, petiolatum
With a leaf stalk, as in *Helichrysum petiolare*

petraeus
pet-RAY-us
petraea, petraeum
Connected with rocky regions, as in *Quercus petraea*

Petrea
peh-TRAY-uh
Named after Robert James Petre (1713–42), British peer and horticulturist (*Verbenaceae*)

Petrocallis [2]

pet-roh-KAL-iss

From Greek *petros*, meaning "rock," and *kallos*, meaning "beautiful," an apt description of this attractive alpine (*Brassicaceae*)

Petrocosmea

pet-roh-KOS-mee-uh

From Greek *petros*, meaning "rock," and *kosmo*, meaning "ornamental," an allusion to its beauty and preferred habitat (*Gesneriaceae*)

Petrophytum

pet-roh-FY-tum

From Greek *petros*, meaning "rock," and *phyton*, meaning "plant," because these prostrate shrubs occupy rocky outcrops (*Rosaceae*)

Petrorhagia

pet-roh-RAJ-ee-uh

From Greek *petros*, meaning "rock," and *rhagas*, meaning "fissure," a native of rock crevices (*Caryophyllaceae*)

Petroselinum [3]

pet-roh-SEL-in-um

From Greek *petros*, meaning "rock," and *selinon*, meaning "celery"; the British word "parsley" (for *P. crispum*) has the same root (*Apiaceae*)

Petunia

puh-TEW-nee-uh

From Tupi-Guarani (Brazil) *petun*, vernacular name for the related tobacco (*Solanaceae*)

Peucedanum

pew-SED-uh-num

From Greek *peuke*, meaning "pine," a reference to the bitter taste (*Apiaceae*)

Phacelia

fuh-SEE-lee-uh

From Greek *phakelos*, meaning a "bundle," a reference to the clustered flowers (*Boraginaceae*)

phaeacanthus

fay-uh-KAN-thus

phaeacantha, phaeacanthum

With gray thorns, as in *Opuntia phaeacantha*

Phaedranassa

fee-druh-NAS-uh

From Greek *phaidros*, meaning "bright," and *anassa*, meaning "queen," an allusion to the attractive blooms with bright colors (*Amaryllidaceae*)

phaeus

FAY-us

phaea, phaeum

Dusky, as in *Geranium phaeum*

Phaius

FY-us

From Greek *phaios*, meaning "dusky," because some species have flowers of a dark color (*Orchidaceae*)

Phalaenopsis [4]

fal-uh-NOP-sis

From Greek *phalaina*, meaning "moth," and *opsis* ("resembling"), because the flowers look like moths (*Orchidaceae*)

3

Petroselinum crispum

4

Phalaenopsis × intermedia

Philadelphus

Phalaris
fuh-LAR-iss
From Greek *phalara*, meaning "white crested," a reference to the inflorescence (*Poaceae*)

Phaseolus
faz-ee-OH-lus
From Greek *phaselos*, meaning "bean" (*Fabaceae*)

Phellodendron
fel-oh-DEN-dron
From Greek *phellos*, meaning "cork," and *dendron*, meaning "tree," a reference to the bark (*Rutaceae*)

philadelphicus
fil-uh-DEL-fih-kus
philadelphica, philadelphicum
Connected with Philadelphia, as in *Lilium philadelphicum*

Philadelphus [1]
fil-uh-DELF-us
From Greek *philos*, meaning "loving," and *adelphus*, meaning "brother," for reasons unknown; alternatively, named after Ptolemy Philadelphus (309–246 BC), King of Egypt (*Hydrangeaceae*)

Philesia
fy-LEE-zee-uh
Etymology uncertain, but probably from Greek *philos*, meaning "loving," because the flowers are attractive (*Philesiaceae*)

philippensis
fil-lip-EN-sis
philippensis, philippense

philippianus
fil-lip-ee-AH-nus
philippiana, philippianum

—

philippii
fil-LIP-ee-eye

—

philippinensis
fil-ip-ee-NEN-sis
philippinensis, philippinense
From the Philippines, as in *Adiantum philippense*

Phillyrea
fil-uh-RAY-uh
From Greek *philyrea*, the vernacular name for *P. latifolia* (*Oleaceae*)

Philodendron
fil-oh-DEN-dron
From Greek *philos*, meaning "loving," and *dendron*, meaning "tree," a reference to the many species of climbers and epiphytes in this genus (*Araceae*)

Philotheca
fil-oh-THEE-kuh
From Greek *philos*, meaning "loving," and *theke*, meaning "container," probably referring to the capsular fruit (*Rutaceae*)

Phlebodium
fluh-BOH-dee-um
From Greek *phlebos*, meaning "vein," because the fronds have prominent venation (*Polypodiaceae*)

phleoides
flee-OY-deez
Resembling *Phleum* (timothy, cat's tail), as in *Phleum phleoides*

Phleum
FLEE-um
From Greek *phleos*, meaning "woolly reed," probably referring to another grass species (*Poaceae*)

phlogiflorus
flo-GIF-flor-us
phlogiflora, phlogiflorum
With flowers in a flame color or flowers like *Phlox*, as in *Verbena phlogiflora*

Phlomis

FLOH-mis

From Greek *phlogmos*, meaning "flame," because the leaves of some species were used as lamp wicks (*Lamiaceae*)

Phlox

FLOX

From Greek *phlogmos*, meaning "flame"; used as the vernacular name for a plant with red flowers, possibly a *Silene* (*Polemoniaceae*)

Phoebe

FEE-bee

From Greek *phoibos*, meaning "bright" or "radiant" (*Lauraceae*)

phoeniceus

feen-ih-KEE-us

phoenicea, phoeniceum

Purple/red, as in *Juniperus phoenicea*

phoenicolasius

fee-nik-oh-LASS-ee-us

phoenicolasia, phoenicolasium

With purple hairs, as in *Rubus phoenicolasius*

Phoenix

FEE-niks

From Greek *phoinix*, the vernacular name for date palm, *P. dactylifera*; perhaps originating in the Phoenician culture of the Mediterranean (*Arecaceae*)

Phormium

FOR-mee-um

From Greek *phormion*, meaning "mat," because fibers from this plant were used for weaving (*Asphodelaceae*)

Photinia

foh-TIN-ee-uh

From Greek *photeinos*, meaning "shining," alluding to the glossy leaves (*Rosaceae*)

Phragmipedium

frag-mi-PEE-dee-um

From Greek *phragma*, meaning "fence," and *pedilon*, meaning "slipper"; this orchid's flowers have a slipperlike lip and an ovary divided into chambers (*Orchidaceae*)

Phragmites

frag-MY-teez

From Greek *phragma*, meaning "fence" or "hedge," because this reed forms large banks and is also collected for construction (*Poaceae*)

phrygius

FRIJ-ee-us

phrygia, phrygium

Connected with Phrygia, Anatolia, as in *Centaurea phrygia*

Phygelius

fy-JEE-lee-us

From Greek *phyge*, meaning "avoidance," and *helios*, meaning "sun," because plants were said to prefer shade (*Scrophulariaceae*)

Phylica

FY-lik-uh

From Greek *phyllikos*, meaning "leafy," because the foliage is dense (*Rhamnaceae*)

Phyllodoce

fy-loh-DOH-see

Named after a sea nymph in Greek mythology, but because these plants typically occur in arctic or alpine areas, the etymology is obscure (*Ericaceae*)

Phyllostachys

fy-loh-STAK-is

From Greek *phyllon*, meaning "leaf," and *stachys*, meaning "ear of grain," referring to the presence of leaves on flowering shoots (*Poaceae*)

phyllostachyus

fy-lo-STAY-kee-us

phyllostachya, phyllostachyum

With a leaf spike, as in *Hypoestes phyllostachya*

Physalis [2]

fis-AH-lis

From Greek *physa*, meaning "bladder," because the fruit is surrounded by inflated sepals (*Solanaceae*)

Physaria

fy-SAIR-ee-uh

From Greek *physa*, meaning "bladder," because some species have inflated fruit (*Brassicaceae*)

2

Physalis alkekengi

Pieris formosa

Physocarpus
(also **physocarpus**)
fy-soh-KAR-pus
From Greek *physa*, meaning "bladder," and
karpos, meaning "fruit," because the fruit of
some species is inflated (*Rosaceae*)

Physostegia
fy-soh-STEE-jee-uh
From Greek *physa*, meaning "bladder," and
stegos, meaning "cover," because the calyx
covering the fruit is inflated (*Lamiaceae*)

Phyteuma (also **phyteuma**)
fy-TEW-muh
From Greek *phuteuma*, meaning "that which
is planted" (*Campanulaceae*)

Phytolacca
fy-toh-LAK-uh
From Greek *phyton*, meaning "plant," and
Latin *lacca*, a red dye, because the fruit yields
a dye (*Phytolaccaceae*)

Picea
py-SEE-uh
From Latin *picis*, meaning "pitch-producing
pine" (*Pinaceae*)

Picrasma
pi-KRAS-muh
From Greek *picros*, meaning "bitter,"
because the bark yields a bitter flavoring
(*Simaroubaceae*)

picturatus
pik-tur-AH-tus
picturata, picturatum
With variegated leaves, as in *Calathea
picturata*

pictus
PIK-tus
picta, pictum
Painted; highly colored, as in *Acer pictum*

Pieris
PEER-iss
Named after the Muses of Greek mythology,
who were worshipped at Pieria in
Macedonia (*Ericaceae*)

Pilea
py-LEE-uh
From Greek *pilos*, meaning "felt cap,"
because the calyx covers the fruit
(*Urticaceae*)

pileatus
py-lee-AH-tus
pileata, pileatum
With a cap, as in *Lonicera pileata*

Pileostegia
pil-ee-oh-STEE-jee-uh
From Greek *pilos*, meaning "felt cap," and
stegos, meaning "cover," because the corolla
is caplike (*Hydrangeaceae*)

piliferus
py-LIH-fer-us
pilifera, piliferum
With short, soft hairs, as in *Ursinia pilifera*

pillansii
pil-AN-see-eye
Named after Neville Stuart Pillans (1884–
1964), South African botanist,
as in *Watsonia pillansii*

Pilosella
py-loh-SEL-uh
From Latin *pilosa*, meaning "hairy," plus
diminutive *ella*, because these are small,
often woolly plants (*Asteraceae*)

Pilosocereus
py-loss-oh-SER-ee-us
From Latin *pilosa*, meaning "hairy," and
Cereus, a related genus (*Cactaceae*)

pilosus
pil-OH-sus
pilosa, pilosum
With long, soft hairs, as in *Aster pilosus*

pilularis
pil-yoo-LAH-ris
pilularis, pilulare
—

piluliferus
pil-yoo-LIH-fer-us
pilulifera, piluliferum
With globular fruit, as in *Urtica pilulifera*

Pimelea
pim-EE-lee-uh
From Greek *pimele*, meaning "fat," because
the seed is oily (*Thymelaeaceae*)

pimeleoides
py-mee-lee-OY-deez
Resembling *Pimelea*, as in *Pittosporum
pimeleoides*

Pimpinella
pimp-i-NEL-uh
From Latin *pimpinella*, a medicinal herb
(*Apiaceae*)

pimpinellifolius
pim-pi-nel-ih-FOH-lee-us
pimpinellifolia, pimpinellifolium
With leaves like anise (*Pimpinella*), as in
Rosa pimpinellifolia

Pinellia
py-NEE-lee-uh
Named after Giovanni Vincenzo Pinelli
(1535–1601), Italian botanist (*Araceae*)

pinetorum
py-net-OR-um
Connected with pine forests, as in *Fritillaria
pinetorum*

pineus
PY-nee-us
pinea, pineum
Relating to pine (*Pinus*), as in *Pinus pinea*

Pinguicula
pin-GWIK-ew-luh
From Latin *pinguis*, meaning "fat," because
the glistening leaves appear oily
(*Lentibulariaceae*)

pinguifolius
pin-gwih-FOH-lee-us
pinguifolia, pinguifolium
With fat leaves, as in *Hebe pinguifolia*

pinifolius
pin-ih-FOH-lee-us
pinifolia, pinifolium
With leaves like pine (*Pinus*), as in
Penstemon pinifolius

pininana
pin-in-AH-nuh
A dwarf pine, as in *Echium pininana*

pinnatifidus
pin-nat-ih-FY-dus
pinnatifida, pinnatifidum
Cut in the form of a feather, as in *Eranthis
pinnatifida*

pinnatifolius
pin-nat-ih-FOH-lee-us
pinnatifolia, pinnatifolium
With leaves like feathers, as in *Meconopsis
pinnatifolia*

pinnatifrons

pin-NAT-ih-fronz

With fronds like feathers, as in *Chamaedorea pinnatifrons*

pinnatus

pin-NAH-tus

pinnata, pinnatum

With leaves that grow from each side of a stalk; like a feather, as in *Santolina pinnata*

Pinus

PY-nus

The classical Latin name for pine trees (*Pinaceae*)

Piper

PY-per

From Greek *peperi*, meaning "pepper"; black pepper comes from *P. nigrum* (*Piperaceae*)

piperitus

pip-er-EE-tus

piperita, piperitum

With a pepperlike taste, as in *Mentha × piperita*

Piptanthus

pip-TAN-thus

From Greek *pipto*, meaning "to fall," and *anthos*, meaning "flower," because the sepals, petals, and stamens fall as one unit (*Fabaceae*)

pisiferus

pih-SIH-fer-us

pisifera, pisiferum

Bearing peas, as in *Chameacyparis pisifera*

Pisonia

py-SOH-nee-uh

Named after Willem Pies (1611–78), Dutch physician and botanist (*Nyctaginaceae*)

Pistacia

pi-STAS-ee-uh

From Greek *pistakia*, the pistachio tree, *P. vera* (*Anacardiaceae*)

Pistia

PIS-tee-uh

From Greek *pistra*, meaning "water trough," a reference to the aquatic habitat (*Araceae*)

Pisum [1]

PY-sum

From Greek *pison*, the vernacular name for pea, *P. sativum* (*Fabaceae*)

pitardii

pit-ARD-ee-eye

Named after Charles-Joseph Marie Pitard-Briau, twentieth-century French plant collector and botanist, as in *Camellia pitardii*

Pitcairnia

pit-KAIR-nee-uh

Named after William Pitcairn (1711–91), British physician and gardener (*Bromeliaceae*)

pittonii

pit-TON-ee-eye

Named after Josef Claudius Pittoni, nineteenth-century Austrian botanist, as in *Sempervivum pittonii*

Pittosporum

pit-OS-po-rum

From Greek *pitta*, meaning "pitch," and *spora*, meaning "seed," because the seed has a sticky coating (*Pittosporaceae*)

Pityrogramma

pit-ay-roh-GRAM-uh

From Greek *pityros*, meaning "bran," and *gramme*, meaning "line," because the sporangia appear dustlike (*Pteridaceae*)

planiflorus

plan-ih-FLOR-us

planiflora, planiflorum

With flat flowers, as in *Echidnopsis planiflora*

planifolius

plan-ih-FOH-lee-us

planifolia, planifolium

With flat leaves, as in *Iris planifolia*

planipes

PLAN-ee-pays

With a flat stalk, as in *Euonymus planipes*

plantagineus

plan-tuh-JIN-ee-us

plantaginea, plantagineum

Like plantain (*Platago*), as in *Hosta plantaginea*

1

Pisum sativum

Platycarya strobilacea

Platycodon grandifloras

Plantago
plan-TAY-goh
From Latin *planta*, meaning "sole of the foot," because in some species the leaves lie flat along the ground; also, common weeds of well-trodden lawns (*Plantaginaceae*)

planus
PLAH-nus
plana, planum
Flat, as in *Eryngium planum*

platanifolius
pla-tan-ih-FOH-lee-us
platanifolia, platanifolium
With leaves like a plane tree (*Platanus*), as in *Begonia platanifolia*

platanoides
pla-tan-OY-deez
Resembling a plane tree (*Platanus*), as in *Acer platanoides*

Platanus
PLAT-ah-nus
From Greek *platanos*, the vernacular name for *P. orientalis* (*Platanaceae*)

platy-
Used in compound words to denote broad (or sometimes flat)

platycanthus
plat-ee-KAN-thus
platycantha, platycanthum
With broad spines, as in *Acaena platycantha*

platycarpus
plat-ee-KAR-pus
platycarpa, platycarpum
With broad fruit, as in *Thalictrum platycarpum*

Platycarya [2]
pla-tee-KAR-ee-uh
From Greek *platys*, meaning "broad," and *karyon*, meaning "nut," because the seed is winged (*Junglandaceae*)

platycaulis
plat-ee-KAWL-is
platycaulis, platycaule
With a broad stem, as in *Allium platycaule*

Platycerium
pla-tee-SEER-ee-um
From Greek *platys*, meaning "broad," and *keras*, meaning "horn," relating to the antlerlike fronds (*Polypodiaceae*)

Platycladus
pla-tee-KLAY-dus
From Greek *platys*, meaning "broad," and *klados*, meaning "branch," alluding to the flattened, spraylike branchlets (*Cupressaceae*)

Platycodon [3]
pla-tee-KOH-don
From Greek *platys*, meaning "broad," and *codon*, meaning "bell," referring to the inflated, balloonlike flowers (*Campanulaceae*)

1

Plumeria rubra

platyglossus

plat-ee-GLOSS-us

platyglossa, platyglossum

With a broad tongue, as in *Phyllostachys platyglossa*

platypetalus

plat-ee-PET-uh-lus

platypetala, platypetalum

With broad petals, as in *Epimedium platypetalum*

platyphyllos

plat-tih-FIL-los

—

platyphyllus

pla-tih-FIL-us

platyphylla, platyphyllum

With broad leaves, as in *Betula platyphylla*

platypodus

pah-tee-POD-us

platypoda, platypodum

With a broad stalk, as in *Fraxinus platypoda*

platyspathus

plat-ees-PATH-us

platyspatha, platyspathum

With a broad spathe, as in *Allium platyspathum*

platyspermus

plat-ee-SPER-mus

platysperma, platyspermum

With broad seed, as in *Hakea platysperma*

Plectranthus

plek-TRAN-thus

From Greek *plektron*, meaning "spur," and *anthos*, meaning "flower," because some species have spurs at the base of their blooms (*Lamiaceae*)

Pleioblastus

ply-oh-BLAST-us

From Greek *pleios*, meaning "more," and *blastos*, meaning "bud," because this bamboo produces many shoots at each node (*Poaceae*)

Pleione

plee-OH-nee

Named after Pleione, wife of Atlas in Greek mythology (*Orchidaceae*)

Pleiospilos

ply-os-PIE-los

From Greek *pleios*, meaning "more," and *spilos*, meaning "spot," because the succulent leaves mimic pebbles and are spotted but hard to spot (*Aizoaceae*)

pleniflorus

plen-ee-FLOR-us

pleniflora, pleniflorum

With double flowers, as in *Kerria japonica* "Pleniflora"

plenissimus

plen-ISS-i-mus

plenissima, plenissimum

Especially with double flowers, as in *Eucalyptus kochii* subsp. *plenissima*

plenus

plen-US

plena, plenum

Double; full, as in *Felicia plena*

Pleurothallis

plew-roh-THAL-us

From Greek *pleuron*, meaning "rib," and *thallos*, meaning "branch," perhaps alluding to the narrow stems of many species (*Orchidaceae*)

plicatus

ply-KAH-tus

plicata, plicatum

Pleated, as in *Thuja plicata*

plumarius

ploo-MAH-ree-us

plumaria, plumarium

With feathers, as in *Dianthus plumarius*

plumbaginoides

plum-bah-gih-NOY-deez

Resembling *Plumbago*, as in *Ceratostigma plumbaginoides*

Plumbago

plum-BAY-goh

From Greek *plumbago*, meaning "galena," a type of lead ore; the name was used for a herb that cured lead poisoning, although it may not have been this plant (*Plumbaginaceae*)

plumbeus

plum-BEY-us

plumbea, plumbeum

Relating to lead, as in *Alocasia plumbea*

Plumeria [1]

plew-MEER-ee-uh

Named after Charles Plumier (1646–1704), French botanist (*Apocynaceae*)

plumosus

plum-OH-sus

plumosa, plumosum

Feathery, as in *Libocedrus plumosa*

pluriflorus

plur-ee-FLOR-us

pluriflora, pluriflorum

With many flowers, as in *Erythronium pluriflorum*

pluvialis

ploo-VEE-uh-lis

pluvialis, pluviale

Relating to rain, as in *Calendula pluvialis*

Poa

POH-uh

From Greek *poa*, meaning "grass" (*Poaceae*)

pocophorus

po-KO-for-us

pocophora, pocophorum

Bearing fleece, as in *Rhododendron pocophorum*

podagraria

pod-uh-GRAR-ee-uh

From *podagra*, Latin for "gout," as in *Aegopodium podagraria*

Podalyria

pod-uh-LIR-ee-uh

Named after Podalirius, son of Asklepios, Greek god of medicine (*Fabaceae*)

Podocarpus

pod-oh-KARP-us

From Greek *podion*, meaning "foot," and *karpos*, meaning "fruit," because some species have a fleshy stalk below the seed (*Podocarpaceae*)

Podophyllum

pod-oh-FIL-um

From Greek *podion*, meaning "foot," and *phyllon*, meaning "leaf," due to the leaves resembling a duck's webbed foot (*Berberidaceae*)

podophyllus

po-do-FIL-us

podophylla, podophyllum

With leaves having a stout stalk, as in *Rodgersia podophylla*

Podranea

pod-RAH-nee-uh

An anagram of the related genus *Pandorea* (*Bignoniaceae*)

Plumbago

Plumbago as a common name applies to two different groups of plants. The first are members of the genus *Plumbago*, tender shrubs and vines that make excellent conservatory plants. The best known is *P. auriculata*, or Cape leadwort, a sprawling South African plant with vibrant blue blooms. The second group comprises hardy (or half-hardy) shrubs with striking blue flowers and often attractive fall foliage color. *Ceratostigma plumbaginoides*, named for its similarity to *Plumbago*, is a suckering shrub suitable for ground cover, while the more upright *C. willmottianum* is an unassuming shrub to 3 feet tall.

Ceratostigma plumbaginoides

241

poeticus

po-ET-ih-kus

poetica, poeticum

Relating to poets, as in *Narcissus poeticus*

polaris

po-LAH-ris

polaris, polare

Connected with the North Pole, as in *Salix polaris*

Polemonium

pol-uh-MOH-nee-um

From Greek *polemonion*, a medicinal herb associated with Polemon I, or his son Polemon II, rulers of regions in modern-day Turkey; may also derive from Greek *polemos*, meaning "war" (*Polemoniaceae*)

Polianthes

poh-lee-ANTH-eez

From Greek *polios*, meaning "gray," and *anthos*, meaning "flower," a somewhat deprecating etymology for these attractive blooms; sometimes interpreted as deriving from Greek *polys*, meaning "many" (*Asparagaceae*)

polifolius

po-lih-FOH-lee-us

polifolia, polifolium

With gray leaves, as in *Andromeda polifolia*

1

Ranunculus polyanthemos

Poliothyrsis

poh-lee-oh-THUR-sis

From Greek *polios*, meaning "gray," and *thyrsus*, meaning "panicle," referring to the panicles of white flowers (*Salicaceae*)

politus

POL-ee-tus

polita, politum

Polished, as in *Saxifraga × polita*

polonicus

pol-ON-ih-kus

polonica, polonicum

Connected with Poland, as in *Cochlearia polonica*

poly-

Used in compound words to denote many

polyacanthus

pol-lee-KAN-thus

polyacantha, polyacanthum

With many thorns, as in *Acacia polyacantha*

polyandrus

pol-lee-AND-rus

polyandra, polyandrum

With many stamens, as in *Conophytum polyandrum*

polyanthemos [1]

pol-lee-AN-them-os

—

polyanthus

pol-ee-AN-thus

polyantha, polyanthum

With many flowers, as in *Jasminum polyanthum*

polyblepharus

pol-ee-BLEF-ar-us

polyblephara, polyblepharum

With many fringes or eyelashes, as in *Polystichum polyblepharum*

polybotryus

pol-lee-BOT-ree-us

polybotrya, polybotryum

With many clusters, as in *Acacia polybotrya*

polybulbon [2]

pol-lee-BUL-bun

With many bulbs, as in *Dinema polybulbon*

polycarpus

pol-ee-KAR-pus

polycarpa, polycarpum

With many fruit, as in *Fatsia polycarpa*

polycephalus

pol-ee-SEF-a-lus

polycephala, polycephalum

With many heads, as in *Cordia polycephala*

polychromus

pol-ee-KROW-mus

polychroma, polychromum

With many colors, as in *Euphorbia polychroma*

Polygala

poh-LIG-uh-luh

From Greek *polys*, meaning "many," and *gala*, meaning "milk," because they have been variously promoted to increase milk yield in cattle and humans (*Polygalaceae*)

polygaloides

pol-ee-gal-OY-deez

Resembling milkwort (*Polygala*), as in *Osteospermum polygaloides*

Polygonatum [3]

po-lig-oh-NAY-tum

From Greek *polys*, meaning "many," and *gonia*, meaning "knee," referring to the jointed rhizome (*Asparagaceae*)

polygonoides

pol-ee-gon-OY-deez

Resembling *Polygonum*, as in *Alternanthera polygonoides*

Polygonum

po-LIG-oh-num

From Greek *polys*, meaning "many," and either *gonia*, meaning "knee," referring to the jointed stems, or *gone*, meaning "seed," which is profuse (*Polygonaceae*)

Polylepis

pol-ee-LEE-pis

From Greek *polys*, meaning "many," and *lepis*, meaning "scale," alluding to the peeling, multilayer bark (*Rosaceae*)

polymorphus

pol-ee-MOR-fus

polymorpha, polymorphum

With many or variable forms, as in *Acer polymorphum*

polypetalus

pol-ee-PET-uh-lus

polypetala, polypetalum

With many petals, as in *Caltha polypetala*

Dinema polybulbon

Polygonatum hirtum

polyphyllus
pol-ee-FIL-us
polyphylla, polyphyllum
With many leaves, as in *Paris polyphylla*

polypodioides
pol-ee-pod-ee-OY-deez
Resembling *Polypodium*, as in *Blechnum polypodioides*

Polypodium
po-lee-POH-dee-um
From Greek *polys*, meaning "many," and *podion*, meaning "foot,"
referring to the branched rhizome (*Polypodiaceae*)

polyrhizus
pol-ee-RY-zus
polyrhiza, polyrhizum
—

polyrrhizus
polyrrhiza, polyrrhizum
With many roots, as in *Allium polyrrhizum*

Polyscias
po-lee-SKY-us
From Greek *polys*, meaning "many," and *skias*, meaning "canopy,"
because the leaves are often compound and numerous (*Araliaceae*)

polysepalus
pol-ee-SEP-a-lus
polysepala, polysepalum
With many sepals, as in *Nuphar polysepala*

Polyspora
po-lee-SPOR-uh
From Greek *polys*, meaning "many," and *spora*, meaning "seed," so
presumably the fruit contains numerous seed, although possibly a
reference to the fruit, which has five valves in *Polyspora*, but only three
in related *Camellia* (*Theaceae*)

polystachyus
pol-ee-STAK-ee-us
polystachya, polystachyum
With many spikes, as in *Ixia polystachya*

polystichoides
pol-ee-stik-OY-deez
Resembling *Polystichum*, as in *Woodsia polystichoides*

Polystichum
pol-IS-ti-kum
From Greek *polys*, meaning "many," and *stichos*, meaning "row,"
because the sori are in rows under the fronds (*Dryopteridaceae*)

polytrichus
pol-ee-TRY-kus
polytricha, polytrichum
With many hairs, as in *Thymus politrichus*

pomeridianus
pom-er-id-ee-AHN-us
pomeridiana, pomeridanium
Flowering in the afternoon, as in *Carpanthea pomeridiana*

pomiferus
pom-IH-fer-us
pomifera, pomiferum
Bearing apples, as in *Maclura pomifera*

pomponius
pomp-OH-nee-us
pomponia, pomponium
With a tuft or pompon, as in *Lilium pomponium*

ponderosus
pon-der-OH-sus
ponderosa, ponderosum
Heavy, as in *Pinus ponderosa*

Pontederia
pon-tuh-DEER-ee-uh
Named after Giulio Pontedera (1688–1757), Italian botanist (*Pontederiaceae*)

ponticus
PON-tih-kus
pontica, ponticum
Connected with Pontus, Asia Minor, as in *Daphne pontica*

populifolius
pop-yoo-lih-FOH-lee-us
populifolia, populifolium
With leaves like poplar (*Populus*), as in *Cistus populifolius*

populneus
pop-ULL-nee-us
populnea, populneum
Relating to poplar (*Populus*), as in *Brachychiton populneus*

Populus
POP-ew-lus
From Latin *arbor populi*, meaning "tree of the people," although the reason is unknown (*Salicaceae*)

porophyllus
po-ro-FIL-us
porophylla, porophyllum
With leaves with (apparent) holes, as in *Saxifraga porophylla*

porphyreus
por-FY-ree-us
porphyrea, porphyreum
Purple-red, as in *Epidendrum porphyreum*

porrifolius
po-ree-FOH-lee-us
porrifolia, porrifolium
With leaves like leek (*Allium porrum*), as in *Tragopogon porrifolius*

porrigens
por-RIG-enz
Spreading, as in *Schizanthus porrigens*

portenschlagianus
port-en-shlag-ee-AH-nus
portenschlagiana, portenschlagianum
Named after Franz von Portenschlag-Leydermayer (1772–1822), Austrian naturalist, as in *Campanula portenschlagiana*

Portulaca
por-tew-LAH-kuh
From Latin *portula*, meaning "little door," because the fruiting capsule initially opens like a door (*Portulacaceae*)

Portulacaria
por-tew-lak-AIR-ee-uh
Resembling the related genus *Portulaca* (*Didiereaceae*)

poscharskyanus
po-shar-skee-AH-nus
poscharskyana, poscharskyanum
Named after Gustav Poscharsky (1832–1914), German horticulturist, as in *Campanula poscharskyana*

Potamogeton
pot-am-oh-GEE-ton
From Greek *potamos*, meaning "river," and *geiton*, meaning "neighbor," referring to the aquatic habit (*Potamogetonaceae*)

potamophilus
pot-am-OH-fil-us
potamophila, potomaphilum
Loving rivers, as in *Begonia potamophila*

potaninii
po-tan-IN-ee-eye
Named after Grigory Nikolaevich Potanin (1835–1920), Russian plant collector, as in *Indigofera potaninii*

potatorum
poh-tuh-TOR-um
Relating to drinking and brewing, as in *Agave potatorum*

Potentilla [1]
poh-ten-TIL-uh
From Latin *potens*, meaning "powerful," plus diminutive *ella*, alluding to medicinal properties (*Rosaceae*)

Poterium
pot-EER-ee-um
From Greek *poterion*, meaning "goblet," referring to the cuplike hypanthium at the base of the flower (*Rosaceae*)

pottsii
POT-see-eye
Named after John Potts or C. H. Potts, nineteenth-century British horticulturists and plant collectors, as in *Crocosmia pottsii*

powellii
pow-EL-ee-eye
Named after John Wesley Powell (1834–1902), American explorer, as in *Crinum × powellii*

praealtus
pray-AL-tus
praealta, praealtum
Especially tall, as in *Aster praealtus*

praecox
pray-koks
Particularly early, as in *Stachyurus praecox*

praemorsus
pray-MOR-sus
praemorsa, praemorsum
With the appearance of bitten tips, as in *Banksia praemorsa*

praeruptorum
pray-rup-TOR-um
Growing in rough ground, as in *Peucedanum praeruptorum*

praestans
PRAY-stanz
Distinguished, as in *Tulipa praestans*

1

Potentilla × *macnabiana*

praetextus

pray-TEX-tus

praetexta, praetextum

With a border, as in *Oncidium praetextum*

prasinus

pra-SEE-nus

prasina, prasinum

The color of leeks, as in *Dendrobium prasinum*

pratensis

pray-TEN-sis

pratensis, pratense

From the meadow, as in *Geranium pratense*

Pratia

PRAH-tee-uh

Named after L. Prat-Bernon (?–1817), French midshipman who died on Louis de Freycinet's circumnavigation of the world (*Campanulaceae*)

prattii

PRAT-tee-eye

Named after Antwerp E. Pratt, nineteenth-century British zoologist, as in *Anemone prattii*

pravissimus

prav-ISS-ih-mus

pravissima, pravissimum

Crooked, as in *Acacia pravissima*

Primula

PRIM-yew-luh

From Latin *primus*, meaning "first," because primrose flowers are some of the first to appear in spring (*Primulaceae*)

primula [2]

PRIM-yew-luh

First flowering, as in *Rosa primula*

primuliflorus

prim-yew-LIF-flor-us

primuliflora, primuliflorum

With flowers like primrose (*Primula*), as in *Rhododendron primuliflorum*

primulifolius

prim-yew-lih-FOH-lee-us

primulifolia primulifolium

With leaves like primrose (*Primula*), as in *Campanula primulifolia*

Primulina

prim-yew-LEE-nuh

Resembling the genus *Primula* (*Gesneriaceae*)

primulinus

prim-yew-LEE-nus

primulina, primulinum

—

primuloides

prim-yew-LOY-deez

Like primrose (*Primula*), as in *Paphiopedilum primulinum*

princeps

PRIN-keps

Most distinguished, as in *Centaurea princeps*

pringlei

PRING-lee-eye

Named after Cyrus Guernesey Pringle (1838–1911), American botanist and plant collector, as in *Monarda pringlei*

Prinsepia

prin-SEEP-ee-uh

Named for James Prinsep (1799–1840), British scholar and colonial administrator (*Rosaceae*)

prismaticus

priz-MAT-ih-kus

prismatica, prismaticum

In the shape of a prism, as in *Rhipsalis prismatica*

Pritchardia

prit-CHARD-ee-uh

Named after William Thomas Pritchard (1829–1907), British consul in Fiji (*Arecaceae*)

proboscideus

pro-bosk-ee-DEE-us

proboscidea, proboscideum

Shaped like a snout, as in *Arisarum proboscideum*

procerus

PRO-ker-us

procera, procerum

Tall, as in *Abies procera*

procumbens

pro-KUM-benz

Prostrate, as in *Gaultheria procumbens*

procurrens

pro-KUR-enz

Spreading underground, as in *Geranium procurrens*

prodigiosus

pro-dij-ee-OH-sus

prodigiosa, prodigiosum

Wonderful; enormous; prodigious, as in *Tillandsia prodigiosa*

productus

pro-DUK-tus

producta, productum

Lengthened, as in *Costus productus*

2

Primula auricula

A
B
C
D
E
F
G
H
I
J
K
L
M
N
O
P
—
Q
R
S
T
U
V
W
X
Y
Z
i

prolifer
PRO-leef-er
—
proliferus
pro-LIH-fer-us
prolifera, proliferum
Increasing by the production of side shoots, as in *Primula prolifera*

prolificus
pro-LIF-ih-kus
prolifica, prolificum
Producing many fruit, as in *Echeveria prolifica*

propinquus
prop-IN-kwus
propinqua, propinquum
Related to, near, as in *Myriophyllum propinquum*

Prosartes
pro-SAR-teez
From Greek *prosarto*, meaning "to hang," due to the pendulous ovules inside the ovary of some species (*Liliaceae*)

Prostanthera
pros-TAN-thu-ruh
From Greek *prostheke*, meaning "appendix," and *anthera*, because the anthers have spurlike appendages (*Lamiaceae*)

prostratus
prost-RAH-tus
prostrata, prostratum
Growing flat on the ground, as in *Veronica prostrata*

Protea
PRO-tee-uh
Named after the Greek god Proteus, who could change his shape; an allusion to the great diversity of form in this genus (*Proteaceae*)

protistus
pro-TISS-tus
protista, protistum
The first, as in *Rhododendron protistum*

provincialis
pro-VIN-ki-ah-lis
provincialis, provinciale
Connected with Provence, France, as in *Arenaria provincialis*

pruinatus
proo-in-AH-tus
pruinata, pruinatum
—
pruinosus
proo-in-NOH-sus
pruinosa, pruinosum
Glistening like frost, as in *Cotoneaster pruinosus*

Prumnopitys
prum-NOP-it-iss
From Greek *prymnos*, meaning "hindmost," and *pitys*, meaning "pine," because the resin duct is below the midvein in the leaf (*Podocarpaceae*)

Prunella
proo-NEL-uh
From Latin *prunum*, meaning "plum," referring to flower color, although many other etymologies have been proposed for this name (*Lamiaceae*)

prunelloides
proo-nel-LOY-deez
Resembling self-heal (*Prunella*), as in *Haplopappus prunelloides*

prunifolius
proo-ni-FOH-lee-us
prunifolia, prunifolium
With leaves like plum (*Prunus*), as in *Malus prunifolia*

Prunus [1, 2]
PROO-nus
From Latin *prunum*, meaning "plum," although this genus also includes cherries, peaches, and almonds (*Rosaceae*)

przewalskianus
prez-WAL-skee-ah-nus
przewalskiana, przewalskianum
—
przewalskii
prez-WAL-skee
Named after Nicolai Przewalski, nineteenth-century Russian naturalist, as in *Ligularia przewalskii*

pseud-
Used in compound words to denote false

pseudacorus
soo-DA-ko-rus
Deceptively like *Acorus* or sweet flag, as in *Iris pseudacorus*

Pseuderanthemum
soo-der-ANTH-uh-mum
From Greek *pseudes*, meaning "false," plus *Eranthemum*, a related genus (*Acanthaceae*)

pseudocamellia
soo-doh-kuh-MEE-lee-uh
Deceptively like a camellia, as in *Stewartia pseudocamellia*

Prunus armeniaca

Prunus domestica

pseudochrysanthus

soo-doh-kris-AN-thus

pseudochrysantha, pseudochrysanthum

Resembling a *chrysanthus* species in the same genus, as in *Rhododendron pseudochrysanthum*, which means resembling *R. chrysanthum*

Pseudocydonia

soo-doh-sy-DOH-nee-uh

From Greek *pseudes*, meaning "false," plus *Cydonia*, a related genus (*Rosaceae*)

pseudodictamnus

soo-do-dik-TAM-nus

Deceptively like *Dictamnus*, as in *Ballota pseudodictamnus*

Pseudofumaria

soo-doh-foo-MAIR-ee-uh

From Greek *pseudes*, meaning "false," plus *Fumaria*, a related genus (*Papaveraceae*)

Pseudogynoxys

soo-doh-jy-NOKS-iss

From Greek *pseudes*, meaning "false," plus *Gynoxys*, a related genus (*Asteraceae*)

Pseudolarix

soo-doh-LAR-iks

From Greek *pseudes*, meaning "false," plus *Larix*, a related genus (*Pinaceae*)

pseudonarcissus

soo-doh-nar-SIS-us

Deceptively like *Narcissus*; in *N. pseudonarcissus*, it means like *N. poeticus*

Pseudopanax

soo-doh-PAN-aks

From Greek *pseudes*, meaning "false," plus *Panax*, a related genus (*Araliaceae*)

Pseudosasa

soo-doh-SAS-uh

From Greek *pseudes*, meaning "false," plus *Sasa*, a related genus (*Poaceae*)

Pseudotsuga

soo-doh-SOO-guh

From Greek *pseudes*, meaning "false," plus *Tsuga*, a related genus (*Pinaceae*)

Pseudowintera

soo-doh-WIN-ter-uh

From Greek *pseudes*, meaning "false," plus *Wintera* (=*Drimys*), a related genus (*Winteraceae*)

Psidium

SID-ee-um

From Greek *psidion*, meaning "pomegranate," because the fruit of many species resemble its namesake (*Myrtaceae*)

psilostemon

sigh-loh-STEE-mon

With smooth stamens, as in *Geranium psilostemon*

Psilotum

sy-LOH-tum

From Greek *psilos*, meaning "naked," because the stems are leafless (*Psilotaceae*)

psittacinus

sit-uh-SIGN-us

psittacina, psittacinum

—

psittacorum

sit-a-KOR-um

Like a parrot, relating to parrots, as in *Hippeastrum psittacinum*

Psychotria

sy-KOH-tre-uh

From Greek *psychro*, meaning "cold," and *trophe*, meaning "food," together, meaning to refresh, as the plant has medicinal properties (*Rubiaceae*)

ptarmica

TAR-mik-uh

ptarmica, ptarmicum

Ancient Greek name for a plant (probably sneezewort) that caused sneezing, as in *Achillea ptarmica*

Ptelea

TEE-lee-uh

From Greek *ptelea*, the vernacular name for elm (*Ulmus*), because the winged fruit is similar (*Rutaceae*)

Pteridium

te-RID-ee-um

From Greek *pteridion*, meaning "small fern" (*Dennstaedtiaceae*)

pteridoides

ter-id-OY-deez

Resembling *Pteris*, as in *Coriaria pteridioides*

Pteris

TE-ris

From Greek *pteris*, meaning "fern," derived from *pteron*, meaning "wing," because fern fronds can resemble bird wings (*Pteridaceae*)

PTERIS AND PTERIDOLOGY

The Greek word *pteris* means "fern," and it is a component of many generic names (*Pteris*, *Dryopteris*, *Cystopteris*). Indeed, the study of ferns is known as pteridology. Fern Latin names often refer to details of fern morphology that are confusing to gardeners, so here is a quick primer. Ferns reproduce using spores, produced by sporangia. In most garden ferns, sporangia are arranged in distinct clusters called sori (sing. sorus) on the leaf (frond) underside, though they can be distributed across the leaf or on separate, fertile fronds. In many ferns, the sori are protected by a membrane (indusium). The location and arrangement of sporangia and indusia are often critical for fern identification.

Pterocarya

te-roh-KAR-ee-uh

From Greek *pteron*, meaning "wing," and *karyon*, meaning "nut," because the fruit is winged (*Juglandaceae*)

Pteroceltis

te-roh-KEL-tis

From Greek *pteron*, meaning "wing," and *Celtis*, a related genus (*Cannabaceae*)

pteroneurus

ter-OH-new-rus

pteroneura, pteroneurum

With nerves that have wings, as in *Euphorbia pteroneura*

Pterostyrax
tair-oh-STY-raks
From Greek *pteron*, meaning "wing," and *Styrax*, a related genus (*Styracaceae*)

Ptilotus
ty-LOH-tus
From Greek *ptilon*, meaning "feather," because the hairy flowers are arranged in featherlike spikes (*Amaranthaceae*)

pubens
PEW-benz
—

pubescens
pew-BESS-enz
Downy, as in *Primula* × *pubescens*

pubigerus
pub-EE-ger-us
pubigera, pubigerum
Producing down, as in *Schefflera pubigera*

pudicus
pud-IH-kus
pudica, pudicum
Shy, as in *Mimosa pudica*

pugioniformis
pug-ee-oh-nee-FOR-mis
pugioniformis, pugioniforme
Shaped like a dagger, as in *Celmisia pugioniformis*

pulchellus
pul-KELL-us
pulchella, pulchellum
—

pulcher
PUL-ker
pulchra, pulchrum
Pretty, beautiful, as in *Correa pulchella*

pulcherrimus
pul-KAIR-ih-mus
pulcherrima, pulcherrimum
Very beautiful, as in *Dierama pulcherrimum*

pulegioides
pul-eg-ee-OY-deez
Like *Mentha pulegium* (pennyroyal), as in *Thymus pulegioides*

pulegium
pul-ee-GEE-um
Latin for pennyroyal, reputed to be a flea repellent, as in *Mentha pulegium*

pullus
PULL-us
pulla, pullum
Having a dark color, as in *Campanula pulla*

Pulmonaria
pul-mon-AIR-ee-uh
From Latin *pulmo*, meaning "lung," because the mottled leaves resemble lungs and were therefore said to be effective at treating chest conditions (*Boraginaceae*)

Pulsatilla
pul-su-TIL-uh
From Latin *pulsare*, meaning "to pulsate," a reference to the way the flowers or seed heads move in the wind; however, it could also originate as a corruption of Hebrew *paschal*, meaning "passion," a reference to Easter flowering (*Ranunculaceae*)

Pultenaea
pul-ten-AY-uh
Named after Richard Pulteney (1730–1801), British physician and botanist (*Fabaceae*)

Pulmonaria

Many medieval herbalists believed that God would guide them to the correct plant to cure a specific illness by shaping the plant like the ailing organ. This philosophy, known as the doctrine of signatures, was initially developed by Greek physicians but gained widespread acceptance. Liverwort (*Hepatica*), birthwort (*Aristolochia*), eyebright (*Euphrasia*), and spleenwort (*Asplenium*) all derive their English and Latin names from their supposed medicinal properties. Lungworts (*Pulmonaria*) have crescent-shape leaves mottled with white, said to resemble ulcerated lungs. However, the presence of toxic pyrrolizidine alkaloids suggests these herbs may be harmful.

Pulmonaria officinalis

pulverulentus
pul-ver-oo-LEN-tus
pulverulenta, pulverulentum
Appearing to be covered in dust, as in *Primula pulverulenta*

pulvinatus
pul-vin-AH-tus
pulvinata, pulvinatum
Like a cushion, as in *Echeveria pulvinata*

pumilio
poo-MIL-ee-oh
Small, dwarf; as in *Edraianthus pumilio*

pumilus
POO-mil-us
pumila, pumilum
Dwarf, as in *Trollius pumilus*

punctatus
punk-TAH-tus
punctata, punctatum
With spots, as in *Anthemis punctata*

pungens
PUN-genz
With a sharp point, as in *Elymus pungens*

Punica
PEW-ni-kuh
From Latin *punicum malum*, meaning "apple of Carthage" (*Lythraceae*)

puniceus
pun-IK-ee-us
punicea, puniceum
Red-purple, as in *Clianthus puniceus*

purpurascens
pur-pur-ASS-kenz
Becoming purple, as in *Bergenia purpurascens*

purpuratus
pur-pur-AH-tus
purpurata, purpuratum
Made purple, as in *Phyllostachys purpurata*

purpureus
pur-PUR-ee-us
purpurea, purpureum
Purple, as in *Digitalis purpurea*

purpusii
pur-PUSS-ee-eye
Named after Carl Purpus (1851–1941) or his brother Joseph Purpus (1860–1932), German plant collectors, as in *Lonicera × purpusii*

Purshia
PUR-shee-uh
Named after Friedrich Traugott Pursch (1774–1820), German botanist (*Rosaceae*)

Puschkinia
push-KIN-ee-uh
Named after Apollos Apollosovich Musin-Pushkin (1760–1805), Russian chemist and botanist (*Asparagaceae*)

pusillus
pus-ILL-us
pusilla, pusillum
Particularly small, as in *Soldanella pusilla*

pustulatus
pus-tew-LAH-tus
pustulata, pustulatum
Appearing to be blistered, as in *Lachenalia pustulata*

Puya
POO-yuh
From Mapuche (Chile) word, meaning "pointed," and the vernacular name for *P. chilensis* (*Bromeliaceae*)

pycnacanthus
pik-na-KAN-thus
pycnacantha, pycnacanthum
Densely spined, as in *Coryphantha pycnacantha*

pycnanthus
pik-NAN-thus
pycnantha, pycnanthum
With densely crowded flowers, as in *Acer pycnanthum*

Pycnostachys
pik-noh-STAK-is
From Greek *pyknos*, meaning "dense," and *stachys*, meaning "ear of grain," because the flowers are tightly clustered (*Lamiaceae*)

pygmaeus
pig-MAY-us
pygmaea, pygmaeum
Dwarf; pygmy, as in *Erigeron pygmaeus*

Pyracantha
py-ruh-KAN-thuh
From Greek *pyr*, meaning "fire," and *akantha*, meaning "thorn," referring to the sharp thorns and scarlet fruit (*Rosaceae*)

pyramidalis
peer-uh-mid-AH-lis
pyramidalis, pyramidale
Shaped like a pyramid, as in *Ornithogalum pyramidale*

pyrenaeus
py-ren-AY-us
pyrenaea, pyrenaeum
—

pyrenaicus
py-ren-AY-ih-kus
pyrenaica, pyrenaicum
Connected with the Pyrenees, as in *Fritillaria pyrenaica*

pyrifolius
py-rih-FOH-lee-us
pyrifolia, pyrifolium
With leaves like pear (*Pyrus*), as in *Salix pyrifolia*

pyriformis
py-rih-FOR-mis
pyriformis, pyriforme
Shaped like a pear, as in *Rosa pyriformis*

Pyrostegia
py-roh-STEE-jee-uh
From Greek *pyr*, meaning "fire," and *stegos*, meaning "cover," because the flame-color flowers can coat this climber (*Bignoniaceae*)

Pyrrosia
py-ROH-see-uh
From Greek *pyrrhos*, meaning "color of flames," a reference to the tawny frond undersides in some species (*Polypodiaceae*)

Pyrus
PY-rus
From Latin *pirum*, the vernacular name for pear, *P. communis* (*Rosaceae*)

quadr-
Used in compound words to denote four

quadrangularis
kwad-ran-gew-LAH-ris

quadrangularis, quadrangulare

—

quadrangulatus
kwad-ran-gew-LAH-tus

quadrangulata, quadrangulatum

With four angles, as in *Passiflora quadrangularis*

quadratus
kwad-RAH-tus

quadrata, quadratum

In fours, as in *Restio quadratus*

quadriauritus
kwad-ree-AWR-ry-tus

quadriaurita, quadriauritum

With four ears, as in *Pteris quadriaurita*

quadrifidus
kwad-RIF-ee-dus

quadrifida, quadrifidum

Cut into four, as in *Calothamnus quadrifidus*

quadrifolius
kwad-rih-FOH-lee-us

quadrifolia, quadrifolium

With four leaves, as in *Marsilea quadrifolia*

quadrivalvis
kwad-rih-VAL-vis

quadrivalvis, quadrivalve

With four valves, as in *Nicotiana quadrivalvis*

quamash
KWA-mash

Nex Perce (Native American) word for *Camassia*, especially *C. quamesh*

quamoclit
KWAM-oh-klit

Old generic name, possibly, meaning kidney bean, as in *Ipomoea quamoclit*

quercifolius
kwer-se-FOH-lee-us

quercifolia, quercifolium

With leaves like oak (*Quercus*), as in *Hydrangea quercifolia*

Quercus
KWER-kus

From classical Latin name for English oak, *Q. robur*; the name may derive from an unknown early European language or from Greek *kerkhaleos* ("rough," like the bark), or *kakhrus*, meaning "acorn" (*Fagaceae*)

Quesnelia
kes-NEE-lee-uh

Named after D. Quesnel (dates unknown), who grew plants then known as *Billbergia quesneliana* in Paris (*Bromeliaceae*)

quin-
Used in compound words to denote five

quinatus
kwi-NAH-tus

quinata, quinatum

In fives, as in *Akebia quinata*

quinoa
KEEN-oh-a

A Spanish word for *Chenopodium quinoa*, from Quechua, *kinua*

quinqueflorus
kwin-kway-FLOR-rus

quinqueflora, quinqueflorum

With five flowers, as in *Enkianthus quinqueflorus*

quinquefolius
kwin-kway-FOH-lee-us

quinquefolia, quinquefolium

With five leaves, often referring to leaflets, as in *Parthenocissus quinquefolia*

quinquevulnerus
kwin-kway-VUL-ner-us

quinquevulnera, quinquevulnerum

With five wounds (for example, marks), as in *Aerides quinquevulnerum*

Quisqualis
KWIS-kwuh-lis

From Malay *udani*, converted to Dutch *hoedanig*, meaning "how, what?," then translated to Latin *quis*, meaning "who," and *qualis*, meaning "what" (*Combretaceae*)

GENUS SPOTLIGHT

Quercus

The English oak (*Quercus robur*) is one of Europe's most familiar trees. The common name is misleading, because England actually has two native oaks (the other is *Q. petraea*), while *Q. robur*'s range stretches east to the Caucasus. One of the easiest ways to distinguish the two British oaks is to look at their acorns. In *Q. robur*, they sit at the end of a stalk (or peduncle) and an alternative common name is pedunculate oak, while in *Q. petraea*, the acorns lack a stalk and so are said to be sessile, thus "sessile oak."

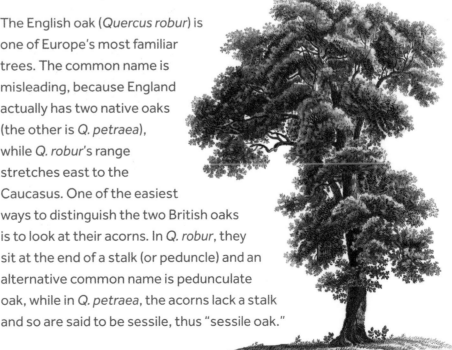

Quercus robur

racemiflorus
ray-see-mih-FLOR-us
racemiflora, racemiflorum
—

racemosus
ray-see-MOH-sus
racemosa, racemosum
With flowers that appear in racemes, as in *Nepeta racemosa*

raddianus
rad-dee-AH-nus
raddiana, raddianum
Named after Giuseppe Raddi (1770–1829), Italian botanist, as in *Adiantum raddianum*

Radermachera
ray-der-MAK-er-uh
Named after Jacob Cornelis Matthieu Radermacher (1741–83), Dutch botanist and artist (*Bignoniaceae*)

radiatus
rad-ee-AH-tus
radiata, radiatum
With rays, as in *Pinus radiata*

radicans
RAD-ee-kanz
With stems that take root, as in *Campsis radicans*

radicatus
rad-ee-KAH-tus
radicata, radicatum
With conspicuous roots, as in *Papaver radicatum*

radicosus
ray-dee-KOH-sus
radicosa, radicosum
With many roots, as in *Silene radicosa*

radiosus
ray-dee-OH-sus
radiosa, radiosum
With many rays, as in *Masdevallia radiosa*

radula
RAD-yoo-luh
From Latin *radula*, a scraper, as in *Silphium radula*

ramentaceus
ra-men-TA-see-us
ramentacea, ramentaceum
Covered with scales, as in *Begonia ramentacea*

ramiflorus
ram-ee-FLOR-us
ramiflora, ramiflorum
With flowers on the older branches, as in *Romulea ramiflora*

Ramonda
ray-MON-duh
Named after Louis-Francois Èlisabeth Ramon de Carbonnières (1755–1827), French politician, botanist, and geologist (*Gesneriaceae*)

ramondioides
ram-on-di-OY-deez
Resembling *Ramonda*, as in *Conandron ramondoides*

ramosissimus
ram-oh-SIS-ih-mus
ramosissima, ramosissimum
Much branched, as in *Lonicera ramosissima*

ramosus
ram-OH-sus
ramosa, ramosum
Branched, as in *Anthericum ramosum*

ramulosus
ram-yoo-LOH-sus
ramulosa, ramulosum
Twiggy, as in *Celmisia ramulosa*

ranunculoides
ra-nun-kul-OY-deez
Resembling buttercup (*Ranunculus*), as in *Anemone ranunculoides*

Ranunculus [1, 2]
ruh-NUNK-ew-lus
From Latin *rana*, meaning "frog," and diminutive *unculus*, because some species, especially water-crowfoots, are aquatic (*Ranunculaceae*)

Ranzania
ran-ZAH-nee-uh
Named after Ono Ranzan (1729–1810), Japanese botanist and herbalist (*Berberidaceae*)

Raoulia
rah-OOL-ee-uh
Named after Étienne Fiacre Louis Raoul (1815–52), French naval surgeon and naturalist (*Asteraceae*)

Raphanus
RAF-uh-nus
From Greek *raphanos*, the vernacular name for radish, *R. sativus* (*Brassicaceae*)

Ranunculus aconitifolius

Ranunculus asiaticus

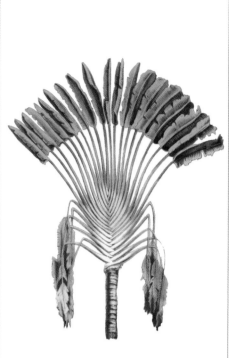

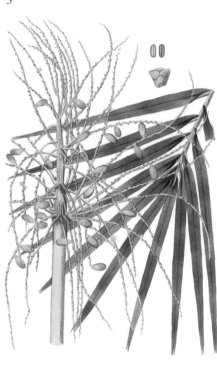

Raphia farinifera

Ravenala madagascariensis

Phoenix reclinata

Raphia [1]
RAF-ee-uh
From Malagasy (Madagascar) *rofia*, the vernacular name for
R. farinifera (*Arecaceae*)

rariflorus
rar-ee-FLOR-us
rariflora, rariflorum
With scattered flowers, as in *Carex rariflora*

Ravenala [2]
rav-uh-NAH-luh
From Malagasy (Madagascar) *ravinala*, meaning "forest leaves"
(*Strelitziaceae*)

re-
Used in compound words to denote back or again

Rebutia
ruh-BEW-tee-uh
Named after Pierre Rebut (c. 1828–98), French horticulturist
(*Cactaceae*)

reclinatus [3]
rek-lin-AH-tus
reclinata, reclinatum
Bent backward, as in *Phoenix reclinata*

rectus
REK-tus
recta, rectum
Upright, as in *Phygelius × rectus*

recurvatus
rek-er-VAH-tus
recurvata, recurvatum

recurvus
re-KUR-vus
recurva, recurvum
Curved backward, as in *Beaucarnea recurvata*

redivivus
re-div-EE-vus
rediviva, redivivum
Revived; brought back to life (for example, after drought) as in
Lunaria rediviva

reductus
red-UK-tus
reducta, reductum
Dwarf, as in *Sorbus reducta*

reflexus
ree-FLEKS-us
reflexa, reflexum
—

refractus
ray-FRAK-tus
refracta, refractum
Bent sharply backward, as in *Correa reflexa*

refulgens
ref-FUL-genz
Shining brightly, as in *Bougainvillea refulgens*

Reseda luteola

regalis
re-GAH-lis
regalis, regale
Regal; of exceptional merit, as in *Osmunda regalis*

reginae
ree-JIN-ay-ee
Relating to a queen, as in *Strelitzia reginae*

reginae-olgae
ree-JIN-ay-ee OL-gy
Named after Queen Olga of Greece (1851–1926), as in *Galanthus reginae-olgae*

regius
REE-jee-us
regia, regium
Royal, as in *Juglans regia*

rehderi
RAY-der-eye
—

rehderianus
ray-der-ee-AH-nus
rehderiana, rehderianum
Named after Alfred Rehder (1863–1949), German-born dendrologist who worked at the Arnold Arboretum, Massachusetts, as in *Clematis rehderiana*

Rehderodendron
ray-der-oh-DEN-dron
Named after Alfred Rehder (1863–1949), German-born American botanist (*Styracaceae*)

Rehmannia
ray-MAN-ee-uh
Named after Joseph Rehmann (?–1831), German physician (*Orobanchaceae*)

rehmannii
ray-MAN-ee-eye
Named after Joseph Rehmann (?–1831), German physician, or Anton Rehmann (1840–1917), Polish botanist, as in *Zantedeschia rehmannii*

reichardii
ri-KAR-dee-eye
Named after Johann Jakob Reichard, (1743–1782), German botanist, as in *Erodium reichardii*

reichenbachiana
rike-en-bak-ee-AH-nuh
—

reichenbachii
ry-ken-BAHK-ee-eye
Named after Heinrich Gottlieb Ludwig Reichenbach (1793–1879) or Heinrich Gustav Reichenbach, as in *Echinocereus reichenbachii*

Reineckea
ry-NEK-ee-uh
Named after Joseph Heinrich Julius Reinecke (1799–1871), German horticulturist (*Asparagaceae*)

Reinwardtia
rine-WARD-tee-uh
Named after Caspar Georg Carl Reinwardt (1773–1854), Dutch botanist (*Linaceae*)

religiosus
re-lij-ee-OH-sus
religiosa, religiosum
Relating to religious ceremonies; sacred, as in *Ficus religiosa*, under which the Buddha attained enlightenment

remotus
ree-MOH-tus
remota, remotum
Scattered, as in *Carex remota*

Renanthera
ray-NAN-thu-ruh
From Latin *renis*, meaning "kidney," and anther, because some species have anthers of this shape (*Orchidaceae*)

renardii
ren-AR-dee-eye
Named after Charles Claude Renard (1809–86), as in *Geranium renardii*

reniformis
ren-ih-FOR-mis
reniformis, reniforme
Shaped like a kidney, as in *Begonia reniformis*

repandus
REP-an-dus
repanda, repandum
With wavy margins, as in *Cyclamen repandum*

repens
REE-penz
With a creeping habit, as in *Gypsophila repens*

replicatus
rep-lee-KAH-tus
replicata, replicatum
Doubled; folded back, as in *Berberis replicata*

reptans
REP-tanz
With a creeping habit, as in *Ajuga reptans*

requienii
re-kwee-EN-ee-eye
Named after Esprit Requien (1788–1851), French naturalist, as in *Mentha requienii*

Reseda [4]
RES-uh-duh
From Latin *resedare*, meaning "to assuage" or "to heal," due to its medicinal properties (*Resedaceae*)

resiniferus
res-in-IH-fer-us
resinifera, resiniferum
—

resinosus
res-in-OH-sus
resinosa, resinosum
Producing resin, as in *Euphorbia resinifera*

Rhinanthus

Yellow rattle (*Rhinanthus minor*) gets its common name from the dry, papery seed pods, which contain numerous small seed that rattle in the breeze. These seed has an important role to play in meadows, boosting the diversity of wildflowers by reining in the grasses. *Rhinanthus* is a parasite and, while it can survive on its own, it prefers to connect to the root system of a vigorous grass and steal water and nutrients. This weakens the grass, allowing other nonparasitic wildflowers, such as orchids, to successfully compete and thrive.

Rhinanthus minor

Restio
RES-tee-oh
From Latin *restis*, meaning "rope," because the plant is a source of cordage (*Restionaceae*)

reticulatus
reh-tick-yoo-LAH-tus
reticulata, reticulatum
Netted, as in *Iris reticulata*

retortus
re-TOR-tus
retorta, retortum
—

retroflexus
re-troh-FLEKS-us
retroflexa, retroflexum
—

retrofractus
re-troh-FRAK-tus
retrofracta, retrofractum
Twisted or turned backward, as in *Helichrysum retortum*

retusus
re-TOO-sus
retusa, retusum
With a rounded and notched tip, *Coryphantha retusa*

reversus
ree-VER-sus
reversa, reversum
Reversed, as in *Rosa* × *reversa*

revolutus
re-vo-LOO-tus
revoluta, revolutum
Rolled backward (for example, of leaves), as in *Cycas revoluta*

rex
REKS
King; with outstanding qualities, as in *Begonia rex*

rhamnifolius
ram-nih-FOH-lee-us
rhamnifolia, rhamnifolium
With leaves like buckthorn (*Rhamnus*), as in *Rubus rhamnifolius*

rhamnoides
ram-NOY-deez
Resembling buckthorn (*Rhamnus*), as in *Hippophae rhamnoides*

Rhamnus
RAM-nus
From Greek *rhamnos*, the name of a spiny shrub (*Rhamnaceae*)

Rhaphidophora
raf-id-oh-FOR-uh
From Greek *rhaphis*, meaning "needle," and *phoreus*, meaning "bearer," because sharp crystals can be found within the tissues (*Araceae*)

Rhaphiolepis
raf-ee-oh-LEE-pis
From Greek *rhaphis*, meaning "needle," and *lepis*, meaning "scale," referring to the slender bracts below the flower clusters (*Rosaceae*)

Rhapidophyllum
rap-id-oh-FIL-um
From *Rhapis*, a related genus, and *phyllon*, meaning leaf; the two genera have similar foliage (*Arecaceae*)

Rhapis
RAH-pis
From Greek *rhapis*, meaning "rod," an allusion to the cane-ike stems (*Arecaceae*)

Rheum
REE-um
From Greek *rheon*, the vernacular name for rhubarb (*Polygonaceae*)

Rhinanthus

ry-NAN-thus

From Greek *rhis*, meaning "nose," and *anthos*, meaning "flower," referring to the snoutlike upper petal (*Orobanchaceae*)

Rhipsalis

RIP-suh-lis

From Greek *rhips*, meaning "wickerwork," a reference to the often entangled, many-branched stems (*Cactaceae*)

rhizophyllus

ry-zo-FIL-us

rhizophylla, rhizophyllum

With leaves that take root, as in *Asplenium rhizophyllum*

Rhodanthe

roh-DAN-thee

From Greek *rhodon*, meaning "rose," and *anthos*, meaning "flower' (*Asteraceae*)

Rhodanthemum

roh-DAN-thuh-mum

From Greek *rhodon*, meaning "rose," and *anthos*, meaning "flower" (*Asteraceae*)

rhodanthus

rho-DAN-thus

rhodantha, rhodanthum

With rose-color flowers, as in *Mammillaria rhodantha*

Rhodiola

roh-dee-OH-luh

From Greek *rhodon*, meaning "rose," because the roots of *R. rosea* have a floral scent (*Crassulaceae*)

Rhodochiton

roh-doh-KY-ton

From Greek *rhodon*, meaning "rose," and *chiton*, meaning "tunic," referring to the red calyx (*Plantaginaceae*)

Rhodocoma

roh-doh-KOH-muh

From Greek *rhodon*, meaning "rose," and *kome*, meaning "hair," an allusion to the rufous hairlike inflorescences (*Restionaceae*)

Rhododendron [1]

roh-doh-DEN-dron

From Greek *rhodon*, meaning "rose," and *dendron*, meaning "tree"; the name rhododendron was applied to oleander (*Nerium oleander*) by the Greeks (*Ericaceae*)

Rhodohypoxis

roh-doh-hy-POX-iss

From Greek *rhodon*, meaning "rose," and *Hypoxis*, a related genus that typically has yellow flowers (*Hypoxidaceae*)

rhodopensis

roh-doh-PEN-sis

rhodopensis, rhodopense

From the Rhodope Mountains, Bulgaria, as in *Haberlea rhodopensis*

Rhodophiala

roh-doh-FEE-ah-luh

From Greek *rhodon*, meaning "rose," and *phiali*, meaning "drinking vessel," referring to the shape and color of the flowers (*Amaryllidaceae*)

Rhodotypos

roh-doh-TY-pos

From Greek *rhodon*, meaning "rose," and *typos*, meaning "model," because the flowers are roselike (*Rosaceae*)

rhoeas

RE-as

Ancient Greek *rhoias*, name for *Papaver rhoeas*

Rhoicissus

roh-i-SIS-us

From Latin *rhoicus*, referring to the genus *Rhus*, and Greek *kissos*, meaning "ivy," because the leaves of this climber vaguely resemble those of *Rhus* (*Vitaceae*)

rhombifolius

rom-bih-FOH-lee-us

rhombifolia, rhombifolium

With diamond-shape leaves, as in *Cissus rhombifolia*

rhomboideus

rom-BOY-dee-us

rhomboidea, rhomboideum

Shaped like a diamond, as in *Rhombophyllum rhomboideum*

Rhopalostylis

roh-pay-loh-STY-lis

From Greek *rhopalos*, meaning "cudgel," and *stylis*, meaning "style," because in the center of male flowers is an infertile, clublike style (*Arecaceae*)

1

Rhododendron cuffeanum

Rhus
ROOS
From Greek *rhous*, the vernacular name for *R. coriaria* (*Anacardiaceae*)

Rhyncholaelia
rin-koh-LAY-lee-uh
From Greek *rhynchos*, meaning "snout," and *Laelia*, a related genus, referring to the beak that separates the ovary from the rest of the flower (*Orchidaceae*)

rhytidophyllus
ry-ti-do-FIL-us
rhytidophylla, rhytidophyllum
With wrinkled leaves, as in *Viburnum rhytidophyllum*

Ribes
RY-beez
From Arabic *ribas*, meaning "acidic" pertaining to the sharp-tasting fruit; *ribas* also refers to the unrelated rhubarb, *Rheum × hybridum* (*Grossulariaceae*)

richardii
rich-AR-dee-eye
Named after various persons with the forename or surname Richard; thus *Cortaderia richardii* commemorates the French botanist Achille Richard (1794–1852)

richardsonii
rich-ard-SON-ee-eye
Named after Sir John Richardson, nineteenth-century Scottish explorer, as in *Heuchera richardsonii*

Richea
RICH-ee-uh
Named after Claude Antoine Gaspard Riche (1762–97), French botanist and entomologist (*Ericaceae*)

Ricinus [1]
RIS-in-us
From Latin *ricinus*, meaning "tick," because the seed resembles this parasite (*Euphorbiaceae*)

rigens
RIG-enz
–

rigidus
RIG-ih-dus
rigida, rigidum
Rigid; inflexible; stiff, as in *Verbena rigida*

rigescens
rig-ES-enz
Rather rigid, as in *Diascia rigescens*

ringens
RIN-jenz
Gaping; open, as in *Arisaema ringens*

riparius
rip-AH-ree-us
riparia, riparium
Of riverbanks, as in *Ageratina riparia*

ritro
RIH-tro
Probably from the Greek for globe thistle, *rhytros*, as in *Echinops ritro*

ritteri
RIT-ter-ee
ritterianus rit-ter-ee-AH-nus
–

ritteriana, ritterianum
Named after Friedrich Ritter (1898–1989), German cactus collector, as in *Cleistocactus ritteri*

rivalis
riv-AH-lis
rivalis, rivale
Growing by the side of streams, as in *Geum rivale*

riversleaianum
riv-ers-lee-i-AY-num
Named after Riverslea Nursery, Hampshire, England, as in *Geranium × riversleaianum*

riviniana
riv-in-ee-AH-nuh
Named after Augustus Quirinus Rivinus (August Bachmann; 1652–1723), German physician and botanist, as in *Viola riviniana*

rivularis
riv-yoo-LAH-ris
rivularis, rivulare
Loving brooks, as in *Cirsium rivulare*

Robinia
rob-IN-ee-uh
Named after Jean Robin (1550–1629), French botanist and horticulturist (*Fabaceae*)

robur
ROH-bur
Oak, as in *Quercus robur*

robustus
roh-BUS-tus
robusta, robustum
Growing strongly; sturdy, as in *Eremurus robustus*

rockii
ROK-ee-eye
Named after Joseph Francis Charles Rock (1884–1962), Austrian-born American plant hunter, as in *Paeonia rockii*

Rodgersia
ro-JER-zee-uh
Named after John Rodgers (1812–82), American naval officer (*Saxifragaceae*)

roebelenii
roh-bel-EN-ee-eye
Named after Carl Roebelen (1855–1927), orchid collector, as in *Phoenix roebelenii*

Rohdea
ROH-dee-uh
Named after Michael Rohde (1782–1812), German physician and botanist (*Asparagaceae*)

romanus
roh-MAHN-us
romana, romanum
Roman, as in *Orchis romana*

1

Ricinus communis

romieuxii

rom-YOO-ee-eye

Named after Henri Auguste Romieux (1857–1937), French botanist, as in *Narcissus romieuxii*

Romneya

ROM-nay-uh

Named after John Thomas Romney Robinson (1792–1882), Irish astronomer (*Papaveraceae*)

Romulea

rom-ew-LEE-uh

Named after Romulus, legendary founder of Rome, Italy, because the plants were said to be common in the Roman countryside (*Iridaceae*)

Rondeletia

ron-duh-LEE-tee-uh

Named after Guillaume Rondelet (1507–66), French physician and botanist (*Rubiaceae*)

Rosa [2]

ROH-zuh

The classical Latin name for rose (*Rosaceae*)

rosa-sinensis

RO-sa sy-NEN-sis

The rose of China, as in *Hibiscus rosa-sinensis*

rosaceus

ro-ZAY-see-us

rosacea, rosaceum

Roselike, as in *Saxifraga rosacea*

Roscoea [3]

ros-KOH-ee-uh

Named after William Roscoe (1753–1831), British lawyer, banker, and politician (*Zingiberaceae*)

roseus

RO-zee-us

rosea, roseum

Coloured like rose (*Rosa*), as in *Lapageria rosea*

rosmarinifolius

rose-ma-rih-nih-FOH-lee-us

rosmarinifolia, rosmarinifolium

With leaves like rosemary (*Rosmarinus*), as in *Santolina rosmarinifolia*

Rosmarinus

roz-MA-rin-us

From Latin *roz*, meaning "dew," and *marinus*, meaning "sea," alluding to its coastal habitat (*Lamiaceae*)

rostratus

ro-STRAH-tus

rostrata, rostratum

With a beak, as in *Magnolia rostrata*

Rosularia

roz-ew-LAIR-ee-uh

From Latin *rosula*, meaning "small rosette," the typical leaf arrangement (*Crassulaceae*)

2

Rosa × centifolia

3

Roscoea species

rotatus

ro-TAH-tus

rotata, rotatum

Shaped like a wheel, as in *Phlomis rotata*

Rothmannia

roth-MAN-ee-uh

Named after Göran Rothman (1739–78), Swedish physician (*Rubiaceae*)

rothschildianus

roths-child-ee-AH-nus

rothschildiana, rothschildianum

Named after Lionel Walter Rothschild (1868–1937), or other members of the House of Rothschild, as in *Paphiopedilum rothschildianum*

rotundatus

roh-tun-DAH-tus

rotundata, rotundatum

Rounded, as in *Carex rotundata*

rotundifolius

ro-tun-dih-FOH-lee-us

rotundifolia, rotundifolium

With leaves that are round, as in *Prostanthera rotundifolia*

rotundus

ro-TUN-dus

rotunda, rotundum

Rounded, as in *Cyperus rotundus*

rowleyanus

ro-lee-AH-nus

Named after Gordon Douglas Rowley, (b. 1921) British botanist and succulent expert, as in *Senecio rowleyanus*

roxburghii

roks-BURGH-ee-eye

Named after William Roxburgh (1751–1815), Superintendent of Calcutta Botanic Garden, as in *Rosa roxburghii*

roxieanum

rox-ee-AY-num

Named after Roxie Hanna, nineteenth-century British missionary, as in *Rhododendron roxieanum*

Roystonea

roy-STOH-nee-uh

Named after Roy Stone (1836–1905), American soldier (*Arecaceae*)

1

Rubus idaeus

rubellus

roo-BELL-us

rubella, rubellum

Pale red, becoming red, as in *Peperomia rubella*

rubens

ROO-benz

—

ruber

ROO-ber

rubra, rubrum

Red, as in *Plumeria rubra*

rubescens

roo-BES-enz

Becoming red, as in *Salvia rubescens*

Rubia

ROO-bee-uh

From Latin *ruber*, meaning "red," because a red dye can be extracted from the roots of madder, *R. tinctoria* (*Rubiaceae*)

rubiginosus

roo-bij-ih-NOH-sus

rubiginosa, rubiginosum

Rusty, as in *Ficus rubiginosa*

rubioides

roo-bee-OY-deez

Resembling madder (*Rubia*), as in *Bauera rubioides*

rubri-

Used in compound words to denote red

rubricaulis

roo-bri-KAW-lis

rubricaulis, rubricaule

With red stems, as in *Actinidia rubricaulis*

rubriflorus

roo-brih-FLOR-us

rubiflora, rubiflorum

With red flowers, as in *Schisandra rubriflora*

Rubus [1]

ROO-bus

From Latin *rubus*, meaning "bramble," possibly deriving from *ruber*, meaning "red," the color of fruit in raspberry, *R. idaeus*, and others (*Rosaceae*)

Rudbeckia

rood-BEK-ee-uh

Named after father and son Olaus Johannes Rudbeck (1630–1702) and Olaus Olai Rudbeck (1660–1740), Swedish scientists (*Asteraceae*)

rudis

ROO-dis

rudis, rude

Coarse, growing on uncultivated ground, as in *Persicaria rudis*

Rumex scutatus

Ruta graveolens

Ruellia
roo-EE-lee-uh
Named after Jean Ruel (1474–1537),
French physician (*Acanthaceae*)

rufinervis
roo-fi-NER-vis
rufinervis, rufinerve
With red veins, as in *Acer rufinerve*

rufus
ROO-fus
rufa, rufum
Red, as in *Prunus rufa*

rugosus
roo-GOH-sus
rugosa, rugosum
Wrinkled, as in *Rosa rugosa*

Rumex [2]
ROO-meks
From Latin *rumo*, meaning "to suck,"
because dock leaves were sucked to quench
the thirst (*Polygonaceae*)

Rumohra
roo-MOH-ruh
Named after Carl Friedrich von Rumohr
(1785–1843), German writer and historian
(*Dryopteridaceae*)

rupestris
rue-PES-tris
rupestris, rupestre
Of rocky places, as in *Leptospermum rupestre*

rupicola
roo-PIH-koh-luh
Growing on cliffs and ledges, as in
Penstemon rupicola

rupifragus
roo-pee-FRAG-us
rupifraga, rupifragum
Breaking rocks, as in *Papaver rupifragum*

ruscifolius
rus-kih-FOH-lee-us
ruscifolia, ruscifolium
With leaves like butcher's broom (*Ruscus*), as
in *Sarcococca ruscifolia*

Ruscus
RUS-kus
From Latin *ruscum*, meaning "butcher's
broom," because the sharp-tipped stems were
used to clean butcher's blocks (*Asparagaceae*)

russatus
russ-AH-tus
russata, russatum
Russet, as in *Rhododendron russatum*

Russelia
rus-EL-ee-uh
Named after Alexander Russell (1715–68),
Scottish physician and naturalist
(*Plantaginaceae*)

russellianus
russ-el-ee-AH-nus
russelliana, russellianum
Named after John Russell, 6th Duke of
Bedford (1766–1839), author of numerous
botanical and horticultural works, as in
Miltonia russelliana

rusticanus
rus-tik-AH-nus
rusticana, rusticanum
—

rusticus
RUS-tih-kus
rustica, rusticum
Relating to the country, as in *Armoracia
rusticana*

Ruta [3]
ROO-tuh
From Latin *ruta*, meaning "disagreeable," in
reference to the pungent smell of the foliage
(*Rutaceae*)

ruta-muraria
ROO-tuh-mur-AY-ree-uh
Literally wall rue, as in *Asplenium ruta-
muraria*

ruthenicus
roo-THEN-ih-kus
ruthenica, ruthenicum
Connected with Ruthenia, a historial area
consisting of parts of Russia and eastern
Europe, as in *Fritillaria ruthenica*

rutifolius
roo-tih-FOH-lee-us
rutifolia, rutifolium
With leaves like rue (*Ruta*), as in *Corydalis
rutifolia*

rutilans
ROO-til-lanz
Reddish, as in *Parodia rutilans*

sabatius

sa-BAY-shee-us

sabatia, sabatium

Connected with Savona, Italy, as in
Convolvulus sabatius

saccatus

sak-KAH-tus

saccata, saccatum

Like a bag, or saccate, as in *Lonicera saccata*

saccharatus

sak-kah-RAH-tus

saccharata, saccharatum

—

saccharinus

sak-kah-EYE-nus

saccharina, saccharinum

Sweet or sugared, as in *Pulmonaria
saccharata*

Saccharum [1]

SAK-uh-rum

From Greek *sakkharon*, meaning "sugar,"
extracted from sugarcane, *S. officinarum*
(*Poaceae*)

sacciferus

sak-IH-fer-us

saccifera, sacciferum

Bearing bags or sacks, as in *Dactylorhiza
saccifera*

sachalinensis

saw-kaw-lin-YEN-sis

sachalinensis, sachalinense

From the island Sakhalin, off the coast of
Russia, as in *Abies sachalinensis*

Sadleria

sad-LEER-ee-uh

Named after Joseph Sadler (1791–1849),
Hungarian physician and botanist
(*Blechnaceae*)

Sageretia

sag-uh-REE-tee-uh

Named after Augustin Sageret (1763–
1851), French botanist (*Rhamnaceae*)

Sagina

suh-JY-nuh

From Latin *sagina*, meaning "food' or
"fodder," because a related genus (*Spergula*)
was fed to livestock (*Caryophyllaceae*)

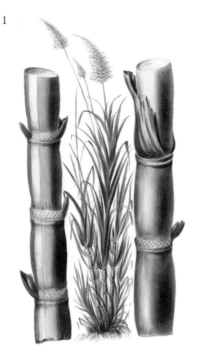

Saccharum officinarum

sagittalis

saj-ih-TAH-lis

sagittalis, sagittale

—

sagittatus

saj-ih-TAH-tus

sagittata, sagittatum

Shaped like an arrow, as in *Genista sagittalis*

Sagittaria

saj-i-TAIR-ee-uh

From Latin *sagitta*, meaning "arrow," because
the leaves resemble arrowheads
(*Alismataceae*)

sagittifolius

saj-it-ih-FOH-lee-us

sagittifolia, sagittifolium

With leaves shaped like arrows, as in
Sagittaria sagittifolia

Saintpaulia

saint-PAWL-ee-uh

Named after Walter von Saint Paul-Illaire
(1860–1940), German colonial
administrator and botanist (*Gesneriaceae*)

salicarius

sa-lih-KAH-ree-us

salicaria, salicarium

Like willow (*Salix*), as in *Lythrum salicaria*

Salix alba

salicariifolius

sa-lih-kar-ih-FOH-lee-us

salicariifolia, salicariifolium

—

salicifolius

sah-lis-ih-FOH-lee-us

salicifolia, salicifolium

With leaves like willow (*Salix*), as in
Magnolia salicifolia

salicinus

sah-lih-SEE-nus

salicina, salicinum

Like willow (*Salix*), as in *Prunus salicina*

salicornioides

sal-eye-korn-ee-OY-deez

Resembling glasswort (*Salicornia*), as in
Hatiora salicornioides

salignus

sal-LIG-nus

saligna, salignum

Like willow (*Salix*), as in *Podocarpus salignus*

salinus

sal-LY-nus

salina, salinum

Of salty regions, as in *Carex salina*

Salix [2]

SAY-liks

The classical Latin name for willow
(*Salicaceae*)

Sambucus nigra

Sambucus ebulus

Sandersonia aurantiaca

Salpiglossis
sal-pi-GLOS-iss
From Greek *salpinx*, meaning "trumpet," and *glossa*, meaning "tongue," because the tubular flowers have tonguelike lobes (*Solanaceae*)

saluenensis
sal-WEN-en-sis
saluenensis, saluenense
From the Salween River, China, as in *Camellia saluenensis*

Salvia
SAL-vee-uh
From Latin *salvia*, the vernacular name for sage, *S. officinalis* (*Lamiaceae*)

salviifolius
sal-vee-FOH-lee-us
salviifolia, salviifolium
With leaves like salvia, as in *Cistus salviifolius*

Salvinia
sal-VIN-ee-uh
Named after Anton Maria Salvini (1633–1729), Italian botanist (*Salviniaceae*)

sambucifolius
sam-boo-kih-FOH-lee-us
sambucifolia, sambucifolium
With leaves like elder (*Sambucus*), as in *Rodgersia sambucifolia*

sambucinus
sam-byoo-ki-nus
sambucina, sambucinum
Like elder (*Sambucus*), as in *Rosa sambucina*

Sambucus [1, 2]
sam-BEW-kus
From classical Latin name for elder, *S. nigra*; a sambuca was a harp supposedly made of elder wood (*Adoxaceae*)

samius
SAM-ee-us
samia, samium
Connected with the isle of Samos, Greece, as in *Phlomis samia*

sanctus
SANK-tus
sancta, sanctum
Holy, as in *Rhododendron sanctum*

sanderi
SAN-der-eye
—
sanderianus
san-der-ee-AH-nus
sanderiana, sanderianum
Named after Henry Frederick Conrad Sander (1847–1920), German-born British plant collector, nurseryman and orchid expert, as in *Dracaena sanderiana*

Sandersonia [3]
san-der-SOH-nee-uh
Named after John Sanderson (1820–81), Scottish journalist and botanist (*Colchicaceae*)

Sanguinaria
san-gwin-AIR-ee-uh
From Latin *sanguis*, meaning "blood," because the roots exude red sap (*Papaveraceae*)

sanguineus
san-GWIN-ee-us
sanguinea, sanguineum
Blood-red, as in *Geranium sanguineum*

Sanguisorba
san-gwi-SOR-buh
From Latin *sanguis*, meaning "blood," and *sorbeo*, meaning "to absorb"; the plant was said to stop bleeding (*Rosaceae*)

Sanicula
san-IK-ew-luh
From Latin *sanus*, meaning "healthy," alluding to medicinal properties (*Apiaceae*)

Sansevieria
san-sev-EER-ee-uh
Named after Raimondo di Sangro, Prince of Sansevero (1710–71), Italian soldier and scientist (*Asparagaceae*)

Santolina
san-toh-LEE-nuh
From Latin *sanctus*, meaning "holy," and *linum*, meaning "flax," a name applied to one species that had medicinal value (*Asteraceae*)

Sanvitalia
san-vi-TAY-lee-uh
Named after Federico Sanvitali (1704–61), Italian mathematician (*Asteraceae*)

sapidus
sap-EE-dus
sapida, sapidum
With a pleasant taste, as in *Rhopalostylis sapida*

Sapindus [4]

sa-PIN-dus

From Latin *sapo*, meaning "soap," and *indicus*, meaning "from India"; the fruit is rich in saponins and has been used as soap (*Sapindaceae*)

Saponaria

sa-pon-AIR-ee-uh

From Latin *sapo*, meaning "soap," because the crushed foliage of soapwort, *S. officinalis*, produces a lather (*Caryophyllaceae*)

saponarius

sap-oh-NAIR-ee-us

saponaria, saponarium

Soapy, as in *Sapindus saponaria*

sarcocaulis

sar-koh-KAW-lis

sarcocaulis, sarcocaule

With a fleshy stem, as in *Crassula sarcocaulis*

Sarcocaulon

sar-koh-KOW-lon

From Greek *sarkos*, meaning "fleshy," and *caulon*, meaning "stem"; these are succulent shrubs (*Geraniaceae*)

Sarcococca

sar-koh-KOK-uh

From Greek *sarkos*, meaning "fleshy," and *kokkos*, meaning "berry' (*Buxaceae*)

sarcodes

sark-OH-deez

Flesh-like, as in *Rhododendron sarcodes*

sardensis

saw-DEN-sis

sardensis, sardense

Of Sardis (Sart), Turkey, as in *Chionodoxa sardensis*

sargentianus

sar-jen-tee-AH-nus

sargentiana, sargentianum

—

sargentii

sar-JEN-tee-eye

Named after Charles Sprague Sargent (1841–1927), dendrologist and director of the Arnold Arboretum, Harvard, Massachusetts, as in *Sorbus sargentiana*

sarmaticus

sar-MAT-ih-kus

sarmatica, sarmaticum

From Sarmatia, historic territory now partly in Poland, partly in Russia, as in *Campanula sarmatica*

sarmentosus

sar-men-TOH-sus

sarmentosa, sarmentosum

Producing runners, as in *Androsace sarmentosa*

Sarmienta

sar-me-EN-tuh

Named after Martín Sarmiento (1695–1772), Spanish cleric and naturalist (*Gesneriaceae*)

sarniensis [5]

sarn-ee-EN-sis

sarniensis, sarniense

From the island of Sarnia (Guernsey), as in *Nerine sarniensis*

Sarracenia [6]

sa-ruh-SEE-nee-uh

Named after Michel Sarrazin (1659–1734), French-born Canadian botanist and physician (*Sarraceniaceae*)

Saruma

suh-ROO-muh

An anagram of the related genus *Asarum* (*Aristolochiaceae*)

Sasa

SA-suh

From Japanese, the vernacular name for these bamboos (*Poaceae*)

sasanqua

suh-SAN-kwuh

From the Japanese name for *Camellia sasanqua*

4

Sapindus saponaria

5

Nerine sarniensis

6

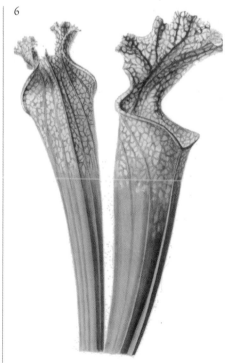

Sarracenia leucophylla

Castanea sativa

Saururus cernuus

Sassafras

SAS-uh-fras

From Spanish/French *sasafras*, probably derived from a Native American name (*Lauraceae*)

sativus [1]

sa-TEE-vus

sativa, sativum

Cultivated, as in *Castanea sativa*

Satureja

sat-ew-REE-uh

The classical Latin name for this herb, possibly derived from Arabic (*Lamiaceae*)

saundersii

son-DER-see-eye

Commemorates various eminent Saunders, for example, Sir Charles Saunders (1857–1935), as in *Pachypodium saundersii*

Sauromatum

saw-roo-MAY-tum

From Greek *sauros*, meaning "lizard," because the mottled spathe resembles reptile skin (*Araceae*)

Saururus [2]

saw-ROO-rus

From Greek *sauros*, meaning "lizard," and *oura*, meaning "tail," alluding to the shape of the flower spike (*Saururaceae*)

Saussurea

saw-SOO-ree-uh

Named after father and son Horace Bénédict (1740–99) and Nicolas Théodore (1767–1845) de Saussure, Swiss naturalists (*Asteraceae*)

saxatilis

saks-A-til-is

saxatilis, saxatile

Of rocky places, as in *Aurinia saxatilis*

Saxegothaea

saks-uh-GOTH-ee-uh

Named after Prince Albert of Saxe-Coburg and Gotha (1819–61), German-born husband to Queen Victoria (*Podocarpaceae*)

saxicola

saks-IH-koh-luh

Growing in rocky places, as in *Juniperus saxicola*

Saxifraga [3]

saks-i-FRAH-gah

From Latin *saxum*, meaning "stone," and *frangere*, meaning "to break," because these alpines often grow in rock crevices (*Saxifragaceae*)

saxorum

saks-OR-um

Of the rocks, as in *Streptocarpus saxorum*

saxosus

saks-OH-sus

saxosa, saxosum

Of rocky places, as in *Gentiana saxosa*

scaber

SKAB-er

scabra, scabrum

Rough, as in *Eccremocarpus scaber*

Scabiosa

ska-bee-OH-suh

From Latin *scabies*, meaning "itch," referring to the use of this plant to cure skin ailments (*Caprifoliaceae*)

scabiosifolius

skab-ee-oh-sih-FOH-lee-us

scabiosifolia, scabiosifolium

With leaves like scabious (*Scabiosa*), as in *Salvia scabiosifolia*

scabiosus

skab-ee-OH-sus

scabiosa, scabiosum

Scabrous, or relating to scabies, as in *Centaurea scabiosa*

Scadoxus

ska-DOKS-us

From Greek *skia*, meaning "shade," and *doxa*, meaning "glory," because the attractive bulbs grow in shady forests (*Amaryllidaceae*)

Scaevola

skuh-VOH-luh

From Greek *scaeva*, meaning "left-handed," because the flowers look like a hand (*Goodeniaceae*)

scalaris

skal-AH-ris

scalaris, scalare

Like a ladder, as in *Sorbus scalaris*

scandens

SKAN-denz

Climbing, as in *Cobaea scandens*

3

Saxifraga hieracifolia

scaposus

ska-POH-sus

scaposa, scaposum

With leafless flowering stems (scapes) as in *Aconitum scaposum*

scariosus

skar-ee-OH-sus

scariosa, scariosum

Shriveled, as in *Liatris scariosa*

sceptrum

SEP-trum

Like a scepter, as in *Isoplexis sceptrum*

schafta

SHAF-tuh

A Caspian vernacular name for *Silene schafta*

Schefflera

SHEF-lu-ruh

Named after Jacob Christoph Scheffler (1698–1742), German physician and botanist (*Araliaceae*)

schidigera

ski-DEE-ger-ruh

Bearing a spine or splinter, as in *Yucca schidigera*

schillingii

shil-LING-ee-eye

Named after Tony Schilling (b. 1935), British plantsman, as in *Euphorbia schillingii*

Schima

SHEE-muh

From Greek *skia*, meaning "shade," alluding to the dense canopy of evergreen leaves (*Theaceae*)

Schinus

SHY-nus

From Greek *schinos*, the vernacular name for mastic tree, *Pistacia lentiscus*, to which it resembles (*Anacardiaceae*)

Schisandra

sky-ZAN-druh

From Greek *schisis*, meaning "splitting," and *andros*, meaning "male"; the anthers are cleft in two (*Schisandraceae*)

Schizachyrium

sky-ZAK-ree-um

From Greek *schisis*, meaning "splitting," and *achryon*, meaning "chaff," because the lemmas (bracts in the flower spikelet) are divided (*Poaceae*)

Schizanthus

sky-ZAN-thus

From Greek *schisis*, meaning "splitting," and *anthos*, meaning "flower," because the petals are split into lobes (*Solanaceae*)

schizopetalus

ski-zo-pe-TAY-lus

schizopetala, schizopetalum

With cut petals, as in *Hibiscus schizopetalus*

Schizophragma

sky-zoh-FRAG-muh

From Greek *schisis*, meaning "splitting," and *phragma*, meaning "fence," alluding to the interior walls of the fruit, which is split into fibers (*Hydrangeaceae*)

schizophyllus

skits-oh-FIL-us

schizophylla, schizophyllum

With cut leaves, as in *Syagurus schizophylla*

Schlumbergera [1]

shloom-BERG-uh-ruh

Named after Frédéric Schlumberger (1823–93), French horticulturist (*Cactaceae*)

schmidtianus

shmit-ee-AH-nus

schmidtiana, schmidtianum

—

schmidtii

SHMIT-ee-eye

Commemorates various eminent botanists called Schmidt, as in *Artemisia schmidtiana*

schoenoprasum

skee-no-PRAY-zum

Epithet for chives (*Allium schoenoprasum*), meaning "rush leek' in Greek

schottii

SHOT-ee-eye

Can commemorate various naturalists called Schott, for example Arthur Carl Victor Schott (1814–75), as in *Yucca schottii*

schubertii

shoo-BER-tee-eye

Named after Gotthilf von Schubert (1780–1860), German naturalist, as in *Allium schubertii*

schumannii

shoo-MAHN-ee-eye

Named after Dr Karl Moritz Schumann (1851–1904), German botanist, as in *Abelia schumannii*

Sciadopitys

skee-ah-DO-pit-is

From Greek *skia*, meaning "shade," which can also be interpreted as "umbrella' or "canopy," plus *pitys*, meaning "pine" (*Sciadopityaceae*)

Scilla

SIL-uh

From Greek *skilla*, meaning "squill" or 'sea onion," possibly referring to the Mediterranean squill, *Drimia maritima* (*Asparagaceae*)

scillaris

sil-AHR-is

scillaris, scillare

Like *Scilla*, as in *Ixia scillaris*

scillifolius

sil-ih-FOH-lee-us

scillifolia, scillifolium

With leaves like *Scilla*, as in *Roscoea scillifolia*

scilloides

sil-OY-deez

Resembling *Scilla*, as in *Puschkinia scilloides*

scilloniensis

sil-oh-nee-EN-sis

scilloniensis, scilloniense

From the Isles of Scilly, England, as in *Olearia* × *scilloniensis*

Scirpus

SKIR-pus

The classical Latin name for a type of rush, possibly *Schoenoplectus lacustris* (*Juncaceae*)

sclarea

SKLAR-ee-uh

From *clarus*, "clear," as in *Salvia sclarea*

sclerophyllus

skler-oh-FIL-us

sclerophylla, sclerophyllum

With hard leaves, as in *Castanopsis sclerophylla*

Scoliopus

skoh-lee-OH-pus

From Greek *skolios*, meaning "crooked," and *pous*, meaning "footed," because the pedicels holding each flower are twisted (*Liliaceae*)

scolopendrius

skol-oh-PEND-ree-us

scolopendria, scolopendrium

From the Greek word for *Asplenium scolopendrium*, from a supposed likeness of the underside of its fronds to a millipede or centipede (Greek *skolopendra*)

scolymus

SKOL-ih-mus

From the Greek for an edible kind of thistle or artichoke, as in *Cynara scolymus*

scoparius

sko-PAIR-ee-us

scoparia, scoparium

Like broom, as in *Cytisus scoparius*

1

Schlumbergera truncata

Scorzonera hispanica

Scrophularia vernalis

Sedum aizoon

Scopolia
skuh-POH-lee-uh
Named after Giovanni Antonio Scopoli (1723–88), Italian physician and naturalist (*Solanaceae*)

scopulorum
sko-puh-LOR-um
Of crags or cliffs, as in *Cirsium scopulorum*

scorodoprasum
skor-oh-doh-PRAY-zum
Greek name for a plant between leek and garlic, as in *Allium scorodoprasum*

scorpioides
skor-pee-OY-deez
Resembling a scorpion's tail, as in *Myosotis scorpioides*

Scorzonera [2]
skor-ZON-uh-ruh
From French *scorzonère*, meaning "viper's grass," possibly due to the medicinal use of the roots in treating snake bites (*Asteraceae*)

scorzonerifolius
skor-zon-er-ih-FOH-lee-us
scorzonerifolia, scorzonerifolium
With leaves like *Scorzonera*, as in *Allium scorzonerifolium*

scoticus
SKOT-ih-kus
scotica, scoticum
Connected with Scotland, as in *Primula scotica*

scouleri
SKOOL-er-ee
Named after Dr. John Scouler (1804–71), Scottish botanist, as in *Hypericum scouleri*

Scrophularia [3]
skrof-ew-LAIR-ee-uh
From Latin *scrofula*, or tuberculosis of the lymph nodes, which when swollen are said to resemble this plant's rhizomes (*Scrophulariaceae*)

scutatus
skut-AH-tus
scutata, scutatum
—

scutellaris
skew-tel-AH-ris
scutellaris, scutellare
—

scutellatus
skew-tel-LAH-tus
scutellata, scutellatum
Shaped like a shield or platter, as in *Rumex scutatus*

Scutellaria
skew-tuh-LAIR-ee-uh
From Latin *scutella*, a small dish; the calyx holding the fruit resembles a dishlike skullcap (*Lamiaceae*)

Secale
SEK-uh-lee
From classical Latin name for rye, *S. cereale*, or derived from Celtic *sega*, meaning "sickle" (*Poaceae*)

secundatus
see-kun-DAH-tus
secundata, secundatum
—

secundiflorus
sek-und-ee-FLOR-us
secundiflora, secundiflorum
—

secundus
se-KUN-dus
secunda, secundum
With leaves or flowers growing on only one side of a stalk, as in *Echeveria secunda*

Sedum [4]
SEE-dum
From Latin *sedeo*, meaning "to sit," alluding to the compact habit of many species (*Crassulaceae*)

seemannianus
see-mahn-ee-AH-nus

seemanniana, seemannianum

—

seemannii
see-MAN-ee-eye

Named after Berthold Carl Seemann
(1825–71), German plant collector, as in
Hydrangea seemannii

segetalis
seg-UH-ta-lis

segetalis, segetale

—

segetum
seg-EE-tum

Of cornfields, as in *Euphorbia segetalis*

Selaginella
sel-aj-uh-NEL-uh

From Latin *selago*, the savin juniper
(*Juniperus sabina*), because the foliage bears
a resemblance, plus diminutive *ella*; *selago*
also refers to the similar-looking genus
Lycopodium (*Selaginellaceae*)

selaginoides
sel-aj-ee-NOY-deez

Resembling clubmoss (*Selaginella*), as in
Athrotaxis selaginoides

Selago
suh-LAY-goh

An ancient name for *Lycopodium*, to which
this plant bears a superficial resemblance;
alternatively, the name may derive from
Celtic *sel*, meaning "sight," and *jach*, meaning
"salutary," suggesting a herb beneficial to the
sight (*Scrophulariaceae*)

Selenicereus
suh-len-i-SER-ee-us

Named after Selene, goddess of the moon
in Greek mythology, plus *Cereus*, a related
genus; these cacti flower at night (*Cactaceae*)

Selinum
suh-LEE-num

The Greek *selinon*, the vernacular name
for the related celery, *Apium graveolens*
(*Apiaceae*)

selloanus
sel-lo-AH-nus

selloana, selloanum

Named after Friedrich Sellow (Sello),
nineteenth-century German explorer and
plant collector, as in *Cortaderia selloana*

Semele
SEM-uh-lee

Named after Semele, mother of Dionysus
in Greek mythology (*Asparagaceae*)

Semiaquilegia
sem-ee-ak-wi-LEE-juh

From Latin *semi*, meaning "half," plus
Aquilegia, a related genus; these plants lack
the spurs typical in *Aquilegia* blooms
(*Ranunculaceae*)

Semiarundinaria
sem-ee-ah-roon-di-NAIR-ee-uh

From Latin *semi*, meaning "half," plus
Arundinaria, a related genus (Poaceae)

semperflorens
sem-per-FLOR-enz

Ever blooming, as in *Grevillea* ×
semperflorens

sempervirens
sem-per-VY-renz

Evergreen, as in *Lonicera sempervirens*

sempervivoides
sem-per-vi-VOY-deez

Resembling houseleek (*Sempervivum*), as in
Androsace sempervivoides

GENUS SPOTLIGHT

Sempervivum

Houseleeks are succulent plants with neat leaf rosettes and
strange, spidery flowers. As succulents, they are adept at
surviving with minimal moisture and so are popular choices for
rock gardens and dry-stone walls. They also grow well on tiled
roofs, where legend has it they will prevent lightning strikes.
With a name like "houseleek," you might expect them to cause
your roof to leak, but the "leek" comes from Anglo-Saxon *laec*,
meaning "plant." This word was also applied to members of the
onion family, such as garlic (*garlaec*) and leeks, but suggests
that *Sempervivum* was a good plant to grow on your house.

Sempervivum tectorum

Sempervivum

sem-per-VY-vum

From Latin *semper*, meaning "always," and *vivere*, meaning "to live," because these succulents cling to life tenaciously (*Crassulaceae*)

Senecio

suh-NEE-see-oh

From Latin *senex*, meaning "old man," because the bristles around the fruit are white (*Asteraceae*)

senegalensis

sen-eh-gal-EN-sis

senegalensis, senegalense

From Senegal, Africa, as in *Persicaria senegalensis*

senescens

sen-ESS-enz

Seeming to grow old (i.e. white or grey), as in *Allium senescens*

senilis

SEE-nil-is

senilis, senile

With white hair, as in *Rebutia senilis*

Senna

SEN-uh

From Arabic *sana*, the vernacular name (*Fabaceae*)

sensibilis

sen-si-BIL-is

sensibilis, sensibile

—

sensitivus

sen-si-TEE-vus

sensitiva, sensitivum

Sensitive to light or touch, as in *Onoclea sensibilis*

sepium

SEP-ee-um

Growing along hedges, as in *Calystegia sepium*

sept-

Used in compound words to denote seven

septemfidus

sep-TEM-fee-dus

septemfida, septemfidum

With seven divisions, as in *Gentiana septemfida*

Sequoiadendron giganteum

septemlobus

sep-tem-LOH-bus

septemloba, septemlobum

With seven lobes, as in *Primula septemloba*

septentrionalis

sep-ten-tree-oh-NAH-lis

septentrionalis, septentrionale

From the north, as in *Beschorneria septentrionalis*

Sequoia

se-KWOY-uh

Named after Sequoyah (c. 1767–1843), Native American silversmith and creator of a syllabary for the Cherokee language (*Cupressaceae*)

Sequoiadendron [1]

se-kwoy-uh-DEN-dron

From the related genus *Sequoia*, plus Greek *dendron*, meaning "tree" (*Cupressaceae*)

sericanthus

ser-ee-KAN-thus

sericantha, sericanthum

With silky flowers, as in *Philadelphus sericanthus*

sericeus

ser-IK-ee-us

sericea, sericeum

Silky, as *Rosa sericea*

Serissa

suh-RIS-uh

From an Indian language name for *S. japonica* (*Rubiaceae*)

serotinus

se-roh-TEE-nus

serotina, serotinum

With flowers or fruit late in the season, as in *Iris serotina*

serpens

SUR-penz

Creeping, as in *Agapetes serpens*

serpyllifolius

ser-pil-ly-FOH-lee-us

serpyllifolia, serpyllifolium

With leaves like wild or creeping thyme (*Thymus serpyllum*), as in *Arenaria serpyllifolia*

serpyllum

ser-PIE-lum

From the Greek word for a kind of thyme, as in *Thymus serpyllum*

serratifolius

sair-rat-ih-FOH-lee-us

serratifolia, serratifolium

With leaves that are serrated or saw-toothed, as in *Photinia serratifolia*

Serratula [1]

suh-RAT-ew-luh

From Latin *serrula*, meaning "little saw," a reference to the serrated leaves (*Asteraceae*)

serratus

sair-AH-tus

serrata, serratum

With small-toothed leaf margins, as in *Zelkova serrata*

serrulatus

ser-yoo-LAH-tus

serrulata, serrulatum,

With small serrations at the leaf margins, as in *Enkianthus serrulatus*

Seseli

SE-suh-lee

From Greek *seseli*, the name for a plant in this family, in English cicely (*Apiaceae*)

Sesleria

sez-LEER-ee-uh

Named after Leonardo Sesler (?–1785), Italian physician and botanist (*Poaceae*)

sesquipedalis

ses-kwee-ped-AH-lis

sesquipedalis, sesquipedale

Eighteen inches long, as in *Angraecum sesquipedale*

sessili-

Used in compound words to denote stalkless

sessiliflorus

sess-il-ee-FLOR-us

sessiliflora, sessililforum

With stalkless flowers, as in *Libertia sessiliflora*

sessilifolius

ses-il-ee-FOH-lee-us

sessilifolia, sessilifolium

With stalkless leaves, as in *Uvularia sessilifolia*

sessilis [2]

SES-sil-is

sessilis, sessile

Without a stalk, as in *Trillium sessile*

1

Serratula tinctoria

2

Trillium sessile

Setaria viridis

Deutzia setchuenensis

setaceus
se-TAY-see-us
setacea, setaceum
With bristles, as in *Pennisetum setaceum*

Setaria [3]
suh-TAIR-ee-uh
From Latin *seta*, meaning "bristle," because the inflorescence is bristly (*Poaceae*)

setchuenensis [4]
sech-yoo-en-EN-sis
setchuenensis, setchuenense
From Sichuan province, China, as in *Deutzia setchuenensis*

seti-
Used in compound words to denote bristled

setiferus
set-IH-fer-us
setifera, setiferum
With bristles, as in *Polystichum setiferum*

setifolius
set-ee-FOH-lee-us
setifolia, setifolium
With bristly leaves, as in *Lathyrus setifolius*

setiger
set-EE-ger
—

setigerus
set-EE-ger-us
setigera, setigerum
Bearing bristles, as in *Gentiana setigera*

setispinus
set-i-SPIN-us
setispina, setispinum
With bristly spines, as in *Thelocactus setispinus*

setosus
set-OH-sus
setosa, setosum
With many bristles, as in *Iris setosa*

setulosus
set-yoo-LOH-sus
setulosa, setulosum
With many small bristles, as in *Salvia setulosa*

sex-
Used in compound words to denote six

sexangularis
seks-an-gew-LAH-ris
sexangularis, sexangulare
With six angles, as in *Sedum sexangulare*

1

Silene vulgaris

2

Silphium terebinthinaceum

3

Silybum marianum

sexstylosus
seks-sty-LOH-sus
sexstylosa, sexstylosum
With six styles, as in *Hoheria sexstylosa*

Shepherdia
shep-ERD-ee-uh
Named after John Shepherd (1764–1836),
British botanist (*Elaeagnaceae*)

sherriffii
sher-RIF-ee-eye
Named after George Sherriff (1898–1967),
Scottish plant collector, as in *Rhododendron sherriffii*

Shibataea
shib-uh-TAY-uh
Named after Keita Shibata (1877–1949),
Japanese botanist (*Poaceae*)

shirasawanus
shir-ah-sa-WAH-nus
shirasawana, shirasawanum
Named after Homi (or Miho) Shirasawa
(1868–1947), Japanese botanist, as in *Acer shirasawanum*

Shortia
SHOR-tee-uh
Named after Charles Wilkins Short
(1794–1863), American botanist
(*Diapensiaceae*)

Sibbaldia
si-BAL-dee-uh
Named after Robert Sibbald (1641–1722),
Scottish physician (*Rosaceae*)

Sibiraea
si-bi-REE-uh
Named after Siberia, where the first
specimens were collected (*Rosaceae*)

sibiricus
sy-BEER-ih-kus
sibirica, sibiricum
Connected with Siberia, as in *Iris sibirica*

Sibthorpia
sib-THOR-pee-uh
Named after Humphrey Waldo Sibthorp
(1713–97), British botanist
(*Plantaginaceae*)

sichuanensis
sy-CHOW-en-sis
sichuanensis, sichuanense
From Sichuan province, China, as in
Cotoneaster sichuanensis

siculus
SIK-yoo-lus
sicula, siculum
From Sicily, Italy, as in *Nectaroscordum siculum*

Sida
SEE-duh
From Greek *side*, meaning "water lily," for
reasons unknown (*Malvaceae*)

Sidalcea
si-DAL-see-uh
A combination of two related genera, *Sida*
and *Alcea* (*Malvaceae*)

Sideritis
sid-uh-RY-tiss
From Greek *sideros*, meaning "iron," because
the plant was used to heal wounds from iron
weapons (*Lamiaceae*)

sideroxylon
sy-der-oh-ZY-lon
Wood like iron, as in *Eucalyptus sideroxylon*

sieberi
sy-BER-ee
Named after Franz Sieber (1789–1844),
Prague-born botanist and plant collector,
as in *Crocus sieberi*

sieboldianus
see-bold-ee-AH-nus
sieboldiana, sieboldianum
—

sieboldii
see-bold-ee-eye
Named after Philipp Franz von Siebold
(1796–1866), German doctor who collected
plants in Japan, as in *Magnolia sieboldii*

signatus
sig-NAH-tus
signata, signatum
Well-marked, as in *Saxifraga signata*

sikkimensis
sik-im-EN-sis
sikkimensis, sikkimense
From Sikkim, India, as in *Euphorbia sikkimensis*

Silene [1]
sy-LEE-nee
Named after Silenus, intoxicated companion of Dionysus in Greek mythology; often portrayed as being covered in foam, and some *Silene* produce a foamlike secretion (*Caryophyllaceae*)

siliceus
sil-ee-SE-us
silicea, siliceum
Growing in sand, as in *Astragalus siliceus*

siliquastrum
sil-ee-KWAS-trum
Roman name for a plant with pods, as in *Cercis siliquastrum*

Silphium [2]
SIL-fee-um
From Greek *silphion*, the name for an unknown plant (*Asteraceae*)

silvaticus
sil-VAT-ih-kus
silvatica, silvaticum
—

silvestris
sil-VES-tris
silvestris, silvestre
Growing in woodlands, as in *Polystichum silvaticum*

Silybum [3]
SIL-i-bum
From Greek *silybon*, a type of thistle (*Asteraceae*)

similis
SIM-il-is
similis, simile
Similar; like, as in *Lonicera similis*

Simmondsia
sim-ONDZ-ee-uh
Named after Thomas William Simmonds (1767–1804), British physician and botanist (*Simmondsiaceae*)

simplex
SIM-plecks
Simple; without branches, as in *Actaea simplex*

simplicifolius
sim-plik-ih-FOH-lee-us
simplicifolia, simplicifolium
With simple leaves, as in *Astilbe simplicifolia*

simulans
sim-YOO-lanz
Resembling, as in *Calochortus simulans*

sinensis [4]
sy-NEN-sis
sinensis, sinense
From China, as in *Corylopsis sinensis*

sinicus
SIN-ih-kus
sinica, sinicum
Connected with China, as in *Amelanchier sinica*

Sinningia
sin-INJ-ee-uh
Named after Wilhelm Sinning (ca. 1794–1874), German horticulturist (*Gesneriaceae*)

Sinobambusa
sy-noh-bam-BOO-suh
From Greek *sino*, meaning "China," plus the related genus *Bambusa* (*Poaceae*)

Sinofranchetia
sy-noh-fran-CHET-ee-uh
Named after Adrien René Franchet (1834–1900), French botanist, plus Greek *sino*, meaning "China' (*Lardizabalaceae*)

Sinojackia
sy-noh-JAK-ee-uh
Named after John George Jack (1861–1949), American botanist, plus Greek *sino*, meaning "China" (*Styracaceae*)

Sinowilsonia
sy-noh-wil-SOH-nee-uh
Named after Ernest Henry Wilson (1876–1930), British botanist, plus Greek *sino*, meaning "China" (*Hamamelidaceae*)

sinuatus
sin-yoo-AH-tus
sinuata, sinuatum
With a wavy margin, as in *Salpiglossis sinuata*

siphiliticus
sigh-fy-LY-tih-kus
siphilitica, siphiliticum
Connected with syphilis, as in *Lobelia siphilitica*

4

Wisteria sinensis

Sisyrinchium
(also **sysrinchium**)
sis-uh-RIN-kee-um

From Greek *sisyrinchion*, the name for the related *Moraea sisyrinchium*, possibly derived from *sisyra*, meaning "goat-hair coat," alluding to the shaggy tunic surrounding the corms; alternatively, derived from *sys*, meaning "pig," and *rynkhos*, meaning "snout," referring to pigs digging for the roots (*Iridaceae*)

sitchensis
sit-KEN-sis

sitchensis, sitchense

From Sitka, Alaska, as in *Sorbus sitchensis*

Skimmia
SKIM-ee-uh

From Japanese *miyama shikimi*, the vernacular name for *S. japonica* (*Rutaceae*)

skinneri
SKIN-ner-ee

Named after George Ure Skinner (1804–67), Scottish plant collector, as in *Cattleya skinneri*

smilacinus
smil-las-SY-nus

smilacina, smilacinum

Relating to greenbriar (*Smilax*), as in *Disporum smilacinum*

Smilax
SMY-laks

From Greek *smilax*, a vernacular name for bindweed, or from the nymph Smilax in Greek mythology (*Smilacaceae*)

smithianus
SMITH-ee-ah-nus

smithiana, smithianum

—

smithii
SMITH-ee-eye

May commemorate any of several Smiths, including Sir James Edward Smith (1759–1828), as in *Senecio smithii*

Smyrnium
SMUR-nee-um

From Greek *smyrna*, "to smell of myrrh" (*Apiaceae*)

soboliferus
soh-boh-LIH-fer-us

sobolifera, soboliferum

With creeping rooting stems, as in *Geranium soboliferum*

socialis
so-KEE-ah-lis

socialis, sociale

Forming colonies, as in *Crassula socialis*

Solandra
soh-LAN-druh

Named after Daniel Carlsson Solander (1733–82), Swedish botanist (*Solanaceae*)

Solanum
soh-LAY-num

From classical Latin name for a nightshade (*Solanaceae*)

Soldanella
sol-duh-NEL-uh

From Italian *solda*, meaning "coin," plus diminutive *ella*, because the leaves resemble small coins (*Primulaceae*)

Soleirolia
soh-lay-ROH-lee-uh

Named after Joseph-François Soleirol (1781–1863), French botanist (*Urticaceae*)

Solenostemon
soh-len-oh-STEM-on

From Greek *solen*, meaning "tube," and *stemon*, for "stamen," because four stamens are partly fused into a tube (*Lamiaceae*)

Solidago
so-li-DAY-goh

From Latin *solidus*, meaning "whole," and *ago*, meaning "becoming," alluding to medicinal properties (*Asteraceae*)

solidus
SOL-id-us

solida, solidum

Solid; dense, as in *Corydalis solida*

somaliensis
soh-mal-ee-EN-sis

somaliensis, somaliense

From Somalia, Africa, as in *Cyanotis somaliensis*

somniferus
som-NIH-fer-us

somnifera, somniferum

Inducing sleep, as in *Papaver somniferum*

sonchifolius
son-chi-FOH-lee-us

sonchifolia, sonchifolium

With leaves like sowthistle (*Sonchus*), as in *Francoa sonchifolia*

Sonchus
SON-kuss

From Greek *sonkos*, the vernacular name for a thistle (*Asteraceae*)

Sophora
SOF-o-ruh

From Arabic *sophera*, the vernacular name for a leguminous tree (*Fabaceae*)

Sorbaria
sor-BAIR-ee-uh

Resembling the related genus *Sorbus* (*Rosaceae*)

sorbifolius
sor-bih-FOH-lee-us

sorbifolia, sorbifolium

With leaves like mountain ash (*Sorbus*), as in *Xanthoceras sorbifolium*

Sorbus
SOR-bus

From classical Latin vernacular name for *S. domestica* (*Rosaceae*)

sordidus
SOR-deh-dus

sordida, sordidum

Having a dirty look, as in *Salix × sordida*

Sorghum
SOR-goom

From Italian *sorgo*, the vernacular name for *S. bicolor* (*Poaceae*)

soulangeanus
soo-lan-jee-AH-nus

soulangeana, soulangeanum

Commemorates Étienne Soulange-Bodin (1774–1846), French diplomat and secretary to the Société Royale et Centrale d'Agriculture (now the Académie d'Agriculture de France), who raised *Magnolia × soulangeana*

spachianus
spak-ee-AH-nus

spachiana, spachianum

Named after Édouard Spach (1801–79), French botanist, as in *Genista × spachiana*

Sparaxis
spuh-RAKS-iss

From Greek *sparasso*, meaning "to tear," referring to the lacerated bracts surrounding the flowers (*Iridaceae*)

Solanum melongena

1

2

3

Sparganium erectum

Hylotelephium spectabile

Blechnum spicant

Sparganium [1]
spar-GAY-nee-um
From Greek *sparganon*, meaning "swaddling band," for the straplike leaves (*Typhaceae*)

Sparmannia
spar-MAN-ee-uh
Named after Anders Sparrman (1748–1820), Swedish botanist (*Malvaceae*)

sparsiflorus
spar-see-FLOR-us
sparsiflora, sparsiflorum
With sparse or scattered flowers, as in *Lupinus sparsiflorus*

Spartium
SPAR-tee-um
Both this and the grass genus *Spartina* (*Poaceae*) derive from Greek *spartion*, meaning "cord," and also "broom"; both were weaved into cordage and the latter into brooms, although *Spartium junceum* is known as Spanish broom (*Fabaceae*)

spathaceus
spath-ay-SEE-us
spathacea, spathaceum
With a spathe, spathelike, as in *Salvia spathacea*

Spathiphyllum
spat-i-FIL-um
From Greek *spathe*, the bract surrounding the inflorescence, plus *phyllon*, meaning "leaf," alluding to the white leaflike spathe (*Araceae*)

spathulatus
spath-yoo-LAH-tus
spathulata, spathulatum
Spatulate, with a broader, flattened end, as in *Aeonium spathulatum*

speciosus
spee-see-OH-sus
speciosa, speciosum
Showy, as in *Ribes speciosum*

spectabilis [2]
speck-TAH-bih-lis
spectabilis, spectabile
Spectacular; showy, as in *Hylotelephium spectabile*

Speirantha
spy-RAN-thuh
From Greek *speiro*, meaning "coiled," and *anthos*, meaning "flower," alluding to a twist in the flowers (*Asparagaceae*)

Sphaeralcea
sfair-AL-see-uh
From Greek *sphaera*, meaning "sphere," plus the related genus *Alcea*; the fruit is spherical (*Malvaceae*)

sphaericus
SFAY-rih-kus
sphaerica, sphaericum
Shaped like a sphere, as in *Mammillaria sphaericus*

sphaerocarpos
sfay-ro-KAR-pus
sphaerocarpa, sphaerocarpum
With round fruit, as in *Medicago sphaerocarpos*

sphaerocephalon
sfay-ro-SEF-uh-lon
—

sphaerocephalus
sfay-ro-SEF-uh-lus
sphaerocephala, sphaerocephalum
With a round head, as in *Allium sphaerocephalon*

spicant [3]
SPIK-ant
Word of uncertain origin; possibly a German corruption of *spica*, spike, tuft, as in *Blechnum spicant*

spicatus
spi-KAH-tus
spicata, spicatum
With ears that grow in spikes, as in *Mentha spicata*

spiciformis
spik-ee-FOR-mis
spiciformis, spiciforme
In the shape of a spike, as in *Celastrus spiciformis*

spicigerus
spik-EE-ger-us
spicigera, spicigerum
Bearing spikes, as in *Justicia spicigera*

spiculifolius

spik-yoo-lih-FOH-lee-us

spiculifolia, spiculifolium

Like small spikes, as in *Erica spiculifolia*

Spigelia

spy-JEE-lee-uh

Named after Adriaan van den Spiegel (1578–1625), Flemish botanist and anatomist (*Loganiaceae*)

Spinacia

spin-AY-see-uh

From Persian *ispanakh*, the vernacular name for spinach, *S. oleracea* (*Amaranthaceae*)

spinescens

spy-NES-enz

—

spinifex

SPIN-ee-feks

—

spinosus

spy-NOH-sus

spinosa, spinosum

With spines, as in *Acanthus spinosus*

spinosissimus [4]

spin-oh-SIS-ih-mus

spinosissima, spinosissimum

Particularly spiny, as in *Rosa spinosissima*

spinulosus

spin-yoo-LOH-sus

spinulosa, spinulosum

With small spines, as in *Woodwardia spinulosa*

Spiraea

spy-REE-uh

From Greek *speiro*, meaning "coiled," because the flexible stems can be weaved into garlands (*Rosaceae*)

spiralis

spir-AH-lis

spiralis, spirale

Spiral, as in *Macrozamia spiralis*

splendens

SPLEN-denz

—

splendidus

splen-DEE-dus

splendida, splendidum

Splendid, as in *Fuchsia splendens*

Sprekelia

spruh-KEE-lee-uh

Named after Johann Heinrich von Spreckelsen (1691–1764), German lawyer and civil servant (*Amaryllidaceae*)

sprengeri

SPRENG-er-ee

Named after Carl Ludwig Sprenger (1846–1917), German botanist and plantsman, who bred and introduced many new plants, as in *Tulipa sprengeri*

spurius

SPEW-eee-us

spuria, spurium

False; spurious, as in *Iris spuria*

squalidus

SKWA-lee-dus

squalida, squalidum

Having a dirty look, dingy, as in *Leptinella squalida*

squamatus

SKWA-ma-tus

squamata, squamatum

With small scalelike leaves or bracts, as in *Juniperus squamata*

squamosus

skwa-MOH-sus

squamosa, squamosum

With many scales, as in *Annona squamosa*

squarrosus

skwa-ROH-sus

squarrosa, squarrosum

With spreading or curving parts at the extremities, as in *Dicksonia squarrosa*

stachyoides

stah-kee-OY-deez

Resembling betony (*Stachys*), as in *Buddleja stachyoides*

Stachys

STA-kiss

From Greek *stachys*, meaning "ear of grain," because the flowers are in wheatlike spikes (*Lamiaceae*)

Stachyurus

stak-ee-EW-rus

From Greek *stachys*, meaning "ear of grain," and *oura*, meaning "tail," a reference to the pendulous flower spikes (*Stachyuraceae*)

4

Rosa spinosissima

Staphylea pinnata

Stellaria nemorum

Stephanotis floribunda

stamineus

stam-IN-ee-us

staminea, stamineum

With pronounced stamens, as in *Vaccinium stamineum*

standishii

stan-DEE-shee-eye

Named after John Standish (1814–1875), British nurseryman, who raised plants collected by Robert Fortune, as in *Lonicera standishii*

Stangeria

stan-JEER-ee-uh

Named after William Stanger (1811–54), British surveyor, botanist, and physician (*Stangeriaceae*)

stans

STANZ

Erect; upright, as in *Clematis stans*

Stapelia

stay-PEE-lee-uh

Named after Johannes Bodaeus van Stapel (1602–36), physician and botanist (*Apocynaceae*)

stapeliiformis

sta-pel-ee-ih-FOR-mis

stapeliiformis, stapeliiforme

Like *Stapelia*, as in *Ceropegia stapeliiformis*

Staphylea [1]

staf-uh-LEE-uh

From Greek *staphyle*, meaning "cluster," because the flowers are clustered (*Staphyleaceae*)

Stauntonia

stawn-TOH-nee-uh

Named after George Leonard Staunton (1737–1801), Irish physician and diplomat (*Lardizabalaceae*)

Stellaria [2]

stuh-LAIR-ee-uh

From Latin *stella*, meaning "star," because the flowers are star-shaped (*Caryophyllaceae*)

stellaris

stell-AH-ris

stellaris, stellare

—

stellatus

stell-AH-tus

stellata, stellatum

Starry, as in *Magnolia stellata*

Stenanthium

stuh-NAN-thee-um

From Greek *stenos*, meaning "narrow," and *anthos*, meaning "flower," alluding to the narrow tepals (*Melanthiaceae*)

steno-

Used in compound words to denote narrow

Stenocactus

sten-oh-KAK-tus

From Greek *stenos*, meaning "narrow," and cactus, referring to the narrow stem ribs (*Cactaceae*)

stenocarpus

sten-oh-KAR-pus

stenocarpa, stenocarpum

With narrow fruit, as in *Carex stenocarpa*

Stenocereus

sten-oh-SER-ee-us

From Greek *stenos*, meaning "narrow," and *Cereus*, a related genus (*Cactaceae*)

stenopetalus

sten-oh-PET-al-us

stenopetala, stenopetalum

With narrow petals, as in *Genista stenopetala*

stenophyllus

sten-oh-FIL-us

stenophylla, stenophyllum

With narrow leaves, as in *Berberis* × *stenophylla*

stenostachyus

sten-oh-STAK-ee-us

stenostachya, stenostachyum

With narrow spikes, as in *Buddleja stenostachya*

Stephanandra

stef-uh-NAN-druh

From Greek *stephanos*, meaning "crown," and *andros*, meaning "male," because the stamens are persistent and form a ring around the fruit (*Rosaceae*)

Stephanotis [3]

stef-uh-NOH-tis

From Greek *stephanos*, meaning "crown," and and *otis*, meaning "ear," because the stamens are arranged in a ring and have earlike lobes (*Apocynaceae*)

sterilis

STER-ee-lis

sterilis, sterile

Infertile; sterile, as in *Potentilla sterilis*

Sternbergia

sturn-BUR-gee-uh

Named after Kaspar Maria von Sternberg (1761–1838), Czech botanist and entomologist (*Amaryllidaceae*)

sternianus

stern-ee-AH-nus

sterniana, sternianum

—

sternii

STERN-ee-eye

Named after Sir Frederick Claude Stern (1884–1967), British horticulturist and author with a particular interest in gardening on chalk, as in *Cotoneaster sternianus*

Stevia [4]

STEE-vee-uh

Named after Pedro Jaime Esteve (1500–56), Spanish physician and botanist (*Asteraceae*)

Stewartia [5]

stew-ART-ee-uh

Named after John Stuart (1713–92), Scottish nobleman and British prime minister (*Theaceae*)

Stipa

STEE-puh

From Greek *tuppe*, meaning "tow" or "fiber," because the grass was used to weave rope (*Poaceae*)

stipulaceus

stip-yoo-LAY-see-us

stipulacea, stipulaceum

—

stipularis

stip-yoo-LAH-ris

stipularis, stipulare

—

stipulatus

stip-yoo-LAH-tus

stipulata, stipulatum

With stipules, as in *Oxalis stipularis*

stoechas

STOW-kas

From *stoichas*, meaning in rows, the Greek name for *Lavandula stoechas*

Stokesia

STOHKS-see-uh

Named after Jonathan Stokes (1755–1831), British physician and botanist (*Asteraceae*)

stoloniferus

sto-lon-IH-fer-us

stolonifera, stoloniferum

With runners that take root, as in *Saxifraga stolonifera*

Stranvaesia

stran-VEE-zee-uh

Named after William Thomas Horner Fox-Strangways (1795–1865), British diplomat and politician (*Rosaceae*)

Stratiotes (also stratiotes)

stra-tee-OH-teez

From Greek *stratio*, meaning "soldier," alluding to the leaves shaped like a sword (*Hydrocharitaceae*)

Strelitzia

stre-LIT-zee-uh

Named after Charlotte of Mecklenburg-Strelitz (1744–1818), German botanist and wife to King George III (*Strelitziaceae*)

strepto-

Used in compound words to denote twisted

Streptocarpus

strep-toh-KAR-pus

From Greek *streptos*, meaning "twisted," and *karpos*, meaning "fruit," because the seed capsules are twisted, untwisting slightly to open (*Gesneriaceae*)

streptophyllus

strep-toh-FIL-us

streptophylla, streptophyllum

With twisted leaves, as in *Ruscus streptophyllum*

Streptopus

strep-TOH-pus

From Greek *streptos*, meaning "twisted," and *pous*, meaning "foot," referring to the bent or twisted flower stalk (*Liliaceae*)

Streptosolen

strep-toh-SOH-len

From Greek *streptos*, meaning "twisted," and *solen*, meaning "tube," for the slight twist in the petal tube (*Solanaceae*)

striatus

stree-AH-tus

striata, striatum

With stripes, as in *Bletilla striata*

4

Stevia ovata

5

Stewartia ovata

Streptocarpus

Commonly known as Cape primroses, *Streptocarpus* are tender plants, often grown as houseplants. Most cultivated hybrids have no stems and clusters of elongated tonguelike leaves. This habit is termed "rosulate," but two other distinctive growth forms exist. Unifoliate plants, such as *S. wendlandii*, produce only a single, large leaf, which keeps growing from the base throughout the life of the plant. Many unifoliates die after flowering. In contrast, caulescent species, such as *S. saxorum*, produce distinct stems and numerous, small leaves, and these species can live for many years.

Streptocarpus wendlandii

strictus
STRIK-tus
stricta, strictum
Erect; upright, as in *Penstemon strictus*

strigosus
strig-OH-sus
strigosa, strigosum
With stiff bristles, as in *Rubus strigosus*

striolatus
stree-oh-LAH-tus
striolata, striolatum
With fine stripes or lines, as in *Dendrobium striolatum*

Strobilanthes
stroh-bi-LANTH-eez
From Greek *strobilus*, meaning "cone," and *anthos*, meaning "flower," because in some species the flowers are arranged in cone-shaped inflorescences (*Acanthaceae*)

strobiliferus
stroh-bil-IH-fer-us
strobilifera, strobiliferum
Producing cones, as in *Epidendrum strobiliferum*

strobus [1]
STROH-bus
From Greek *strobos*, a whirling motion (cf. Greek *strobilos*, pinecone), or Latin *strobus*, an incense-bearing tree in Pliny, as in *Pinus strobus*

Strongylodon
strong-EYE-loh-don
From Greek *strongylus*, meaning "round," and *odontos*, meaning "tooth," because the calyx lobes are rounded (*Fabaceae*)

strumosus
stroo-MOH-sus
strumosa, strumosum
With cushionlike swellings. as in *Nemesia strumosa*

struthiopteris
struth-ee-OP-ter-is
Like an ostrich wing, as in *Matteuccia struthiopteris*

stygianus
sty-jee-AH-nuh
stygiana, stygianum
Dark, as in *Euphorbia stygiana*

Stylidium
sty-LID-ee-um
From Greek *stylos*, meaning "column," referring to the unique mechanism at the center of the flower. Male and female parts are fused into a column and, when insects visit the flower, they trigger this column to spring forward and dab pollen on the insect's body, or collect pollen from it (*Stylidiaceae*)

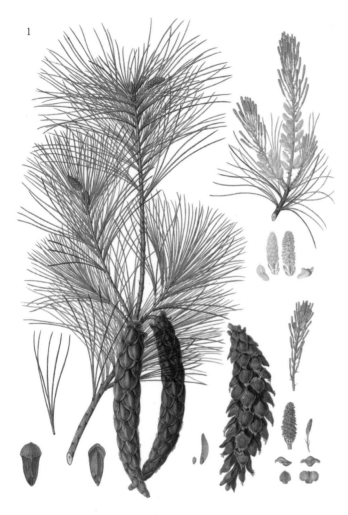

1

Pinus strobus

Stylophorum

sty-loh-FOR-um

From Greek *stylos*, meaning "column," and *phoreus*, meaning "bearer"; these flowers have a style in the flower, unusual in poppies (*Papaveraceae*)

stylosus

sty-LOH-sus

stylosa, stylosum

With pronounced styles, as in *Rosa stylosa*

Styphnolobium

stif-noh-LOH-bee-um

From Greek *stryphnos*, meaning "astringent," and *lobion*, meaning "pod," referring to the sour taste of the fruit (*Fabaceae*)

styracifluus

sty-rak-IF-lu-us

styraciflua, styracifluum

Producing gum, from *styrax*, the Greek name for storax, as in *Liquidambar styraciflua*

Styrax

STY-raks

From Arabic *assthirak*, the vernacular name for *S. officinalis* (*Styracaceae*)

suaveolens

swah-vee-OH-lenz

With a sweet fragrance, as in *Brugmansia suaveolens*

suavis

SWAH-vis

suavis, suave

Sweet; with a sweet scent, as in *Asperula suavis*

sub-

Used in compound words to denote a variety of meanings, such as almost, partly, slightly, somewhat, under

subacaulis

sub-a-KAW-lis

subacaulis, subacaule

Without much stem, as in *Dianthus subacaulis*

subalpinus

sub-al-PY-nus

subalpina, subalpinum

Growing at the lower levels of mountain ranges, as in *Viburnum subalpinum*

subcaulescens

sub-kawl-ESS-enz

With a small stem, as in *Geranium subcaulescens*

subcordatus

sub-kor-DAH-tus

subcordata, subcordatum

Shaped somewhat like a heart, as in *Alnus subcordata*

suberosus

sub-er-OH-sus

suberosa, suberosum

With cork bark, as in *Scorzonera suberosa*

subhirtellus

sub-hir-TELL-us

subhirtella, subhirtellum

Somewhat hairy, as in *Prunus* × *subhirtella*

submersus

sub-MER-sus

submersa, submersum

Submerged, as in *Ceratophyllum submersum*

subsessilis

sub-SES-sil-is

subsessilis, subsessile

Fixed, as in *Nepeta subsessilis*

subterraneus

sub-ter-RAY-nee-us

subterranea, subterraneum

Underground, as in *Parodia subterranea*

subtomentosus

sub-toh-men-TOH-sus

subtomentosa, subtomentosum

Almost hairy, as in *Rudbeckia subtomentosa*

subulatus

sub-yoo-LAH-tus

subulata, subulatum

Awl or needle shape, as in *Phlox subulata*

subvillosus

sub-vil-OH-sus

subvillosa, subvillosum

With somewhat soft hairs, as in *Begonia subvillosa*

Succisa [1]

suk-SY-zuh

From Latin *succis*, meaning "cut off," because the short rhizomes end abruptly, as if cut off (*Caprifoliaceae*)

succulentus

suk-yoo-LEN-tus

succulenta, succulentum

Fleshy; juicy, as in *Oxalis succulenta*

suffrutescens

suf-roo-TESS-enz

—

suffruticosus

suf-roo-tee-KOH-sus

suffruticosa, suffruticosum

Somewhat shrubby, as in *Paeonia suffruticosa*

sulcatus

sul-KAH-tus

sulcata, sulcatum

With furrows, as in *Rubus sulcatus*

sulphureus

sul-FER-ee-us

sulphurea, sulphureum

Sulfur yellow, as in *Lilium sulphureum*

suntensis

sun-TEN-sis

suntensis, suntense

Named after Sunte House, Sussex, England, as in *Abutilon × suntense*

Succisa pratensis

superbiens

soo-PER-bee-enz

—

superbus

soo-PER-bus

superba, superbum

Superb, as in *Salvia × superba*

supinus

sup-EE-nus

supina, supinum

Prostrate, as in *Verbena supina*

surculosus

sur-ku-LOH-sus

surculosa, surculosum

Producing suckers, as in *Dracaena surculosa*

suspensus

sus-PEN-sus

suspensa, suspensum

Hanging, as in *Forsythia suspensa*

sutchuenensis

sech-yoo-en-EN-sis

sutchuenensis, sutchuenense

From Sichuan province, China, as in *Adonis sutchuenensis*

Sutherlandia [2]

suth-er-LAN-dee-uh

Named after James Sutherland (1639–1719), Scottish botanist (*Fabaceae*)

Sutherlandia frutescens

sutherlandii

suth-er-LAN-dee-eye

Named after Dr. Peter Sutherland (1822–1900), who discovered *Begonia sutherlandii*

Swainsona

swayn-SOH-nuh

Named after Isaac Swainson (1746–1812), British physician and horticulturist (*Fabaceae*)

Syagrus

sy-AG-rus

From classical Latin name for a palm, possibly the date, *Phoenix dactylifera* (*Arecaceae*)

Sycopsis

sy-KOP-sis

From Greek *sykon*, meaning "fig," a plant they are said to resemble (*Hamamelidaceae*)

sylvaticus

sil-VAT-ih-kus

sylvatica, sylvaticum

—

sylvester

sil-VESS-ter

—

sylvestris

sil-VESS-tris

sylvestris, sylvestre

—

sylvicola

sil-VIH-koh-luh

Growing in woodlands, as in *Pinus sylvestris*, *Nyssa sylvatica*

Symphoricarpos

sim-for-ee-KAR-pos

From Greek *symphora*, meaning "gathering," and *karpos*, meaning "fruit," because the fruit clusters closely together (*Caprifoliaceae*)

Symphyotrichum

sim-fy-oh-TREE-kum

From Greek *symphysis*, meaning "junction," and *trichos*, meaning "hair"; the botanist who established this genus noticed the hairs around each flower were fused together in a ring, but unfortunately, this is not the case in most species (*Asteraceae*)

Symphytum

sim-FY-tum

From Greek *symphyo*, meaning "grow together," and *phyton*, meaning "plant," alluding to the use of comfrey to heal broken bones (*Boraginaceae*)

Symplocos

sim-PLOK-us

From Greek *symplokos*, meaning "entwined," because the stamens are partly fused together, and to the petals in some species (*Symplocaceae*)

Syneilesis

sy-nuh-LEE-sis

From Greek *syneilesis*, meaning "rolled up," referring to the unfurling leaves in spring (*Asteraceae*)

Syngonium

sin-GOH-nee-um

From Greek *syn*, meaning "together," and *gone*, meaning "gonad," because the ovaries are fused together (*Araceae*)

Synthyris

sin-THY-ris

From Greek *syn*, meaning "together," and *thyris*, meaning "little door," referring to the small valves that release the seed (*Plantaginaceae*)

syriacus

seer-ee-AH-kus

syriaca, syriacum

Connected with Syria, as in *Asclepias syriaca*

Syringa [3]

suh-RING-uh

From Greek *syrinx*, meaning "pipe," because the stems are easily hollowed (*Oleaceae*)

Syzygium

sy-ZEE-jee-um

From Greek *syzygos*, meaning "yoked together," because the first species described had paired leaves (*Myrtaceae*)

szechuanicus

se-CHWAN-ih-kus

szechuanica, szechuanicum

Connected with Szechuan, China, as in *Populus szechuanica*

3

Syringa vulgaris

NAMING A GIANT

The giant redwood (*Sequoiadendron giganteum*) is a marvel of the natural world. Its first Latin name, *Wellingtonia gigantea*, honored the great British war hero and Prime Minister Arthur Wellesley, Duke of Wellington (1769–1852), who had recently died, but unfortunately, *Wellingtonia* had already been published in another plant family (*Sabiaceae*). The naming of an American giant after a British soldier did not go unnoticed. An American botanist christened the tree *Washingtonia californica*, after the first U.S. president, but that too was thwarted, because *Washingtonia* is a conserved name for a palm genus. Ultimately, the big tree became *Sequoiadendron*, recognizing its similarity to the related *Sequoia*.

Tabernaemontana
tab-ur-nay-mon-TAN-uh
Named after Jakob Theodor von Bergzabern (1525–90), German physician and botanist, who Latinized his family name as Tabernaemontanus (*Apocynaceae*)

tabularis
tab-yoo-LAH-ris
tabularis, tabulare
—

tabuliformis
tab-yoo-lee-FORM-is
tabuliformis, tabuliforme
Flat, as in *Blechnum tabulare*

Tacca
TAK-uh
From Indonesian *taka*, the vernacular name (*Dioscoreaceae*)

Tagetes
TAJ-uh-teez
Etymology uncertain, although possibly named after Tages, a god in Etruscan mythology (*Asteraceae*)

tagliabuanus
tag-lee-ah-boo-AH-nus
tagliabuana, tagliabuanum
Commemorates Alberto and Carlo Tagliabue, nineteenth-century Italian nurserymen, as in *Campsis × tagliabuana*

taiwanensis
tai-wan-EN-sis
taiwanensis, taiwanense
From Taiwan, as in *Chamaecyparis taiwanensis*

takesimanus
tak-ess-ih-MAH-nus
takesimana, takesimanum
Connected with the Liancourt Rocks (Takeshima in Japanese), as in *Campanula takesimana*

taliensis
tal-ee-EN-sis
taliensis, taliense
From the Tali Range, Yunnan, China, as in *Lobelia taliensis*

Talinum
TAL-in-um
Possibly from Greek *taleia*, meaning "abundant blooms" (*Talinaceae*)

Tamarix
TAM-ur-iks
From Latin *tamariscus*, the vernacular name (*Tamaricaceae*)

tanacetifolius
tan-uh-kee-tih-FOH-lee-us
tanacetifolia, tanacetifolium
With leaves like tansy (*Tanacetum*), as in *Phacelia tanacetifolia*

Tanacetum
tan-uh-SEE-tum
From Greek *athanasia*, meaning "immortality," because the flowers are everlasting (*Asteraceae*)

Tanakaea
tuh-NAH-kee-uh
Named after Tanaka Yoshio (1838–1916), Japanese civil servant and botanist (*Saxifragaceae*)

tangelo
TAN-jel-oh
A hybrid of tangerine (*Citrus reticula*) and pomelo (*C. maxima*), as in *Citrus × tangelo*

tanguticus
tan-GOO-tih-kus
tangutica, tanguticum
Connected with the Tangut region of Tibet, as in *Daphne tangutica*

Taraxacum [1]
tuh-RAKS-uh-kum
From Arabic *tarakhshaqun*, a bitter herb, possibly chicory, *Cichorium intybus* (*Asteraceae*)

tardiflorus
tar-dee-FLOR-us
tardiflora, tardiflorum
Flowering late in the season, as in *Cotoneaster tardiflorus*

1

Taraxacum officinalis

tardus
TAR-dus
tarda, tardum
Late, as in *Tulipa tarda*

tasmanicus
tas-MAN-ih-kus
tasmanica, tasmanicum
Connected with Tasmania, Australia, as in *Dianella tasmanica*

Tasmannia
tas-MAN-ee-uh
Named after Abel Janszoon Tasman (1603–59), Dutch explorer and the first European to reach Tasmania, where some species occur (*Winteraceae*)

tataricus
tat-TAR-ih-kus
tatarica, tataricum
Connected with the historical region of Tartary (now the Crimea), as in *Lonicera tatarica*

tatsienensis
tat-see-en-EN-sis
tatsienensis, tatsienense
From Tatsienlu, China, as in *Delphinium tatsienense*

tauricus
TAW-ih-kus
taurica, tauricum
Connected with Taurica (now Crimea), as in *Onosma taurica*

taxifolius
taks-ih-FOH-lee-us
taxifolia, taxifolium
With leaves like yew (*Taxus*), as in *Prumnopitys taxifolia*

Taxodium
taks-OH-dee-um
Resembling the conifer genus *Taxus*, despite being deciduous (*Cupressaceae*)

Taxus [2]
TAK-sus
The classical Latin name for yew, *T. baccata* (*Taxaceae*)

tazetta [3]
taz-ET-tuh
Little cup, as in *Narcissus tazetta*

Tecoma
te-KOH-muh
From Nahuatl (Mexico) *tecomaxochitl*, vernacular name for a flower that resembles a clay vessel (*Bignoniaceae*)

Tecophilaea
tek-oh-FIL-ee-uh
Named after Tecophila Billoti (1824–?), Italian botanical artist (*Tecophilaeaceae*)

tectorum [1]
tek-TOR-um
Of house roofs, as in *Sempervivum tectorum*

Telekia [4]
te-LEE-kee-uh
Named after Sámuel Teleki de Szék (1739–1822), Chancellor of Transylvania and book collector (*Asteraceae*)

Tellima [5]
TEL-i-muh
An anagram of the related genus *Mitella* (*Saxifragaceae*)

Telopea
tel-OH-pee-uh
From Greek *telopos*, meaning "seen from afar," alluding to the remarkable flowers (*Proteaceae*)

temulentus
tem-yoo-LEN-tus
temulenta, temulentum
Inebriated, as in *Lolium temulentum*

1

Sempervivum tectorum

2

Taxus baccata

3

Narcissus tazetta

tenax

TEN-aks

Tough; matted, as in *Phormium tenax*

tenebrosus

teh-neh-BROH-sus

tenebrosa, tenebrosum

Connected with dark and shady places, as in *Catasetum tenebrosum*

tenellus

ten-ELL-us

tenella, tenellum

Tender; delicate, as in *Prunus tenella*

tener

TEN-er

tenera, tenerum

Slender; soft, as in *Adiantum tenerum*

tentaculatus

ten-tak-yoo-LAH-tus

tentaculata, tentaculatum

With tentacles, as in *Nepenthes tentaculata*

tenuicaulis

ten-yoo-ee-KAW-lis

tenuicaulis, tenuicaule

With slender stems, as in *Dahlia tenuicaulis*

tenuis

TEN-yoo-is

tenuis, tenue

Slender; thin, as in *Bupleurum tenue*

tenuiflorus

ten-yoo-ee-FLOR-us

tenuiflora, tenuiflorum

With slender flowers, as in *Muscari tenuiflorum*

tenuifolius

ten-yoo-ih-FOH-lee-us

tenuifolia, tenuifolium

With slender leaves, as in *Pittosporum tenuifolium*

tenuissimus

ten-yoo-ISS-ih-mus

tenuissima, tenuissimum

Very slender; thin, as in *Stipa tenuissima*

tequilana

te-kee-lee-AH-nuh

Connected with Tequila (Jalisco), Mexico, as in *Agave tequilana*

terebinthifolius

ter-ee-binth-ih-FOH-lee-us

terebinthifolia, terebinthifolium

With leaves that smell of turpentine, as in *Schinus terebinthifolius*

teres

TER-es

With a cylindrical form, as in *Vanda teres*

Terminalia

ter-min-AY-lee-uh

From Latin *terminus*, because the leaves are clustered at the tips of branches (*Combretaceae*)

terminalis

term-in-AH-lis

terminalis, terminale

Ending, as in *Erica terminalis*

ternatus

ter-NAH-tus

ternata, ternatum

With clusters of three, as in *Choisya ternata*

Ternstroemia

tern-STROH-mee-uh

Named after Christopher Tärnström (1711–46), Swedish botanist (*Pentaphylacaceae*)

terrestris

ter-RES-tris

terrestris, terrestre

From the ground; growing in the ground, as in *Lysimachia terrestris*

tessellatus

tess-ell-AH-tus

tessellata, tessellatum

Checkered, as in *Indocalamus tessellatus*

testaceus

test-AY-see-us

testacea, testaceum

The color or brick, as in *Lilium* × *testaceum*

testicularis

tes-tik-yoo-LAY-ris

testicularis, testiculare

Shaped like testicles, as in *Argyroderma testiculare*

testudinarius

tes-tuh-din-AIR-ee-us

testudinaria, testudinarium

Shaped like a tortoise shell, as in *Durio testudinarius*

tetra-

Used in compound words to denote four

Telekia speciosa

Tellima grandiflora

Tetradium ruticarpum

Teucrium botrys

Tetracentron
tet-ruh-SEN-tron
From Greek *tetra*, meaning "four," and *kentron*, meaning "spur," because the fruit has four prominent protrusions (*Trochodendraceae*)

Tetradium [1]
tet-RAD-ee-um
From Greek *tetra*, meaning "four," allegedly because the flower parts are in fours, but they can also be in fives (*Rutaceae*)

tetragonus
tet-ra-GON-us
tetragona, tetragonum
With four angles, as in *Nymphaea tetragona*

tetrandrus
tet-RAN-drus
tetrandra, tetrandrum
With four anthers, as in *Tamarix tetrandra*

Tetrapanax
tet-ruh-PAN-aks
From Greek *tetra*, meaning "four," and *Panax*, a related genus; flower parts in fours, unlike in *Panax*, where they are in fives (*Araliaceae*)

tetraphyllus
tet-ruh-FIL-us
tetraphylla, tetraphyllum
With four leaves, as in *Peperomia tetraphylla*

tetrapterus
tet-rap-TER-us
tetraptera, tetrapterum
With four wings, as in *Sophora tetraptera*

Tetrastigma
tet-ruh-STIG-muh
From Greek *tetra*, meaning "four," because the stigma has four lobes (*Vitaceae*)

Teucrium [2]
TEW-kree-um
From Greek *teukrion*, the vernacular name, possibly derived from Teucer, 1st King of Troy (*Lamiaceae*)

texanus
tek-SAH-nus
texana, texanum
—

texensis
tek-SEN-sis
texensis, texense
Of or from Texas, as in *Echinocactus texensis*

textilis
teks-TIL-is
textilis, textile
Relating to weaving, as in *Bambusa textilis*

Thalia
THA-lee-uh
Named after Johann Thal (1542–83), German physician and botanist (*Marantaceae*)

thalictroides
thal-ik-TROY-deez
Resembling meadow rue (*Thalictrum*), as in *Anemonella thalictroides*

Theobroma cacao

4

Thevetia peruviana

Thalictrum
tha-LIK-trum
From Greek *thaliktron*, a plant with divided leaves (*Ranunculaceae*)

Thamnocalamus
tham-noh-KAL-uh-mus
From Greek *thamnos*, meaning "shrub," and *kalamos*, meaning "reed," a compact, bushy bamboo (*Poaceae*)

Thelypteris
thel-IP-tur-iss
From Greek *thelys*, meaning "female," and *pteris*, meaning "fern," due to a supposed resemblance to lady fern, *Athyrium filix-femina* (*Thelypteridaceae*)

Theobroma [3]
thee-oh-BROH-muh
From Greeks *theos*, meaning "god," and *broma*, meaning "food," an appropriate name for chocolate, *T. cacao* (*Malvaceae*)

Thermopsis
thur-MOP-sis
From Greek *thermos*, meaning "lupine," a resemblance (*Fabaceae*)

Thevetia [4]
thu-VEE-tee-uh
Named after André Thevet (1502–92), French cleric and explorer (*Apocynaceae*)

thibetanus
ti-bet-AH-nus
thibetana, thibetanum
—

thibeticus
ti-BET-ih-kus
thibetica, thibeticum
Connected with Tibet, as in *Rubus thibetanus*

Thlaspi
THLAS-pee
From Greek *thalo*, meaning "to compress," because the fruit is flat and disklike (*Brassicaceae*)

thomsonii
tom-SON-ee-eye
Named after Dr. Thomas Thomson, nineteenth-century Scottish naturalist and superintendent of the Calcutta Botanic Garden, India, as in *Clerodendrum thomsoniae*

Thrinax
THRI-naks
From Greek *thrinax*, meaning "trident," a reference to the pointed leaflets (*Arecaceae*)

Thuja
THOO-yuh
From classical Greek name for a juniper or other similar conifer (*Cupressaceae*)

Thujopsis
thoo-YOP-sis
Resembling the related genus *Thuja* (*Cupressaceae*)

Thunbergia
thun-BER-jee-uh
Named after Carl Peter Thunberg (1743–1828), Swedish botanist and physician (*Acanthaceae*)

thunbergii
thun-BER-jee-eye
Named after Carl Peter Thunberg (1743–1828), Swedish botanist, as in *Spiraea thunbergii*

thymifolius
ty-mih-FOH-lee-us
thymifolia, thymifolium
With leaves like thyme (*Thymus*), as in *Lythrum thymifolium*

thymoides
ty-MOY-deez
Resembling thyme (*Thymus*), as in *Eriogonum thymoides*

Thymus
TY-muss
From Greek *thymos*, meaning thyme, *T. vulgaris*; derived from *thyein*, meaning "to smoke" or "to cure," because the plant has a smoky scent and may have been used as incense (*Lamiaceae*)

thyrsiflorus
thur-see-FLOR-us
thyrsiflora, thyrsiflorum
With thyrselike flower clusters, a central spike with side branches also bearing flower clusters, as in *Ceanothus thyrsiflorus*

thyrsoideus
thurs-OY-dee-us
thyrsoidea, thyrsoideum
–

thyrsoides
thurs-OY-deez
Like a Bacchic staff, as in *Ornithogalum thyrsoides*

Tiarella
tee-uh-REL-uh
From Greek *tiara*, meaning "crown," plus diminutive *ella*, because the seed capsule resembles a small hat (*Saxifragaceae*)

tiarelloides
tee-uh-rell-OY-deez
Resembling *Tiarella*, as in × *Heucherella tiarelloides*

tibeticus
ti-BET-ih-kus
tibetica, tibeticum
Connected with Tibet, as in *Roscoea tibetica*

Tibouchina
ti-boo-CHEE-nuh
From *tibouche*, the vernacular name in a Guyanese indigeneous language (*Melastomataceae*)

Tigridia
ti-GRID-ee-uh
From Latin *tigris*, meaning "tiger," for the patterning on the petals (*Iridaceae*)

tigrinus
tig-REE-nus
tigrina, tigrinum
With stripes like the Asiatic tiger or with spots like a jaguar (known as "tiger" in South America), as in *Faucaria tigrina*

Tilia
TIL-ee-uh
From Latin *tilia*, the vernacular name for linden, *T. cordata*, or *T. platyphyllos* (*Malvaceae*)

Tillandsia [1]
ty-LAND-zee-uh
Named after Elias Erici Tillandz (1640–93), Swedish physician and botanist (*Bromeliaceae*)

tinctorius
tink-TOR-ee-us
tinctoria, tinctorium
Used as a dye, as in *Genista tinctoria*

tingitanus
ting-ee-TAH-nus
tingitana, tingitanum
Connected with Tangiers, as in *Lathyrus tingitanus*

Titanopsis
ty-tan-OP-sis
From Greek *titanos*, meaning "limestone," and *opsis*, meaning "resembling"; this succulent masquerades as rock (*Aizoaceae*)

titanus
ti-AH-nus
titana, titanum
Enormous, as in *Amorphophallus titanum*

1

Tillandsia cyanea

2

Trachelospermum jasminoides

Tithonia

ty-THOW-nee-uh
Named after Tithonus, prince of Troy in Greek mythology; Tithonus was made immortal by Zeus but not granted eternal youth, and his name may have been used here as some *Tithonia* have white hairs on the foliage; another etymology relates to Eos, goddess of dawn, who chose Tithonus as her consort, the orange flowers resembling the sun (*Asteraceae*)

tobira

TOH-bir-uh
From the Japanese name, as in *Pittosporum tobira*

Tolmiea

TOL-mee-uh
Named after William Fraser Tolmie (1812–86), Scottish surgeon and politician (*Saxifragaceae*)

tomentosus

toh-men-TOH-sus
tomentosa, tomentosum
Especially woolly; matted, as in *Paulownia tomentosa*

tommasinianus

toh-mas-see-nee-AH-nus
tommasiniana, tommasinianum
Named after Muzio Giuseppe Spirito de' Tommasini, nineteenth-century Italian botanist, as in *Campanula tommasiniana*

Toona

TOO-nuh
From Sanskrit *toon* or *tunna*, the vernacular name for *T. ciliata* (*Meliaceae*)

Torenia

to-REE-nee-uh
Named after Olof Torén (1718–53), Swedish cleric and botanist (*Linderniaceae*)

Torreya

TOR-ay-uh
Named after John Torrey (1796–1873), American botanist, chemist and physician (*Taxaceae*)

torreyanus

tor-ree-AH-nus
torreyana, torreyanum
Named after Dr. John Torrey (1796–1873), American botanist, as in *Pinus torreyana*

tortifolius

tor-tih-FOH-lee-us
tortifolia, tortifolium
With twisted leaves, as in *Narcissus tortifolius*

tortilis

TOR-til-is
tortilis, tortile
Twisted, as in *Acacia tortilis*

tortuosus

tor-tew-OH-sus
tortuosa, tortuosum
Particularly twisted, as in *Arisaema tortuosum*

tortus

TOR-tus
torta, tortum
Twisted, as in *Masdevallia torta*

totara

toh-TAR-uh
From the Maori name for this tree, as in *Podocarpus totara*

tournefortii

toor-ne-FOR-tee-eye
Named after Joseph Pitton de Tournefort (1656–1708), French botanist, first to define the genus, as in *Crocus tournefortii*

Townsendia

town-ZEN-dee-uh
Named after David Townsend (1787–1858), American botanist (*Asteraceae*)

toxicarius

toks-ih-KAH-ree-us
toxicaria, toxicarium
Poisonous, as in *Antiaris toxicaria*

Toxicodendron

toks-i-koh-DEN-dron
From Greek *toxikon pharmakon*, meaning "poison," and *dendron*, meaning "tree," a fitting name for poison ivy, *T. radicans* (*Anacardiaceae*)

Trachelium (also trachelium)

tra-KEE-lee-um
From Greek *trachelos*, meaning "neck," because the plant is used to treat ailments of the throat (*Campanulaceae*)

Trachelospermum [2]

trak-uh-loh-SPER-mum
From Greek *trachelos*, meaning "neck," and *sperma*, meaning "seed," referring to the shape of the seed or perhaps the fruit (*Apocynaceae*)

Trachycarpus

tra-kee-KAR-pus
From Greek *trachys*, meaning "rough," and *karpos*, meaning "fruit," because the fruit is irregularly shaped (*Arecaceae*)

trachyspermus

trak-ee-SPER-mus
trachysperma, trachyspermum
With rough seed, as in *Sauropus trachyspermus*

Trachystemon

tra-kee-STEM-on
From Greek *trachys*, meaning "rough," and *stemon*, meaning "stamen"; the stamen filaments are hairy (*Boraginaceae*)

Tradescantia

tra-duh-SKAN-tee-uh
Named after father and son John Tradescant the Elder (ca. 1570–1638) and John Tradescant (1608–62), British botanists (*Commelinaceae*)

tragophylla

tra-go-FIL-uh
Literally goat leaf, as in *Lonicera tragophylla*

Tragopogon

tra-goh-POH-gon

From Greek *tragos*, meaning "goat," and *pogon*, meaning "beard," alluding to the silky, parachute-like hairs attached to the fruit (*Asteraceae*)

transcaucasicus

tranz-kaw-KAS-ih-kus

transcaucasica, transcaucasicum

Connected with Caucasus, Turkey, as in *Galanthus transcaucasicus*

transitorius

tranz-ee-TAW-ree-us

transitoria, transitorum

Short lived, as in *Malus transitoria*

transsilvanicus

tranz-il-VAN-ih-kus

transsilvanica, transsilvanicum

—

transsylvanicus

transsylvanica, transsylvanicum

Connected with Romania, as in *Hepatica transsilvanica*

trapeziformis

tra-pez-ih-FOR-mis

trapeziformis, trapeziforme

With four unequal sides, as in *Adiantum trapeziforme*

Trautvetteria

trout-vet-EER-ee-uh

Named after Ernst Rudolph von Trautvetter (1809–89), Baltic German botanist (*Ranunculaceae*)

traversii

trav-ERZ-ee-eye

Named after William Travers (1819–1903), New Zealand lawyer and plant collector, as in *Celmisia traversii*

tremulus

TREM-yoo-lus

tremula, tremulum

Quivering; trembling, as in *Populus tremula*

tri-

Used in compound words to denote three

triacanthos

try-a-KAN-thos

With three spines, as in *Gleditsia triacanthos*

triandrus

TRY-an-drus

triandra, triandrum

With three stamens, as in *Narcissus triandrus*

triangularis

try-an-gew-LAH-ris

triangularis, triangulare

—

triangulatus

try-an-gew-LAIR-tus

triangulata, triangulatum

With three angles, as in *Oxalis triangularis*

tricho-

Used in compound words to denote hairy

trichocarpus

try-ko-KAR-pus

trichocarpa, trichocarpum

With hairy fruit, as in *Rhus trichocarpa*

trichomanes

try-KOH-man-ees

Relating to a Greek name for fern, as in *Asplenium trichomanes*

trichophyllus

try-koh-FIL-us

tricophylla, tricophyllum

With hairy leaves, as in *Ranunculus trichophyllus*

trichotomus

try-KOH-toh-mus

trichotoma, trichotomum

With three branches, as in *Clerodendrum trichotomum*

tricolor [1]

TRY-kull-lur

With three colors, as in *Tropaeolum tricolor*

tricuspidatus

try-kusp-ee-DAH-tus

tricuspidata, tricuspidatum

With three points, as in *Parthenocissus tricuspidata*

Tricyrtis

try-SUR-tis

From Greek *treis*, meaning "three," and *kyrtos*, meaning "humped," referring to the swollen nectaries at the base of the flower (*Liliaceae*)

Viola tricolor

Trientalis

try-EN-tuh-lis

From Latin *trientalis*, meaning "third of a foot," referring to diminutive stature (*Primulaceae*)

trifasciata

try-fask-ee-AH-tuh

Three groups or bundles, as in *Sansevieria trifasciata*

trifidus

TRY-fee-dus

trifida, trifidum

Cut in three, as in *Carex trifida*

triflorus

TRY-flor-us

triflora, triflorum

With three flowers, as in *Acer triflorum*

trifoliatus

try-foh-lee-AH-tus

trifoliata, trifoliatum

—

trifolius

try-FOH-lee-us

trifolia, trifolium

With three leaves, as in *Gillenia trifoliata*

Trifolium [2]

try-FOH-lee-um

From Greek *triphyllon*, meaning "three-leaved," because clover leaves are made up of three leaflets (*Fabaceae*)

trifurcatus

try-fur-KAH-tus

trifurcata, trifurcatum

With three forks, as in *Artemisia trifurcata*

Trigonella

trig-oh-NEL-uh

From Greek *trigonella*, meaning fenugreek (*T. foenum-graecum*) seed; derived from *trigonon*, meaning "triangle," plus diminutive *ella*, referring to the shape of the leaflets (*Fabaceae*)

trigonophyllus

try-gon-oh-FIL-us

trigonophylla, trigonophyllum

With triangular leaves, as in *Acacia trigonophylla*

Trillium

TRIL-ee-um

From Latin *trilix*, meaning "triple," because the flower parts are in threes (*Melanthiaceae*)

trilobatus

try-lo-BAH-tus

trilobata, trilobatum

—

trilobus

try-LO-bus

triloba, trilobum

With three lobes, as in *Aristolochia trilobata*

trimestris

try-MES-tris

trimestris, trimestre

Of three months, as in *Lavatera trimestris*

trinervis

try-NER-vis

trinervis, trinerve

With three nerves, as in *Coelogyne trinervis*

tripartitus

try-par-TEE-tus

tripartita, tripartitum

With three parts, as in *Eryngium × tripartitum*

tripetalus

try-PET-uh-lus

tripetala, tripetalum

With three petals, as in *Moraea tripetala*

triphyllus

try-FIL-us

triphylla, triphyllum

With three leaves, as in *Penstemon triphyllus*

triplinervis

trip-lin-ner-vis

triplinervis, triplinerve

With three veins, as in *Anaphalis triplinervis*

tripteris

TRIPT-er-is

—

tripterus

TRIPT-er-us

triptera, tripterum

With three wings, as in *Coreopsis tripteris*

Tripterygium

trip-tur-IJ-ee-um

From Greek *tri*, meaning "three," and *pteryx*, meaning "wing," referring to the three-winged fruit (*Celastraceae*)

tristis

TRIS-tis

tristis, triste

Dull; sad, as in *Gladiolus tristis*

Tritelaia

try-tuh-LEE-uh

From Greek *treis*, meaning "three," and *teleios*, meaning "perfect," because the flower parts are in threes (*Asparagaceae*)

triternatus

try-tern-AH-tus

triternata, triternatum

Literally three threes, referring to leaf shape, as in *Corydalis triternata*

Tritonia

try-TOH-nee-uh

From Latin *triton*, meaning "weather vane," because the stamens can point in several directions (*Iridaceae*)

trivialis

tri-vee-AH-lis

trivialis, triviale

Common; ordinary; usual, as in *Rubus trivialis*

2

Trifolium reflexum

Tropaeolum

The nasturtium is a familiar garden annual with colorful flowers and sprawling stems. The flowers and foliage are also edible with a peppery taste. It is worth noting that the Latin name for nasturtium is *Tropaeolum majus*, because *Nasturtium* is also a genus name, but for an entirely different plant. Watercress is *Nasturtium officinale* and is also known for its peppery taste. The shared flavor is the result of glucosinolates, a group of chemicals common to all plant families in the order *Brassicales*, which includes *Brassicaceae* (*Nasturtium*) and *Tropaeolaceae* (*Tropaeolum*). Glucosinolates also provide the pungent taste of horseradish, wasabi, and mustard.

Tropaeolum majus

Trochodendron

trok-oh-DEN-dron

From Greek *trochos*, meaning "wheel," and *dendron*, meaning "tree," alluding to the stamens, which radiate like spokes (*Trochodendraceae*)

Trollius

TROL-ee-us

From German *troll*, meaning "round," the shape of the flowers (*Ranunculaceae*)

Tropaeolum

trop-ee-OH-lum

From Latin *tropaeum*, meaning "trophy"; the leaves of nasturtium (*T. majus*) resemble a shield and the flowers a helmet, and both were hung in a tree as a trophy by the victor after battle (*Tropaeoleaceae*)

truncatus

trunk-AH-tus

truncata, truncatum

Cut square, as in *Haworthia truncata*

tsariensis

sar-ee-EN-sis

tsariensis, tsariense

From Tsari, China, as in *Rhododendron tsariense*

tschonoskii

chon-OSK-ee-eye

Named after Sugawa Tschonoski (1841–1925), Japanese botanist and plant collector, as in *Malus tschonoskii*

Tsuga

TSOO-guh

From Japanese *tsuga*, the vernacular name for hemlock (*Pinaceae*)

tsussimensis

tsoos-sim-EN-sis

tsussimensis, tsussimense

From Tsushima Island, between Japan and Korea, as in *Polystichum tsussimense*

tuberculatus

too-ber-kew-LAH-tus

tuberculata, tuberculatum

—

tuberculosus

too-ber-kew-LOH-sus

tuberculosa, tuberculosum

Covered in lumps, as in *Anthemis tuberculata*

tuberosus

too-ber-OH-sus

tuberosa, tuberosum

Tuberous, as in *Polianthes tuberosa*

tubiferus
too-BIH-fer-us
tubifera, tubiferum
—

tubulosus
too-bul-OH-sus
tubulosa, tubulosum
Shaped like a tube or pipe, as in *Clematis tubulosa*

tubiflorus
too-bih-FLOR-us
tubiflora, tubiflorum
With flowers shaped like a trumpet, as in *Salvia tubiflora*

Tulbaghia
tool-BAG-ee-uh
Named after Ryk Tulbagh (1699–1771), Dutch governor of the
Cape Colony, now South Africa (*Amaryllidaceae*)

Tulipa [1]
TEW-lip-uh
From Turkish *tulbend*, meaning "turban," the shape of the flowers
(*Liliaceae*)

tulipiferus
too-lip-IH-fer-us
tulipifera, tulipiferum
Producing tulips or tuliplike flowers, as in *Liriodendron tulipifera*

tuolumnensis
too-ah-lum-NEN-sis
tuolumnensis, tuolumnense
From Tuolumne County, California, as in *Erythronium tuolumnense*

tupa
TOO-pa
Local name for *Lobelia tupa*

turbinatus
turb-in-AH-tus
turbinata, turbinatum
Swirling around, as in *Aesculus turbinata*

turczaninowii
tur-zan-in-NOV-ee-eye
Named after Nicholai S. Turczaninov (1796–1863), Russian
botanist, as in *Carpinus turczaninowii*

turkestanicus
tur-kay-STAN-ih-kus
turkestanica, turkestanicum
Connected with Turkestan, as in *Tulipa turkestanica*

Tussilago
tew-si-LAH-goh
From Latin *tussis*, meaning "cough"; coltsfoot (*T. farfara*) was used as
an expectorant (*Asteraceae*)

1

Tulipa gesneriana

Tweedia
TWEE-dee-uh
Named after John (James) Tweedie (1775–1862), Scottish
horticulturist (*Apocynaceae*)

tweedyi
TWEE-dee-eye
Named after Frank Tweedy, nineteenth-century American
topographer, as in *Lewisia tweedyi*

Typha
TY-fuh
From Greek *typhe*, meaning "cat's tail," possibly derived from *typhos*,
meaning "smoke," because the tiny seed disperses in a cloud and also
makes excellent tinder (*Typhaceae*)

typhinus
ty-FEE-nus
typhina, typhinum
Like *Typha* (reed mace), as in *Rhus typhina*

Typhonium
ty-FOH-nee-um
From Greek *typhonion*, a type of lavender, an allusion to the shape of
the spadix (*Araceae*)

Ugni

UG-nee

From Mapuche (Chile) *uñi*, the vernacular name for *U. molinae* (*Myrtaceae*)

Ulex

EW-leks

From *ulicis*, the vernacular Latin name for a shrub resembling rosemary (*Fabaceae*)

ulicinus

yoo-lih-SEE-nus

ulicina, ulicinum

Like gorse (*Ulex*), as in *Hakea ulicina*

uliginosus

yoo-li-gi-NOH-sus

uliginosa, uliginosum

From swampy and wet regions, as in *Salvia uliginosa*

ulmaria

ul-MAR-ee-uh

Like elm (*Ulmus*), as in *Filipendula ulmaria*

ulmifolius

ul-mih-FOH-lee-us

ulmifolia, ulmifolium

With leaves like elm (*Ulmus*), as in *Rubus ulmifolius*

Ulmus [1]

UL-mus

The classical Latin name for an elm tree (*Ulmaceae*)

umbellatus [2]

um-bell-AH-tus

umbellata, umbellatum

With umbels, as in *Butomus umbellatus*

Umbellularia

um-bel-ew-LAIR-ee-uh

From Latin *umbellula*, meaning "small umbel," alluding to the flower arrangement (*Lauraceae*)

Umbilicus

um-bil-i-kus

From Latin *umbilicus*, meaning "navel," because the leaves appear to have a "belly button" (*Crassulaceae*)

umbrosus

um-BROH-sus

umbrosa, umbrosum

Growing in shade, as in *Phlomis umbrosa*

uncinatus

un-sin-NA-tus

uncinata, uncinatum

With a hooked end, as in *Uncinia uncinata*

Uncinia

un-SIN-ee-uh

From Latin *uncinus*, meaning "hook," for the hooklike appendage on the fruit (*Cyperaceae*)

1

Ulmus minor

2

Iberis umbellata

Iris unguicularis

Urtica dioica

undatus
un-DAH-tus
undata, undatum
—

undulatus
un-dew-LAH-tus
undulata, undulatum
Wavy; undulating, as in *Hosta undulata*

unedo
YOO-nee-doe
Edible but of doubtful taste, from *unum edo*, "I eat one," as in *Arbutus unedo*

Ungnadia
un-GNAH-dee-uh
Named after David Ungnad von Sonnegg (ca. 1535–1600), Austrian diplomat (*Sapindaceae*)

unguicularis [1]
un-gwee-kew-LAH-ris
unguicularis, unguiculare
—

unguiculatus
un-gwee-kew-LAH-tus
unguiculata, unguiculatum
With claws, as in *Iris unguicularis*

uni-
Used in compound words to denote one

unicolor
YOO-nih-ko-lor
Of one color, as in *Lachenalia unicolor*

uniflorus
yoo-nih-FLOR-us
uniflora, uniflorum
With one flower, as in *Silene uniflora*

unifolius
yoo-nih-FOH-lee-us
unifolia, unifolium
With one leaf, as in *Allium unifolium*

unilateralis
yoo-nih-LAT-uh-ra-lis
unilateralis, unilaterale
One-sided, as in *Penstemon unilateralis*

uplandicus
up-LAN-ih-kus
uplandica, uplandicum
Connected with Uppland, Sweden, as in *Symphytum × uplandicum*

urbanus
ur-BAH-nus
urbana, urbanum
—

urbicus
UR-bih-kus
urbica, urbicum
—

urbius
UR-bee-us
urbia, urbium
From a town, as in *Geum urbanum*

urceolatus
ur-kee-oh-LAH-tus
urceolata, urceolatum
Shaped like an urn, as in *Galax urceolata*

urens
UR-enz
Stinging; burning, as in *Urtica urens*

Urginea
ur-JIN-ee-uh
Named after an Algerian tribe, the Beni-Urgin (*Asparagaceae*)

urophyllus
ur-oh-FIL-us
urophylla, urophyllum
With leaves with a tip like a tail, as in *Clematis urophylla*

Ursinia
ur-SIN-ee-uh
Named after Johann Heinrich Ursinus (1608–67), German theologian (*Asteraceae*)

ursinus
ur-SEE-nus
ursina, ursinum
Like a bear, as in *Eriogonum ursinum*

Urtica [2]
UR-tik-uh
From Latin *urere*, meaning "to burn," referring to the sting of nettles, *U. dioica* (*Urticaceae*)

urticifolius
ur-tik-ih-FOH-lee-us
urticifolia, urticifolium
With leaves like nettle (*Urtica*), as in *Agastache urticifolia*

uruguayensis
ur-uh-gway-EN-sis
uruguayensis, uruguayense
From Uruguay, South America, as in *Gymnocalycium uruguayense*

---

urumiensis
ur-um-ee-EN-sis
urumiensis urumiense
From Urmia, Iran, as in *Tulipa urumiensis*

urvilleanus
ur-VIL-ah-nus
urvilleana, urvilleanum
Named after J. S. C. Dumont d'Urville (1790–1842), French botanist and explorer, as in *Tibouchina urvilleana*

ussuriensis
oo-soo-ree-EN-sis
ussuriensis, ussuriense
From the Ussuri River, Asia, as in *Pyrus ussuriensis*

utahensis
yoo-tah-EN-sis
utahensis, utahense
From Utah, as in *Agave utahensis*

utilis
YOO-tih-lis
utilis, utile
Useful, as in *Betula utilis*

Utricularia [3]
oo-trik-ew-LAIR-ee-uh
From Latin *utriculus*, meaning "small bag," alluding to the pouchlike traps of many bladderworts (*Lentibulariaceae*)

utriculatus
uh-trik-yoo-LAH-tus
utriculata, utriculatum
Like a bladder, as in *Alyssoides utriculata*

---

uva-crispa
OO-vuh-KRIS-puh
Curled grape, as in *Ribes uva-crispa*

uva-ursi
OO-va UR-see
Bear's grape, as in *Arctostaphylos uva-ursi*

uvaria
oo-VAR-ee-uh
Like a bunch of grapes, as in *Kniphofia uvaria*

Uvularia
oo-vew-LAIR-ee-uh
From Latin *uvula*, the structure hanging from the back of the human soft palate, to which the flowers are said to resemble (*Colchicaceae*)

Utricularia alpina

vacciniifolius

vak-sin-ee-FOH-lee-us

vacciniifolia, vacciniifolium

With leaves like blueberry (*Vaccinium*),
as in *Persicaria vacciniifolia*

vaccinioides

vak-sin-ee-OY-deez

Resembling blueberry (*Vaccinium*),
as in *Rhododendron vaccinioides*

Vaccinium

vak-SIN-ee-um

From classical Latin name for bilberry,
V. myrtillus (*Ericaceae*)

vagans

VAG-anz

Widely distributed, as in *Erica vagans*

vaginalis

vaj-in-AH-lis

vaginalis, vaginale

—

vaginatus

vaj-in-AH-tus

vaginata, vaginatum

With a sheath, as in *Primula vaginata*

valdivianus

val-div-ee-AH-nus

valdiviana, valdivianum

Connected with Valdivia, Chile, as in *Ribes
valdivianum*

valentinus

val-en-TEE-nus

valentina, valentinum

Connected with Valencia, Spain, as in
Coronilla valentina

Valeriana

val-eer-ee-AH-nuh

From classical Latin name for valerian,
V. officinalis; from *valere*, meaning "to be
healthy" or "to be strong," alluding to
medicinal properties (*Caprifoliaceae*)

Vancouveria

van-koo-VEER-ee-uh

Named after George Vancouver (1757–98),
British naval officer and explorer
(*Berberidaceae*)

Vanda

VAN-duh

From Sanskrit *vanda*, the name for an
epiphytic plant (*Orchidaceae*)

Vanilla

vuh-NIL-uh

From Spanish *vainilla*, meaning "little pod,"
alluding to the fruit of *V. planifolia*
(*Orchidaceae*)

variabilis

var-ee-AH-bih-lis

variabilis, variabile

—

varians

var-ee-anz

—

variatus

var-ee-AH-tus

variata, variatum

Variable, as in *Eupatorium variabile*

varicosus

var-ee-KOH-sus

varicosa, varicosum

With dilated veins, as in *Oncidium
varicosum*

variegatus

var-ee-GAH-tus

variegata, variegatum

Variegated, as in *Pleioblastus variegatus*

GENUS SPOTLIGHT

Vaccinium

The genus *Vaccinium* comprises evergreen and
deciduous shrubs widely distributed around the northern
hemisphere, although with the odd outlier in tropical
locations, such as Hawaii and Madagascar. *Vaccinium*
is the source of numerous edible berries, including
blueberries, cranberries, lingonberries, and bilberries.
Blueberries and kin prefer an ericaceous soil, the name
deriving from their family *Ericaceae*. Ericaceous soil is
alkaline and popular with plants that grow in nutrient-
poor or peat-rich soil, such as camellias, magnolias,
Japanese maples, and, of course, most members of
family *Ericaceae* (heathers, rhododendrons, and azaleas).

Vaccinium corymbosum

varius

VAH-ree-us

varia, varium

Diverse, as in *Calamagrostis varia*

vaseyi

VAS-ee-eye

Named after George Richard Vasey
(1822–93), American plant collector,
as in *Rhododendron vaseyi*

vedrariensis

ved-rar-ee-EN-sis

vedrariensis, vedrariense

From Verrières-le-Buisson, France, and the
nurseries of Vilmorin-Andrieux & Cie, as in
Clematis × vedrariensis

vegetus

veg-AH-tus

vegeta, vegetum

Vigorous, as in *Ulmus × vegeta*

veitchianus

veet-chee-AH-nus

veitchiana, veitchianum

—

veitchii

veet-chee-EYE

Named after members of the Veitch family,
nurserymen of Exeter and Chelsea in
England, as in *Paeonia veitchii*

Vellozia

vel-OH-zee-uh

Named after José Mariano da Conceição
Vellozo (1742–1811), Brazilian botanist
(*Velloziaceae*)

Veltheimia

vel-THY-me-uh

Named after August Ferdinand Graf von
Veltheim (1741–1801), German geologist
(*Asparagaceae*)

velutinus

vel-oo-TEE-nus

velutina, velutinum

Like velvet, as in *Musa velutina*

venenosus

ven-ee-NOH-sus

venenosa, venenosum

Very poisonous, as in *Caralluma venenosa*

venosus

ven-OH-sus

venosa, venosum

With many veins, as in *Vicia venosa*

1

Veratrum nigrum

ventricosus

ven-tree-KOH-sus

ventricosa, ventricosum

With a swelling on one side, bellylike,
as in *Ensete ventricosum*

venustus

ven-NUSS-tus

venusta, venustum

Handsome, as in *Hosta venusta*

Veratrum [1]

vuh-RAT-rum

From Latin *vere*, meaning "truth," and *ater*,
meaning "black," because some species have
black roots (*Melanthiaceae*)

verbascifolius

ver-bask-ih-FOH-lee-us

verbascifolia, verbascifolium

With leaves like mullein (*Verbascum*), as in
Celmisia verbascifolia

Verbascum

ver-BAS-kum

From Latin *barbascum*, meaning "bearded,"
referring to the hairy stamen filaments
(*Scrophulariaceae*)

Verbena

ver-BEE-nuh

From Latin *verbena*, meaning "sacred
bough," fragrant stems carried in religious
processions, although probably not the
plants called *Verbena* today (*Verbenaceae*)

Verbesina

ver-bes-EE-nuh

Indicating a resemblance to *Verbena*
(*Asteraceae*)

verecundus

ver-ay-KUN-dus

verecunda, verecundum

Modest, as in *Columnea verecunda*

veris

VER-is

Relating to spring; flowering in spring,
as in *Primula veris*

vernalis

ver-NAH-lis

vernalis, vernale

Relating to spring; flowering in spring,
as in *Pulsatilla vernalis*

vernicifluus

ver-nik-IF-loo-us

verniciflua, vernicifluum

Producing varnish, as in *Rhus verniciflua*

vernicosus

vern-ih-KOH-sus

vernicosa, vernicosum

Varnished, as in *Hebe vernicosa*

Vernonia

ver-NOH-nee-uh

Named after William Vernon (ca. 1666–1715), British naturalist (*Asteraceae*)

vernus

VER-nus

verna, vernum

Relating to spring, as in *Leucojum vernum*

Veronica

vuh-RON-i-kuh

Named after Saint Veronica, who gave a cloth to Jesus, so he could mop his brow on the way to Calvary; some species have flowers with markings said to resemble the imprint of Christ's face on that cloth (*Plantaginaceae*)

Veronicastrum

vuh-ron-i-KAS-trum

Somewhat resembling the related genus *Veronica* (*Plantaginaceae*)

verrucosus

ver-oo-KOH-sus

verrucosa, verrucosum

Covered with warts, as in *Brassia verrucosa*

verruculosus

ver-oo-ko-LOH-sus

verruculosa, verruculosum

With small warts, as in *Berberis verruculosa*

versicolor

VER-suh-kuh-lor

With various colors, as in *Oxalis versicolor*

verticillatus [2]

ver-ti-si-LAH-tus

verticillata, verticillatum

With a whorl or whorls, as in *Sciadopitys verticillata*

Verticordia

vur-ti-KOR-dee-uh

From Latin *vertere*, meaning "to turn," and *cordatus*, meaning "good heart," a reference to the Roman goddess Venus, who could turn hearts; her floral emblem was myrtle (*Myrtus communis*), a member of the same family, and *Verticordia* blooms are extremely attractive (*Myrtaceae*)

verus

VER-us

vera, verum

True; standard; regular, as in *Aloe vera*

vescus

VES-kus

vesca, vescum

Thin; feeble, as in *Fragaria vesca*

vesicarius

ves-ee-KAH-ree-us

vesicaria, vesicarium

—

vesiculosus

ves-ee-kew-LOH-sus

vesiculosa, vesiculosum

Like a bladder; with small bladders, as in *Eruca vesicaria*

vespertinus

ves-per-TEE-nus

vespertina, vespertinum

Relating to the evening; flowering in the evening, as in *Moraea vespertina*

Vestia

VES-tee-uh

Named after Lorenz Chrysanth von Vest (1776–1840), Austrian physician and botanist (*Solanaceae*)

vestitus

ves-TEE-tus

vestita, vestitum

Covered; clothed, as in *Sorbus vestita*

vexans

VEKS-anz

Vexatious or troublesome in some respect, as in *Sorbus vexans*

vialii

vy-AL-ee-eye

Named after Paul Vial (1855–1917), as in *Primula vialii*

2

Sciadopitys verticillata

vialis
vee-AH-lis
vialis, viale
From the wayside, as in *Calyptocarpus vialis*

viburnifolius
vy-burn-ih-FOH-lee-us
viburnifolia, viburnifolium
With leaves like *Viburnum*, as in *Ribes viburnifolium*

viburnoides
vy-burn-OY-deez
Resembling *Viburnum*, as in *Pileostegia viburnoides*

Viburnum
vy-BUR-num
From the classical Latin name for *V. lantana* (*Adoxaceae*)

Vicia
VISS-ee-uh
From Latin *vincere*, meaning "to bind," referring to the tendrils (*Fabaceae*)

Victoria
vik-TOR-ee-uh
Named after Queen Victoria (1819–1901), the genus was named in 1837, the year she ascended the throne (*Nymphaeaceae*)

victoriae
vik-TOR-ee-ay
—

victoriae-reginae
vik-TOR-ee-ay re-JEE-nay
Named after Queen Victoria (1819–1901), British monarch, as in *Agave victoriae-reginae*

vigilis
VIJ-il-is
—

vigilans
VIJ-il-anz
Vigilant, as in *Diascia vigilis*

villosus
vil-OH-sus
villosa, villosum
With soft hairs, as in *Photinia villosa*

vilmorinianus
vil-mor-in-ee-AH-nus
vilmoriniana, vilmorinianum
—

vilmorinii
vil-mor-IN-ee-eye
Named after Maurice de Vilmorin (1849–1918), French nurseryman, as in *Cotoneaster vilmorinianus*

viminalis
vim-in-AH-lis
viminalis, viminale
—

vimineus
vim-IN-ee-us
viminea, vimineum
With long, slender shoots, as in *Salix viminalis*

Vinca
VIN-kuh
From Latin *vincere*, meaning "to bind," because the stems were used for making wreaths (*Apocynaceae*)

viniferus
vih-NIH-fer-us
vinifera, viniferum
Producing wine, as in *Vitis vinifera*

Viola
vee-OH-luh
From Latin *viola*, the name for several plants with fragrant flowers (*Violaceae*)

violaceus
vy-oh-LAH-see-us
violacea, violaceum
The color of violet, as in *Hardenbergia violacea*

violescens
vy-oh-LESS-enz
Turning violet, as in *Phyllostachys violescens*

virens
VEER-enz
Green, as in *Penstemon virens*

virescens
veer-ES-enz
Turning green, as in *Carpobrotus virescens*

virgatus
vir-GA-tus
virgata, virgatum
Twiggy, as in *Panicum virgatum*

virginalis
vir-jin-AH-lis
virginalis, virginale
—

virgineus
vir-JIN-ee-us
virginea, virgineum
White; virginal, as in *Anguloa virginalis*

virginianus
vir-jin-ee-AH-nus
virginiana, virginianum
—

virginicus
vir-JIN-ih-kus
virginica, virginicum
—

virgineus
vir-JIN-ee-us
virginea, virgineum
Connected with Virginia, as in *Hamamelis virginiana*

viridi-
Used in compound words to denote green

viridescens
vir-ih-DESS-enz
Turning green, as in *Ferocactus viridescens*

viridiflorus
vir-id-uh-FLOR-us
viridiflora, viridiflorum
With green flowers, as in *Lachenalia viridiflora*

viridis
VEER-ih-dis
viridis, viride
Green, as in *Trillium viride*

viridissimus
vir-id-ISS-ih-mus
viridissima, viridissimum
Very green, as in *Forsythia viridissima*

viridistriatus
vi-rid-ee-stry-AH-tus
viridistriata, viridistriatum
With green stripes, as in *Pleioblastus viridistriatus*

viridulus
vir-ID-yoo-lus
viridula, viridulum
Somewhat green, as in *Tricyrtis viridula*

Viscum

European mistletoe (*Viscum album*) is a parasitic plant that relies on its host tree for water and soil nutrients, although it can manufacture its own food through photosynthesis. Because it can survive only when attached to a host, mature plants must find a way to transport their seed to other suitable hosts, and birds are key. Mistle thrushes (*Turdus viscivorus*) feed on the white berries, providing a courier service. The berry flesh is sticky (viscous, thus *Viscum*), so once the bird eats the berry, a sticky excrement that can adhere to tree branches transfers seed to the perfect site for germination on a new host.

Viscum album

viscidus
VIS-kid-us
viscida, viscidum
Sticky; clammy, as in *Teucrium viscidum*

viscosus
vis-KOH-sus
viscosa, viscosum
Sticky; clammy, as in *Rhododendron viscosum*

Viscum
VIS-kum
The classical Latin name for mistletoe (*V. album*); also used for birdlime, a sticky substance used to catch birds and manufactured from mistletoe berries; the viscid nature of the fruit allows the seed to stick to tree branches (*Santalaceae*)

vitaceus
vee-TAY-see-us
vitacea, vitaceum
Like vine (*Vitis*), as in *Parthenocissus vitacea*

vitellinus
vy-tel-LY-nus
vitellina, vitellinum
The color of egg yolk, as in *Encyclia vitellina*

Vitex
VI-teks
The classical Latin name for chaste tree, *V. agnus-castus* (*Lamiaceae*)

viticella
vy-tee-CHELL-uh
Small vine, as in *Clematis viticella*

vitifolius
vy-tih-FOH-lee-us
vitifolia, vitifolium
With leaves like vine (*Vitis*), as in *Abutilon vitifolium*

Vitis
VY-tis
The classical Latin name for grapevine, *V. vinifera* (*Vitaceae*)

vitis-idaea
VY-tiss-id-uh-EE-uh
Vine of Mount Ida, as in *Vaccinium vitis-idaea*

vittatus
vy-TAH-tus
vittata, vittatum
With lengthwise stripes, as in *Billbergia vittata*

vivax
VY-vaks
Long lived, as in *Phyllostachys vivax*

viviparus
vy-VIP-ar-us
vivipara, viviparum
Producing plantlets; self-propagating, as in *Persicaria vivipara*

volubilis
vol-OO-bil-is
volubilis, volubile
Twining, as in *Aconitum volubile*

vomitorius
vom-ih-TOR-ee-us
vomitoria, vomitorium
Emetic, as in *Ilex vomitoria*

Vriesea
vry-EEZ-ee-uh
Named after Willem Hendrik de Vriese (1806–62), Dutch botanist and physician (*Bromeliaceae*)

vulgaris
vul-GAH-ris
vulgaris, vulgare
—

vulgatus
vul-GAIT-us
vulgata, vulgatum
Common, as in *Aquilegia vulgaris*

Wachendorfia [1]

wak-un-DORF-ee-uh

Named after Evert Jacob van Wachendorff (1703–58), Dutch botanist (*Haemodoraceae*)

wagnerii

wag-ner-EE-eye

—

wagneriana

wag-ner-ee-AH-nuh

—

wagnerianus

wag-ner-ee-AH-nus

Named after Warren Wagner (1920–2000), American botanist, as in *Trachycarpus wagnerianus*

Wahlenbergia

wal-un-BERG-ee-uh

Named after Georg Göran Wahlenburg (1780–1851), Swedish botanist (*Campanulaceae*)

Waldsteinia

wald-STY-nee-uh

Named after Franz de Paula Adam von Waldstein (1759–1823), Austrian soldier and botanist (*Rosaceae*)

walkerae

WAL-ker-ah

Commemorates various Walkers, including Ernest Pillsbury Walker (1891–1969), American zoologist, as in *Chylismia walkeri*

wallerianus

wall-er-ee-AH-nus

walleriana, wallerianum

Named after Horace Waller (1833–96), British missionary, as in *Impatiens walleriana*

wallichianus [2]

wal-ik-ee-AH-nus

wallichiana, wallichianum

Named after Dr. Nathaniel Wallich (1786–1854), Danish botanist and plant hunter, as in *Pinus wallichiana*

walteri

WAL-ter-ee

Named after Thomas Walter, eighteenth-century American botanist, as in *Cornus walteri*

wardii

WAR-dee-eye

Named after Frank Kingdon-Ward (1885–1958), British botanist and plant collector, as in *Roscoea wardii*

warscewiczii

vark-zeh-wik-ZEE-eye

Named after Joseph Warsczewicz (1812–66), Polish orchid collector, as in *Kohleria warscewiczii*

Washingtonia

wash-ing-TOH-nee-uh

Named after George Washington (1732–99), first President of the United States (*Arecaceae*)

1

Wachendorfia paniculata

2

Pinus wallichiana

1

Weigela floribunda

2

Westringia eremicola

3

Rhododendron wightii

watereri
wat-er-EER-eye
Named after Waterers Nurseries, Knaphill, England, as in *Laburnum* × *watereri*

Watsonia
wat-SOH-nee-uh
Named after William Watson (1715–87), British physician and scientist (*Iridaceae*)

webbianus
web-bee-AH-nus
webbiana, webbianum
Named after Philip Barker Webb (1793–1854), British botanist and traveler, as in *Rosa webbiana*

Weigela [1]
WY-juh-luh
Named after Christian Ehrenfried Weigel (1748–1831), German scientist (*Caprifoliaceae*)

Weinmannia
wine-MAN-ee-uh
Named after Johann Wilhelm Weinmann (1683–1741), German apothecary (*Cunoniaceae*)

Westringia [2]
west-RING-ee-uh
Named after Johan Petrus Westring (1753–1833), Swedish physician and botanist (*Lamiaceae*)

weyerianus
wey-er-ee-AH-nus
weyeriana, weyerianum
Named after William van de Weyer, twentieth-century horticulturist, who bred *Buddleja* × *weyeriana*

Wikstroemia
wik-STROH-mee-uh
Named after Johan Emanuel Wikström (1789–1856), Swedish botanist (*Thymelaeaceae*)

Wisteria
wis-TEER-ee-uh
Named after Caspar Wistar (1761–1818), American physician and anatomist; the spelling was deliberately altered from "*Wistaria*" to produce a better sound, but perhaps also to honor his friend Charles Jones Wister (1782–1865), (*Fabaceae*)

wheeleri
WHEE-ler-ee
Named after George Montague Wheeler (1842–1905), American surveyor, as in *Dasylirion wheeleri*

wherryi
WHER-ee-eye
Named after Dr. Edgar Theodore Wherry (1885–1982), American botanist and geologist, as in *Tiarella wherryi*

whipplei
WHIP-lee-eye
Named after Lieutenant Amiel Weeks Whipple (1818–63), American surveyor, as in *Yucca whipplei*

wichurana
whi-choo-re-AH-nuh
Named after Max Ernst Wichura (1817–1866), German botanist, as in *Rosa wichurana*

wightii [3]
WIGHT-ee-eye
Named after Robert Wight (1796–1872), botanist and superintendent of Madras Botanic Garden, as in *Rhododendron wightii*

wildpretii
wild-PRET-ee-eye
Named after Hermann Josef Wildpret, nineteenth-century Swiss botanist, as in *Echium wildpretii*

wilkesianus
wilk-see-AH-nus
wilkesiana, wilkesianum
Named after Charles Wilkes (1798–1877), American naval officer and explorer, as in *Acalypha wilkesiana*

williamsii

wil-yams-EE-eye

Named for various eminent botanists and horticulturists called Williams, including John Charles Williams, nineteenth-century British plant collector, as in *Camellia × williamsii*

willmottianus

wil-mot-ee-AH-nus

willmottiana, willmottianum

—

willmottiae

wil-MOT-ee-eye

Named after Ellen Willmott (1858–1934), British horticulturist, of Warley Place, Essex, as in *Rosa willmottiae*

wilsoniae

wil-SON-ee-ay

—

wilsonii

wil-SON-ee-eye

Named after Dr. Ernest Henry Wilson (1876–1930), British plant hunter, as in *Spiraea wilsonii*. The epithet *wilsoniae* commemorates his wife Helen

wintonensis

win-ton-EN-sis

wintonensis, wintonense

From Winchester; used especially of Hillier Nurseries, Hampshire, England, as in *Halimiocistus × wintonensis*

wisleyensis

wis-lee-EN-sis

wisleyensis, wisleyense

Named after Royal Horticultural Society's RHS Garden Wisley, Surrey, England, as in *Gaultheria × wisleyensis*

wittrockianus

wit-rok-ee-AH-nus

wittrockiana, wittrockianum

Named after Professor Veit Brecher Wittrock (1839–1914), Swedish botanist, as in *Viola × wittrockiana*

Wollemia

wo-LEM-ee-uh

Named after Wollemi National Park in New South Wales, Australia, where *W. nobilis* was discovered in 1994 (*Araucariaceae*)

Woodsia

WOOD-zee-uh

Named after Joseph Woods (1776–1864), British botanist and geologist (*Woodsiaceae*)

Woodwardia

wood-WARD-ee-uh

Named after Thomas Jenkinson Woodward (1745–1820), British botanist (*Blechnaceae*)

woronowii

wor-on-OV-ee-eye

Named after Georg Woronow (1874–1931), Russian botanist and plant collector, as in *Galanthus woronowii*

Worsleya

WORS-lay-uh

Named after Arthington Worsley (1861–1944), British horticulturist and engineer (*Amaryllidaceae*)

wrightii

RITE-ee-eye

Named after Charles Wright, nineteenth-century American botanist and plant collector, as in *Viburnum wrightii*

Wulfenia

wul-FEE-nee-uh

Named after Franz Xaver von Wulfen (1728–1805), Serbian-born Austrian cleric and scientist (*Plantaginaceae*)

GENUS SPOTLIGHT

Wollemia

Wollemia nobilis, a conifer in the monkey-puzzle family (*Araucariaceae*), is native to New South Wales, Australia. Only described in 1995, its discovery the year before created a sensation in the botanical world. It was found in Wollemi National Park, less than 100 miles from Sydney, but how could a whole new conifer genus be discovered so close to a major city? It seems the reason is that the park is large, and the wollemia pines were in a remote location. There were only a few of them, among thousands of other trees, and it took an alert, and intrepid, botanist to discover these trees for what they were. The plants are rare in the wild, too, with less than 100 adult trees, and they are threatened by an introduced fungal pathogen. In order to make sure of their survival, plants have been propagated and the offspring sold, with profits being used to support *Wollemia* conservation.

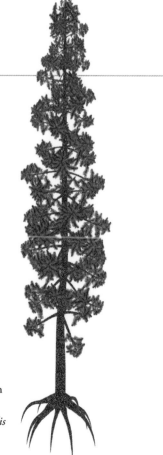

Reconstruction of fossil *Wollemia nobilis*

Xanthoceras sorbifolium

Xerophyllum asphodeloides

xanth-

Used in compound words to denote yellow

xanthinus

zan-TEE-nus

xanthina, xanthinum

Yellow, as in *Rosa xanthina*

xanthocarpus

zan-tho-KAR-pus

xanthocarpa, xanthocarpum

With yellow fruit, as in *Rubus xanthocarpus*

Xanthoceras [1]

zan-tho-SEER-as

From Greek *xanthos*, meaning "yellow," and *keras*, meaning "horn," referring to the hornlike glands between the petals (*Sapindaceae*)

Xanthocyparis

zan-tho-SIP-uh-ris

From Greek *xanthos*, meaning "yellow," and *cyparissos*, meaning "cypress," alluding to the color of the wood (*Cupressaceae*)

xantholeucus

zan-THO-luh-cus

xantholeuca, xantholeucum

Yellow-white, as in *Sobralia xantholeuca*

Xanthorhiza

zan-tho-RY-zuh

From Greek *xanthos*, meaning "yellow," and *rhiza*, meaning "root," because the rhizomes are yellow (*Ranunculaceae*)

Xanthosoma

zan-tho-SOH-muh

From Greek *xanthos*, meaning "yellow," and *soma*, meaning "body," referring either to the stigma or inner tissues (*Araceae*)

Xeranthemum

zeer-ANTH-uh-mum

From Greek *xeros*, meaning "dry," and *anthemon*, meaning "flower," because the flower heads are everlasting when dry (*Asteraceae*)

Xerochrysum

zeer-oh-KRY-zum

From Greek *xeros*, meaning "dry," and *chrysos*, meaning "gold," alluding to the papery yellow bracts (*Asteraceae*)

Xerophyllum [2]

zeer-oh-FIL-um

From Greek *xeros*, meaning "dry," and *phyllon*, meaning "leaf"; the dried leaves are used for weaving (*Melanthiaceae*)

yakushimanus

ya-koo-shim-MAH-nus

yakushimana, yakushimanum

Connected with Yakushima Island, Japan, as in *Rhododendron yakushimanum*

yedoensis

YED-oh-en-sis

yedoensis, yedoense

—

yesoensis

yesoensis, yesoense

—

yezoensis

yezoensis, yezoense

From Tokyo, Japan, as in *Prunus × yedoensis*

Ypsilandra

ip-si-LAN-druh

From Greek *ypsilon*, meaning the letter "Y," and *andros*, meaning "male," because the stamens supposedly have a Y shape (*Melanthiaceae*)

Yucca [1]

YUK-uh

From Caribbean *yuca*, the name for cassava (*Manihot esculenta*) roots, wrongly applied to this genus (*Asparagaceae*)

yuccifolius

yuk-kih-FOH-lee-us

yuccifolia, yuccifolium

With leaves like *Yucca*, as in *Eryngium yuccifolium*

yuccoides

yuk-KOY-deez

Resembling *Yucca*, as in *Beschorneria yuccoides*

yunnanensis

yoo-nan-EN-sis

yunnanensis, yunnanense

From Yunnan, China, as in *Magnolia yunnanensis*

Yushania

yew-SHAN-ee-uh

Named after Yushan, the highest mountain in Taiwan, where the first species was collected (*Poaceae*)

1

Yucca gloriosa

KEEPING IT IN THE FAMILY

The Latin names given to plant families are constructed by choosing one genus from the family, such as *Erica*, then adding the suffix *-aceae*, so *Ericaceae* are plants that resemble *Erica* (heather). Typically, the genus chosen is typical of the family, but not always. *Acanthaceae* are mainly tropical shrubs, but *Acanthus* is a hardy herbaceous perennial. In a few families, the representative genus is no longer recognized, as in *Caprifoliaceae*, *Aquifoliaceae*, and *Cactaceae*, but the family name remains unchanged. *Caprifolium* is a synonym of *Lonicera*, *Aquifolium* is now *Ilex*, while *Cactus* species now belong to several other genera.

A
B
C
D
E
F
G
H
I
J
K
L
M
N
O
P
Q
R
S
T
U
V
W
X
Y
Z
i

zabelianus

zah-bel-ee-AH-nus

zabeliana, zabelianum

Named after Hermann Zabel, nineteenth-century German dendrologist, as in *Berberis zabeliana*

Zaluzianskya

zah-loo-zee-AN-ski-uh

Named after Adam Zaluziansky von Zaluzian (1558–1613), Czech physician (*Scrophulariaceae*)

zambesiacus

zam-bes-ee-AH-kus

zambesiaca, zambesiacum

Connected with the Zambezi River, Africa, as in *Eucomis zambesiaca*

Zamia

ZAY-mee-uh

From Greek *azaniae*, meaning "pinecone," because *Zamia* is a cone-producing cycad (*Zamiaceae*)

Zamioculcas

zay-mee-oh-KUL-kas

From *Zamia*, an unrelated plant of similar appearance, plus Arabic *culcas*, the vernacular name for the related *Colocasia* (*Araceae*)

Zantedeschia

zan-tuh-DESH-ee-uh

Named after Giovanni Zantedeschi (1773–1846), Italian botanist and physician (*Araceae*)

Zanthoxylum

zan-THOKS-uh-lum

From Greek *xanthos*, meaning "yellow," and *xylon*, meaning "wood"; some species have yellow heartwood (*Rutaceae*)

Zea

ZEE-uh

From Greek *zeia*, a type of cereal (*Poaceae*)

zebrinus

zeb-REE-nus

zebrina, zebrinum

With stripes like a zebra, as in *Tradescantia zebrina*

Zelkova

zel-KOH-vuh

From Georgian *dzelkva*, meaning "stone pillar," because the wood is hard and popular for construction (*Ulmaceae*)

Zenobia

zuh-NOH-bee-uh

Named after Septimia Zenobia (ca. AD 240–274), Queen of Palmyra, in present-day Syria (*Ericaceae*)

Zephyranthes

zef-ur-ANTH-eez

From Greek Zephyros, god of the west wind, plus *anthos*, meaning "flower"; presumably alluding to their distribution in the western hemisphere (*Amaryllidaceae*)

GENUS SPOTLIGHT

Zamioculcas

This popular houseplant is often known as the ZZ plant, due to the initials of its Latin name *Zamioculcas zamiifolia*. Both the genus and species names recognize the similarity of this plant to the genus *Zamia*. However, the two plants are not closely related. *Zamioculcas* is an aroid, a member of the *Araceae* family, most notable for their unusual flower structure where tiny flowers cluster around a central rod (spadix), which is then enclosed within a bract (spathe). Other popular family members include peace lily (*Spathiphyllum*) and flamingo flower (*Anthurium*). *Zamia*, however, is a cycad, a nonflowering plant most closely related to conifers.

Zamioculcas zamiifolia

zeyheri

ZAY-AIR-eye

Named after Karl Ludwig Philipp Zeyher (1799–1859), German botanist and plant collector, as in *Philadelphus zeyheri*

zeylanicus

zey-LAN-ih-kus

zeylanica, zeylanicum

From Ceylon (Sri Lanka), as in *Pancratium zeylanicum*

zibethinus

zy-beth-EE-nus

zibethina, zibethinum

Smelling foul, like a civet cat, as in *Durio zibethinus*

Zigadenus

zig-uh-DEE-nus

From Greek *zygos*, meaning "pair" or "yoke," and *aden*, meaning "gland," referring to the pair of glands on each petal of some species (*Melanthiaceae*)

Zingiber

ZIN-jib-ur

From Greek *ziggiberis*, the vernacular name for ginger, *Z. officinale* (*Zingiberaceae*)

Zinnia [1]

ZIN-ee-uh

Named after Johann Gottfried Zinn (1727–59), German botanist and ophthalmologist (*Asteraceae*)

Zizania

zy-ZAY-nee-uh

From Greek *zizanion*, a weedy grass growing among wheat crops (*Poaceae*)

Ziziphus [2]

ZIZ-i-fus

From Persian *zizafun*, the vernacular name for *Z. lotus* (*Rhamnaceae*)

zonalis

zo-NAH-lis

zonalis, zonale

—

zonatus

zo-NAH-tus

zonata, zonatum

With bands, often colored, as in *Cryptanthus zonatus*

Zygopetalum

zy-goh-PET-ah-lum

From Greek *zygos*, meaning "yoke," and *petalon*, meaning "petal," because the flower lip is swollen at the base, appearing to yoke together the other petals (*Orchidaceae*)

1

Zinnia elegans

2

Ziziphus spina-christi

The Gardener's Botanical Picture Credits

Chrysanthemum indicum

Index of common names

This index can be used to find the botanical names of some plants in the book. It is not a definitive index, but if you know the common name, it cross-references to listings for each part of the botanical name.

For example: 'Adam's needle' is a common name for *Yucca filamentosa*. The cross references for that listing direct you to entries in the main part of the book for both *Yucca* (315) and *filamentosa* (132), allowing you to understand the origins and meanings of each part of the binomial name. Some common names lead to more than one botanical name (where the same common name is applied to different species). Common names for a genus will have only one botanical term beneath them, for example: Amaryllis, *Hippeastrum* (155).

A
B
C
D
E
F
G
H
I
J
K
L
M
N
O
P
Q
R
S
T
U
V
W
X
Y
Z

i

A
B
C
D
E
F
G
H
I
J
K
L
M
N
O
P
Q
R
S
T
U
V
W
X
Y
Z

i

A
B
C
D
E
F
G
H
I
J
K
L
M
N
O
P
Q
R
S
T
U
V
W
X
Y
Z

i

A
B
C
D
E
F
G
H
I
J
K
L
M
N
O
P
Q
R
S
T
U
V
W
X
Y
Z

i

A
B
C
D
E
F
G
H
I
J
K
L
M
N
O
P
Q
R
S
T
U
V
W
X
Y
Z

i

A
B
C
D
E
F
G
H
I
J
K
L
M
N
O
P
Q
R
S
T
U
V
W
X
Y
Z

i

A
B
C
D
E
F
G
H
I
J
K
L
M
N
O
P
Q
R
S
T
U
V
W
X
Y
Z

i

A
B
C
D
E
F
G
H
I
J
K
L
M
N
O
P
Q
R
S
T
U
V
W
X
Y
Z

i

A
B
C
D
E
F
G
H
I
J
K
L
M
N
O
P
Q
R
S
T
U
V
W
X
Y
Z

i

A
B
C
D
E
F
G
H
I
J
K
L
M
N
O
P
Q
R
S
T
U
V
W
X
Y
Z

i

A
B
C
D
E
F
G
H
I
J
K
L
M
N
O
P
Q
R
S
T
U
V
W
X
Y
Z

i

A
B
C
D
E
F
G
H
I
J
K
L
M
N
O
P
Q
R
S
T
U
V
W
X
Y
Z

i

Please note that some Latin names are used as the common names so are also listed in the main section. These include:

Alyssum	*Dahlia*	*Hydrangea*
Aster	*Daphne*	*Iris*
Aubrieta	*Forsythia*	*Lobelia*
Begonia	*Freesia*	*Magnolia*
Ceanothus	*Geranium*	*Petunia*
Chrysanthemum	*Hebe*	*Phlox*
Clematis	*Heuchera*	*Rhododendron*
Cosmos	*Hosta*	

A
B
C
D
E
F
G
H
I
J
K
L
M
N
O
P
Q
R
S
T
U
V
W
X
Y
Z

i

Dahlia